Table of Formula Weights

based on 1961 Atomic Weights and Rounded to Two Decimal Places

$AgBr$	187.78
$AgCl$	143.32
Ag_2CrO_4	331.74
AgI	234.77
$AgNO_3$	169.88
Ag_2S	247.80
$AgSCN$	165.95
$Al(C_9H_6ON)_3$ (oxinate)	459.45
Al_2O_3	101.96
$Al(OH)_3$	78.00
$Al_2(SO_4)_3$	342.15
As_2O_3	197.84
As_2O_5	229.84
As_2S_3	246.03
$BaCO_3$	197.35
$BaCl_2$	208.24
$BaCrO_4$	253.34
BaF_2	175.34
BaO	153.34
BaO_2	169.34
$Ba(OH)_2$	171.36
$Ba_3(PO_4)_2$	601.97
$BaSO_4$	233.40
Bi_2S_3	514.16
B_2O_3	69.62
$Ca_3(AsO_4)_2$	398.07
$CaBr_2$	199.90
$CaCO_3$	100.09
CaC_2O_4	128.10
CaF_2	78.08
$Ca(IO_3)_2$	389.89
CaO	56.08
$Ca(OH)_2$	74.10
$Ca(NO_3)_2$	164.10
$Ca_3(PO_4)_2$	310.19
$CaSO_3$	120.14
$CaSO_4$	136.14
$C_6H_5NH_3Cl$	129.59
$C_{12}H_{22}O_{11}$	342.30
$Ce(IO_3)_3$	489.91
CeO_2	172.12
$Ce(SO_4)_2$	332.25
$Ce_2(SO_4)_3$	316.42
$(NH_4)_2Ce(NO_3)_6$	548.22
$(NH_4)_2Ce(SO_4)_3 \cdot 2H_2O$	500.41
CO	28.01
CO_2	44.01
$CO(NH_2)_2$ (urea)	60.05
Co_3O_4	240.80
CrO_3	99.99
Cr_2O_3	151.99
$Cr(OH)_3$	103.01
$Cu(NO_3)_2$	187.54
CuO	79.54
Cu_2O	143.08
$CuSO_4$	159.60
$CuSO_4 \cdot 5H_2O$	249.68
CuS	95.60
Cu_2S	159.14
EDTA, disodium $\cdot 2H_2O$	372.24
$FeCO_3$	115.86
FeO	71.85
Fe_2O_3	159.69
Fe_3O_4	231.54
$Fe(OH)_2$	89.87
$Fe(OH)_3$	106.87
FeS	87.91
$Fe(SCN)_3$	230.09
$FeSO_4 \cdot 7H_2O$	278.02
$FeSO_4 \cdot (NH_4)_2SO_4 \cdot 6H_2O$	392.14
$Fe_2(SO_4)_3$	399.89
$Fe_2(SO_4)_3 \cdot (NH_4)_2SO_4 \cdot 24H_2O$	964.38
H_3AsO_4	141.94
H_3AsO_3	125.94
HBO_2	43.82
H_3BO_3	61.83
HBr	80.92
$HCHO_2$ (formic)	46.03
$HC_2H_3O_2$ (acetic)	60.05
$HC_7H_5O_2$ (benzoic)	122.12
$HC_3H_5O_3$ (lactic)	90.08
$H_2C_4H_2O_4$ (maleic)	116.08
$H_2C_4H_4O_5$ (malic)	134.09
$H_2C_3H_2O_4$ (malonic)	104.06
$H_2C_2O_4 \cdot 2H_2O$ (oxalic)	126.07
$H_2C_4H_4O_6$ (tartaric)	150.09
$H_2C_8H_4O_4$ (phthalic)	166.14
HCl	36.46
$HClO_4$	100.46
HF	20.01
HNO_2	47.01
HNO_3	63.02
HSO_3NH_2 (sulfamic)	97.09
H_2O_2	34.02
H_3PO_4	97.99
H_2S	34.08

(continued inside back cover)

EX LIBRIS
MARK A. DE ROSCH

ANALYTICAL SEPARATIONS AND DETERMINATIONS

Analytical Separations and Determinations

A TEXTBOOK IN QUANTITATIVE ANALYSIS

C. T. KENNER
Southern Methodist University
Dallas, Texas

ASSISTED BY **RICHARD E. O'BRIEN**
El Centro Junior College
Dallas, Texas

The Macmillan Company
New York
Collier-Macmillan Limited
London

The Macmillan Company
866 Third Avenue, New York, New York 10022

Collier-Macmillan Canada, Ltd., Toronto, Ontario

Library of Congress catalog card number: 76-12389.
First printing

PREFACE

Analytical chemistry plays an important role in many areas of present-day research and industry. Research personnel in chemistry, in other physical and biological sciences, and many engineering fields rely heavily upon the information obtained through chemical analysis, and many industrial processes depend upon adequate control analyses for economical performance. The rapid expansion of knowledge in all scientific fields has focused much thought and discussion upon the chemistry curriculum and upon necessary changes, both of omission and addition, which may improve the dissemination of knowledge to the student. The entire chemistry curriculum is in a state of flux, and many modifications and rearrangements are being suggested and tried. In terms of the course in analytical chemistry, normally taught at the sophomore level, some schools have abandoned it completely, some have incorporated it as a laboratory course at the freshman level, some have moved it to the junior year with physical chemistry as a corequisite, and others have made it a senior course with physical chemistry as a prerequisite. The schools which have kept the formal course in quantitative analysis have modified the content in many ways: Some have eliminated all gravimetric determinations; some have installed sophisticated instrumentation such as mass spectrometry, x-ray diffraction, and automatic scanning photometers; but the majority have retained a moderate number of gravimetric and volumetric procedures and have introduced simple instrumentation together with some of the recent developments in separation chemistry such as chromatography, extraction, and ion exchange. Some institutions have retained the formal course as a service course for other departmental and preprofessional majors and placed the course for chemistry majors at the junior or senior level with physical chemistry as a prerequisite.

Since many of the present quantitative analysis texts are written for the chemistry major, the explanations of the basic principles often presuppose a knowledge of physical chemistry and calculus. As a consequence many analytical students who have not had physical chemistry or calculus and do not intend to take them find themselves at a disadvantage with many texts.

We believe there is a definite need for a formal course in analytical chemistry, both for chemistry majors and others, and that it should be given as soon as the student is capable of assimilating the knowledge. In some cases this may be at the freshman level for outstanding students with excellent chemistry backgrounds. In most cases, it will be at the sophomore or junior level. We also believe that the text should not presuppose a knowledge of physical chemistry and calculus, especially when given at the freshman or sophomore level.

The beginning course in analysis must not only teach the basic principles and techniques of analytical determinations, it should also emphasize the application of the quantitative approach to the solution of problems, which is the basis of all chemical progress. As a consequence, we believe that the course should include the fundamental principles and techniques of gravimetric and volumetric analysis together with an introduction to separation theory and instrumental analysis. This text endeavors to follow these principles and does not presuppose a knowledge of either physical chemistry or calculus. It can be used as the basis of a one-semester terminal course or as an introductory course for chemistry and other preprofessional majors who will take a later course or courses at the advanced level.

This text differs from most texts for analytical chemistry in that it does not include laboratory directions for individual experiments. The theoretical basis of typical determinations of each of the main classes of analytical methods is included so that the text can be used with practically any laboratory manual or set of experiments. Chapter 17, which discusses the most widely used basic techniques and provides clear illustrative diagrams and pictures, serves as a guide to the development of good quantitative habits.

The author feels that most instructors prefer to design their own experiments and that they usually modify the experiments in any text by issuance of mimeographed instructions. Also, periodic change of texts is rendered difficult if the same experiments are not included in both. The author has used his own laboratory manual or set of instructions for several years and finds that the students prefer an inexpensive manual so that they will not have to bring their expensive texts into the laboratory where they may be damaged through no fault of their own.

Even though there is no formal separation into parts, the text is actually divided into four sections. The first section, which includes the first five chapters, deals with introductory material and the fundamental calculations involved in all types of analysis. Section two includes Chapters 6 through 11 and covers the basic theory and practice of precipitation, neutralization, and oxidation-reduction methods. Section three covers the theory and practice of separations and simple electrochemical and photometric instrumentation. The last two chapters discuss the reliability of analytical results and the techniques required for quantitative laboratory work.

The discussions of colorimetry, photometry, chromatography, and ion exchange are more detailed than in many texts at this level since these subjects are extremely important to preprofessional students in all fields, especially those who will receive no further training in analytical chemistry.

I wish to express my appreciation to the many students who have aided in the preparation of this text through their comments and class notes during the development stage. Thanks are expressed to Mr. Richard E. O'Brien, who helped write one chapter and whose comments and criticisms have helped clarify some of the material. I am deeply grateful to Mrs. Shirley McLean, who typed the entire manuscript with a minimum of errors. My thanks also go to my wife, Bess, for her forbearance during the preparatory stages and for sharing in the task of proofreading the manuscript.

C. T. K.

Dallas

CONTENTS

DETAILED CONTENTS

CHAPTER 1

Introduction

A. Types and Methods of Analysis

Analytical chemistry deals with the determination of the composition of materials and substances. The word *substance* in chemistry is usually used to refer to a pure element or compound, while a *material* is considered to be a mixture. Thus the compound sodium chloride is referred to as a substance, but sodium chloride as it occurs in nature and as it is mined from salt domes is referred to as a material since it contains substances other than sodium chloride as impurities. As a result a chemist practically never analyzes a substance but deals primarily with materials as they occur in nature or as they are manufactured or prepared from other materials.

The two main classes of analysis are qualitative and quantitative. The determination of the actual substances or constituents present in a material, without regard to the amount of each present, is known as qualitative analysis, while the determination of the actual amount of a constituent present is considered to be quantitative analysis. Thus qualitative analysis is concerned only with what materials or substances are present, while quantitative analysis deals with how much of each constituent determined is present. The result of a quantitative analysis is usually reported as a number, often the percentage by weight.

Analyses may be classified in several different ways. One method of classification is concerned with the number and type of constituents determined and reported. An ultimate or complete analysis accounts for each constituent present and should represent 100% of the sample. If all the constituents are not determined, the analysis is called a partial analysis, and if only one constituent is determined, it is often referred to as single-constituent analysis. Since the properties of materials are often affected by one or a few of the major, and sometimes the minor, constituents, partial analyses are run much more often than complete analyses.

Often similar materials may be determined and reported together if it is not necessary to know the value for each individually. Such analyses are called proximate analyses. An example is the determination of iron and aluminum in rocks. Both of these elements precipitate upon addition of ammonia to form hydrous oxides which are filtered, washed, and ignited to a mixture of Fe_2O_3 and Al_2O_3, which is weighed. The report would show the combined percentages of Fe_2O_3 and Al_2O_3. Since this particular precipitate is known as the R_2O_3 precipitate, the report often simply shows per cent R_2O_3. If it is necessary to know either or both of the metals individually, other methods would have to be used to determine one or both.

Analyses may also be classified on the basis of the type of material being analyzed. Thus we refer to inorganic analysis and to organic analysis. Similarly we speak of rock analysis, nonferrous alloy analysis, steel analysis, petroleum product analysis, fuel analysis, etc., to indicate the type of material being analyzed.

Analyses may also be classified as to the amount of sample taken for analysis. In the beginning course in quantitative analysis, and in many industrial, research, and government laboratories, the majority of samples are 0.1 g or more and are classified as macroanalyses. If the sample weight is between 1 and 10 mg (0.010 to 0.001 g), the analysis is said to be a microanalysis. Sample weights between micro and macro are known as semimicroanalyses. Recently, methods which use a sample weight of less than 1 mg have been developed. Such methods are known as ultramicroanalyses.

The classification of analytical methods is based upon the final measurement step in the procedure and includes gravimetric, volumetric, and physicochemical methods. Methods in which the final determination is the accurate measurement of a weight of a substance of known purity are known as gravimetric methods, while those in which the final measurement deals with a volume are called volumetric methods. If the final volume measurement is made with a burette, the procedure is called a titration, and the method is classified as a titrimetric analysis. Neither gravimetric nor volumetric methods measure individual properties of the substance being determined or measured. Methods based upon the measurement of some physical or chemical property are known as physicochemical methods. Such methods often require special instruments to make the measurement and are known as instrumental methods. Physicochemical or instrumental methods include the measurement of optical properties (transmittance, absorbance, reflectance, emission, rotation), electrical properties (conductance, potential, resistance, etc.), radioactivity, and crystalline structure.

B. Steps in an Analysis

To analyze a material, the following steps are usually required: (1) The sample must be obtained; (2) the method of analysis must be selected; (3) the sample must be measured (usually weighed); (4) the sample must be dissolved; and finally (5) the constituent being determined must be separated and measured. Each of these will be discussed separately.

1. Obtaining the Sample

The sample used in the actual analytical determination must be a representative sample; that is, the determination sample must have the same average composition as the larger sample from which it is obtained and which it is supposed to represent. While representative samples seldom create a problem in beginning courses in analytical chemistry, they are very important for laboratories which run analyses on large lots of material in order to check purchase specifications. The weight of the determination sample varies depending upon the method being used, but in the majority of cases it is 1 g or less. Yet this 1 g may have to represent as much as a trainload of coal or a boatload of ore. It is imperative that the determination sample which represents large amounts of material actually be a representative sample; otherwise the analysis, no matter how carefully run, will be erroneous and may result in substantial loss or gain to the companies involved.

Correct methods of sampling large lots of materials are specified by various industrial and/or scientific associations which are concerned with the use of the materials. Such associations include the American Society for Testing Materials (ASTM), the Association of Official Analytical Chemists (AOAC), the American Oil Chemists Society (AOCS), the Association of the Petroleum Industry (API), and the National Bureau of Standards (NBS). Details of standard methods of sampling can be found in many reference books and in articles and books published by the associations involved.

In general the sampling of large lots of material is performed in three steps. The first sample taken is called the gross sample and is usually obtained during the unloading process. Some specified portion is retained, such as every twenty-fifth shovel, or some sort of riff is used on the unloading chute to separate a certain fraction of the material. The gross sample may range as high as several tons of material and must be obtained by a procedure which ensures that it is representative of the large lot. The gross sample must be reduced to an amount suitable for transmission to the laboratory, usually 25 lb or less.

Reduction of the amount of the gross sample may be performed by the long pile and alternate shovel method or by the cone and quarter method. In the long pile method, the gross sample is arranged in a long pile during the unloading process. It is then separated into two equal piles using a shovel and throwing alternate shovels to opposite sides. One-half is discarded and the other half again separated in the same manner. This is continued until the remaining sample is of the proper weight. In the cone and quarter method, the gross sample is piled into a cone and the top flattened. The cone is then separated into quarters. Two opposite quarters are discarded and the other two opposite quarters retained and formed into a second cone. The process is repeated until the sample is of the correct amount to send to the laboratory.

During the reduction of the amount of the gross sample, it is often necessary to reduce the particle size to retain the proper proportion of large and small particles, to ensure uniformity, and to ensure that the sample sent to the laboratory is representative. Various types of crushers, grinders, ball mills, etc., are used to reduce the particle size of the sample.

After receipt of the laboratory sample, the analyst must also be sure that the sample used in the actual analysis (called the determination sample) is uniform and is representative of the large lot. The analyst must often use the same methods (on a smaller scale, of course) for reduction of the amount and particle size of the sample. Sometimes the sample is such that it cannot be reduced to particles which are small enough to ensure uniform mixing. In such cases, the analyst will use a series of standard sieves to separate the sample into fractions of relatively equal particle size. The fraction of the total weight of the sample represented by each fraction is then calculated. The necessary portion of each fraction is then weighed individually and mixed to obtain the determination sample.

Standard sieves are made in the form of wire screens and are classified on the basis of the number of openings (meshes) per square inch. The larger the number of openings, the smaller is the mesh size and, as a consequence, the smaller a particle has to be to pass through a sieve. A particle which is retained by a number 40 screen would pass through any screen of a smaller number. Particle sizes of solids are usually specified in terms of the smallest number screen through which the particles will pass and the largest number screen upon which they will be retained. A sample which is classified a 40/80 will pass a number 40 sieve and be retained on a number 80 sieve. If only one sieve number is given in the particle size specifications, it means that all the particles will pass that size screen. All the particles of a sample which is referred to as 100 mesh will pass through a number 100 screen. Standard screens vary in size from number 1 to number 300.

Some materials must be reported on an "as-received" basis and also on a "dry" basis. As a consequence, the analyst will take some rather large representative portion and weigh and dry it at 100–110°C to remove the adventitious moisture. Adventitious moisture is moisture that is loosely held by the solid and which may be removed by simple heating to a temperature slightly above the boiling temperature of water. The determination sample itself must be dry, and as a result, all that portion of the laboratory sample which is to be used for analytical determinations is dried for at least 1 hr at a temperature between 100 and 110°C and is stored in a desiccator (see Fig. 17–11, Chapter 17) until the individual samples are weighed out. The results are then calculated on both the dry basis and the as-received basis.

2. Selection of the Method of Determination

Before deciding on the actual sample weight to be weighed, it is necessary to select the method of determination to be used. The first and primary consideration in the selection is the accuracy with which the constituent must be determined. The method selected must be capable of this requisite accuracy since it is not economical to select a method which will not give adequate results or to select a method which is capable of accuracy beyond that which is required, especially if the more accurate method requires more time or more expensive equipment.

The method must be selected also on the basis of the percentage of the constituent in the sample and the size of sample which is available. Methods which are satisfactory for major components are usually not satisfactory to determine the same constituent if it is present in trace quantities. Also, a method which is satisfactory for a relatively large sample is usually not satisfactory for a sample size in the milligram range.

If several methods are available, the next consideration is the time involved in the analysis. This factor can be very important in control work in metallurgical and manufacturing plants when the product must be delayed until the analysis is complete. For instance, a steel mill prepares large quantities of special steels at one time, and usually must keep the steel molten while components are being added. After the addition, the steel must be analyzed before the molten material can be poured into the forms desired. A few minutes of time saved in such an analysis may save the company a large sum of money over a comparatively short period of time. As a consequence, steel mills will often pay over $100,000 for an instrument which can save a few minutes in analysis time while the steel must be kept in the molten state.

For situations in which time is not critical, the actual time spent by the analyst on the determination must be considered. If one determination requires 1 hr of the analyst's time over a period of 2 days and another requires 2 hr during one 8 hr shift, the slower method will be selected if time is not important.

Other factors which are usually considered include costs other than analyst time, difficulty of procedure, availability and stability of reagents, special equipment needed and whether or not it can be justified by the number of analyses run, and other similar considerations.

3. Measurement of the Sample

a. SELECTION OF PROPER SAMPLE WEIGHT. After selection of the method, the actual sample weight to be utilized must be chosen. In general, it is desirable to weigh the largest possible sample in order to decrease the relative error in weighing. However, the time and diffculties involved in later steps of the analysis must also be considered since these usually increase with sample weight. The optimum weight to be used will be the largest weight possible which will not cause undue difficulties in the later steps of the analysis. Only an experienced analyst can determine what the optimum weight will be for a given analysis since the decision requires a knowledge of the procedure and the experimental difficulties which will be encountered in the course of the determination. Optimum weights are seldom expressed to more than two significant figures (0.50, 1.0, 10, etc.) since a sample weight within 5 or 10% of the optimum weight is usually satisfactory.

To select the actual weight of the sample to be weighed it must be decided whether speed of weighing or ease of calculation is of primary importance. Weighing an exact weight of sample to the nearest part per thousand (ppt) can be a tedious and time-consuming operation if the sample weight is 1 g or less since this means that a deviation of only 1 mg or less can be allowed. It is much faster to obtain an accurate weight of a sample if a range is allowed on either side of the optimum weight. As an example, weighing a sample of exactly 1.000 g takes much longer than weighing a sample the weight of which can be anywhere between 0.950 and 1.050 g as long as the actual weight is measured to one ppt.

However, calculations are simplified by using weights which are exact fractions or multiples of 1 since most calculations involve dividing by the weight of the sample in at least one step. Thus, it is much easier to divide by 1.000 than by 0.979 or by 1.036. An even simpler method of calculation is involved if the sample weight chosen is the exact factor weight or some exact fraction or multiple of the factor weight. The factor weight in a given determination is that weight of sample which will allow the percentage of the constituent being determined to be calculated by multiplying the weight of the precipitate produced (or the volume of a solution used) by 100 to obtain the percentage directly. If the factor weight deviates widely from the optimum weight, some exact fraction or multiple of the factor weight can be used so that the multiplier will be 10, 50, 200, etc. Methods of calculating factor weights will be discussed later. Ease of calculation is not considered as necessary now as it was formerly since most modern laboratories are equipped with electric calculators.

b. WEIGHING BY DIFFERENCE. If the sample changes weight upon exposure to the atmosphere, it is usually weighed by difference. When weighing a sample by difference, the material is weighed in a closed container, the sample to be analyzed is poured out of this container, and the container is weighed again. The weight of sample taken is then calculated from the difference in the weights before and after removal of the determination sample. It is needless to point out that it is essentially impossible to weigh a sample of an exact predetermined weight by this method and that a range of weight must be used.

c. ALIQUOTING PROCEDURE. Sometimes it is preferable to weigh a relatively large sample so that it can be dissolved and diluted to a definite, accurately known volume. A known accurate fraction of this large volume is then used for the determination sample. This procedure is known as aliquoting, and the sample used is called an aliquot. An aliquot may be defined as an accurately known fraction of a larger volume from which it is taken. The aliquoting procedure may be used if the determination sample is too small to allow accurate weighing or if the sample changes weight upon exposure. In either case, the larger sample size can be weighed more accurately and more rapidly, especially if a range of weight is allowable. The aliquoting procedure to be used must be planned before the sample is weighed in order that the aliquots used will contain the optimum weight of sample.

The analyst usually has available various sizes of volumetric (transfer) pipettes and various sizes of volumetric flasks. The volumetric flasks are calibrated to contain (TC) and thus can be used to prepare an accurately known volume of a solution. The volumetric pipettes are usually calibrated to deliver (TD) and are used to transfer an accurately known volume of a solution. The actual amount of sample to be weighed will depend upon the volumetric flask and the volumetric pipette used and upon the optimum weight of the determination sample. These values are related by the equation

$$\text{wt. of sample to be weighed} = \text{optimum weight of determination sample} \times \frac{\text{volume of flask}}{\text{volume of pipette}} \quad (1)$$

As an example, calculate the weight of sample to be weighed if the optimum weight is 0.25 g, the sample is to be diluted to 500.0 ml, and a 25.00-ml aliquot is to be used. Substitution in Eq. (1) gives

$$\text{wt. of sample} = 0.25\text{ g} \times \frac{500.0\text{ ml}}{25.00\text{ ml}} = 5.0\text{ g}$$

In weighing the 5-g sample, a range of 5–10 per cent may be used without seriously affecting the analysis so that the actual sample weight could be any value between 4.500 and 5.500 g, as long as it was known accurately.

It should be noted that the calculation of the actual weight of the determination sample is the reverse of the calculation in Eq. (1). For example, assume that the actual weight of the sample in the example given above turned out to be 5.012 g, and calculate the weight of the actual determination sample. The actual weight of that determination sample will be the weight of solute in the aliquot or will equal the weight weighed multiplied by the fraction of the total weight found in the aliquot. This relationship is expressed by the equation

$$\text{wt of determination sample} = \text{weight weighed} \times \frac{\text{volume of aliquot}}{\text{volume of solution}} \qquad (2)$$

The calculation then becomes

$$\text{wt of determination sample} = 5.012 \text{ g} \times \frac{25.00 \text{ ml}}{500.0 \text{ ml}} = 0.2506 \text{ g}$$

4. Solution of the Sample

In practically all analyses, the first step after weighing the sample is to dissolve the weighed sample. The solvent used should dissolve the sample completely in the least amount of time possible and should cause the least possible interference with the later steps of the analysis. The general rule is that the solvent should be as gentle as possible and should dissolve the sample at a reasonable rate as close to room temperature as possible. Solvents are usually tried in the following order.

a. WATER. Samples which are composed of soluble salts may be dissolved in water at room temperature or at elevated temperatures. If the sample is composed of a mixture of soluble and insoluble substances, the residue from the water solution may be dissolved in a stronger solvent (such as a strong acid or a strong base) or may be fused with an acidic or basic flux. A simple solubility table is given in Appendix 3.

b. NONOXIDIZING ACIDS. If the sample is not soluble in water, the next solvent to be tried is a nonoxidizing acid such as hydrochloric acid. Other nonoxidizing acids include dilute sulfuric acid and dilute perchloric acid. Concentrated sulfuric and perchloric acids can act as oxidizing acids at elevated temperatures. Hydrofluoric acid may be used as a nonoxidizing acid for many silicate samples.

c. OXIDIZING ACIDS. Oxidizing acids such as nitric, concentrated sulfuric, and perchloric acids at elevated temperatures, and aqua regia (which is a mixture of hydrochloric and nitric acids) are often used.

d. FUSION. If the sample is not soluble in acids, it usually must be fused with an acidic or basic flux. The most popular acidic flux is potassium bisulfate ($KHSO_4$), which decomposes to potassium pyrosulfate ($K_2S_2O_7$)

when melted. Basic fluxes include sodium carbonate or a mixture of sodium and potassium carbonates. If simple acidic or basic fusion will not render the sample soluble, oxidizing fluxes can be used. Sodium peroxide is often used as an oxidizing basic flux. Sodium carbonate plus potassium chlorate is also used as an oxidizing flux. Fusion is accomplished by mixing the sample intimately with the flux and heating until the mixture is molten. The melt is then cooled and dissolved in water or in acid.

In many cases, the sample is dissolved as completely as possible in an acid and the insoluble residue is then fused with a basic flux such as sodium carbonate.

e. SOLUBILITY IN ORGANIC SOLVENTS. Some samples may be soluble in organic solvents such as chloroform, carbon tetrachloride, alcohols, ethers, etc. These samples are usually organic materials which are insoluble in inorganic solvents such as water and acids.

The methods usually used to dissolve various types of materials are summarized in Table 1-1.

TABLE 1.1. Methods Used to Dissolve Inorganic Substances

Substances	Solvent
Very active metals	Water
Moderately active metals	Nonoxidizing acids
Inactive metals	Oxidizing acids
Soluble salts	Water
Carbonates, hydroxides, oxides, phosphates, sulfides	Nonoxidizing acids
Silicates	Hydrofluoric acid or fusion
Various ores	Acidic, basic, or oxidizing fusion

In some cases the type of determination being performed must be considered in the choice of solvent. For example, in the analysis of certain types of brasses and bronzes, tin may be determined either by a gravimetric or a volumetric method. For the gravimetric method, nitric acid is chosen as the solvent since metastannic acid, which is insoluble in nitric acid and can be filtered and ignited, is formed. A mixture of hydrochloric acid and nitric acid is used for the volumetric tin procedure since the hydrochloric acid will keep the tin in solution and the nitric acid will dissolve the copper.

5. Separation and Measurement of the Constituent(s) Being Determined

After the sample has been dissolved, the constituent or constituents being determined must be separated from interfering substances before actual measurement. Separation methods include precipitation, extraction, chromatography, ion exchange, distillation, and electrodeposition. Most of these are discussed in later chapters. In many cases, an interfering substance can be *masked* to prevent the interference. Masking in analytical chemistry means the conversion to a form which will not enter into the reaction being performed. This usually involves the conversion to a soluble complex or to a different state of oxidation.

C. Problems

1. Arrange the following list of samples in order of decreasing particle size: 20/50, 4/8, 60/100, 300, 10/30, 100/200.
2. Define and state the difference between (a) qualitative and quantitative analysis, (b) gross and determination sample, (c) micro- and ultramicro-analysis, (d) micro- and macroanalysis, (e) complete and partial analysis, (f) oxidizing and nonoxidizing acids, (g) acidic and basic flux, (h) aliquot and total volume, (i) organic and inorganic analysis, (j) representative sample and laboratory sample, and (k) optimum weight sample and factor weight sample.
3. Calculate the sample weight to be weighed and the range of weight which would be allowed for the following aliquoting procedures:

Optimum weight	Volumetric flask used (ml)	Transfer pipette used (ml)
(a) 0.015	1000	25
(b) 0.50	250	50 (*Ans.* 2.500; 2.250–2.750 g)
(c) 0.25	100	25
(d) 0.33	500	100
(e) 0.0035	500	10

4. Calculate the weight of the determination sample for the following aliquoting procedures:

Grams of sample weighed	Volumetric flask used (ml)	Volumetric pipette used (ml)
(a) 1.036	100	25
(b) 2.356	500	25 (*Ans.* 0.1178)
(c) 8.579	1000	20
(d) 2.624	250	10
(e) 0.1039	500	50

CHAPTER 2

The Analytical Balance

A. History of Weighing

Nomadic societies had little need for tools of measurement. However, with the advent of agrarian groups, length measurements became necessary to denote field boundaries, and weight measurements became necessary for the exchange of grains[1] and other products of land cultivation.

Although some agrarian societies had scholars of science, especially mathematicians, there existed little exchange between these men of science and the common man of the field. As a consequence, the system of weights and measures which evolved only provided a degree of precision which satisfied the comparisons necessary for everyday commerce.

Allied to the development of modern science is the development of technology. The empirical consideration of physical reality must use reliable and suitable tools of investigation. A hoe would not be practical for the preparation of a 40-acre field, nor would a tractor be suitable for the preparation of a four square foot backyard garden. Tools of measurement are continually being developed and perfected to expand the body of scientific knowledge.

Our chief consideration as analytical chemists, at this stage, is the analytical balance. The principle of the analytical balance can be found in the ancient Greeks' laws of mechanics. Balances of rough construction features were used for hundreds of years for weight comparisons. However,

---

[1] Grain continues to be a unit in the English system of weights; it is derived from the weight of a grain of wheat; 437.5 grains = 1 oz. (Avoirdupois).

the birth of this instrument as a reliable and suitable tool for precise investigative work occurred in the eighteenth century when Antoine Lavoisier (1743–1794) recognized the necessity of precision in weighing for the clear understanding of the behavior of matter.

Lavoisier's experiments with the oxidation of metals required very precise weighings, a precision not possible on available balances. Lavoisier perfected the design of the analytical balance, and the accurate data he obtained led him to a successful theory of combustion. Lavoisier left his scientific descendants with a very important legacy because chemistry has been equated with the quantification of data since his time. As a result of his far-reaching influence on the direction of chemistry, Lavoisier is often referred to as the Father of Modern Chemistry.

B. Mass and Weight

The analytical balance is so designed that objects of unknown mass are compared with objects of known mass until any variation in the two masses is less than the limits of detection of the balance. The concept of mass is sometimes obscure to the beginning student. Mass is the quantity of matter contained by an object and is constant. It is a measure of the inertia of a body, that is, the tendency of a body to remain at rest or in motion in a straight line.

By contrast, weight is a force measurement and is the result of the earth exerting a gravitational force on an object. It varies with geographical location since the force of attraction depends on the distance of the object from the center of the earth.

These two terms, mass and weight, might be illustrated in the following manner. When the astronauts of the Apollo 11 space project embarked from Cape Kennedy, they possessed several pencils, one of which was not used. This pencil had a certain mass and weight on the earth, but from the time of blast-off its weight changed continually until it finally reached zero, i.e., weightlessness. On the moon's surface its weight was approximately one-sixth the value on earth due to the much less massive moon. During the same period, the mass, or quantity of matter, was constant.

IN SUMMARY. Mass is constant over the universe and depends only on the quantity of matter in the object; weight is not constant over the universe and depends on the quantity of matter and the relationship of the object to a much larger body that exerts a gravitational attraction on it.

Having drawn this careful distinction between mass and weight, we should add that they are frequently used interchangeably since the analytical balance is designed to compare an unknown weight with a known mass under conditions of equal gravitational attraction. The geographical location, and also the gravitational attraction, of the object and the known mass is essentially the same during the comparison, and, as a result, the weights are equal when the masses are equal. For convenience, the term *weight* is used as being identical with the term *mass*, and the distinction between the two is made only when necessary for clarity.

C. Theory of Weighing

There are two types of balance used in most analytical laboratories today. The equal-arm or double-pan balance used by Lavoisier was the most popular type of blance until recently but is rapidly being replaced in many laboratories by the substitution or single-pan balance. The beams of these two designs are shown in Fig. 2-1, and pictures of the complete balances are shown in Fig. 2-2 and 2-3.

The theory of the balance is the theory of a lever of the first class in which the fulcrum lies between the ends of the lever. The theory is explained using the equal-arm balance first since the explanation of the substitution balance is easier to understand if the principle of an equal-arm lever is first understood.

With the double-pan balance the object to be weighed is placed on the left-hand pan, represented by M_x in Fig. 2-4, and known weights, M_c, are placed on the right. M_x and M_c would in reality represent not only the object and weights, respectively, but also the stirrups and pans suspended from the terminal knife edges. For simplicity, we shall assume that the two terminal suspension systems are matched, allowing us to consider only the additional objects on the pans.

The forces acting on the two lever arms under load conditions are

$$F_x = M_x g \qquad F_c = M_c g \tag{1}$$

where g is the acceleration due to gravity.

AOB represents the equilibrium position of the balance. In this position the moments are equal, and using the law of moments we may write

$$F_x L_x = F_c L_c \tag{2}$$

Substituting eq. (1) into (2) yeields

$$M_x g L_x = M_c g L_c \tag{3}$$

Since $L_x = L_c$ by design and g is constant, this becomes

$$M_x = M_c$$

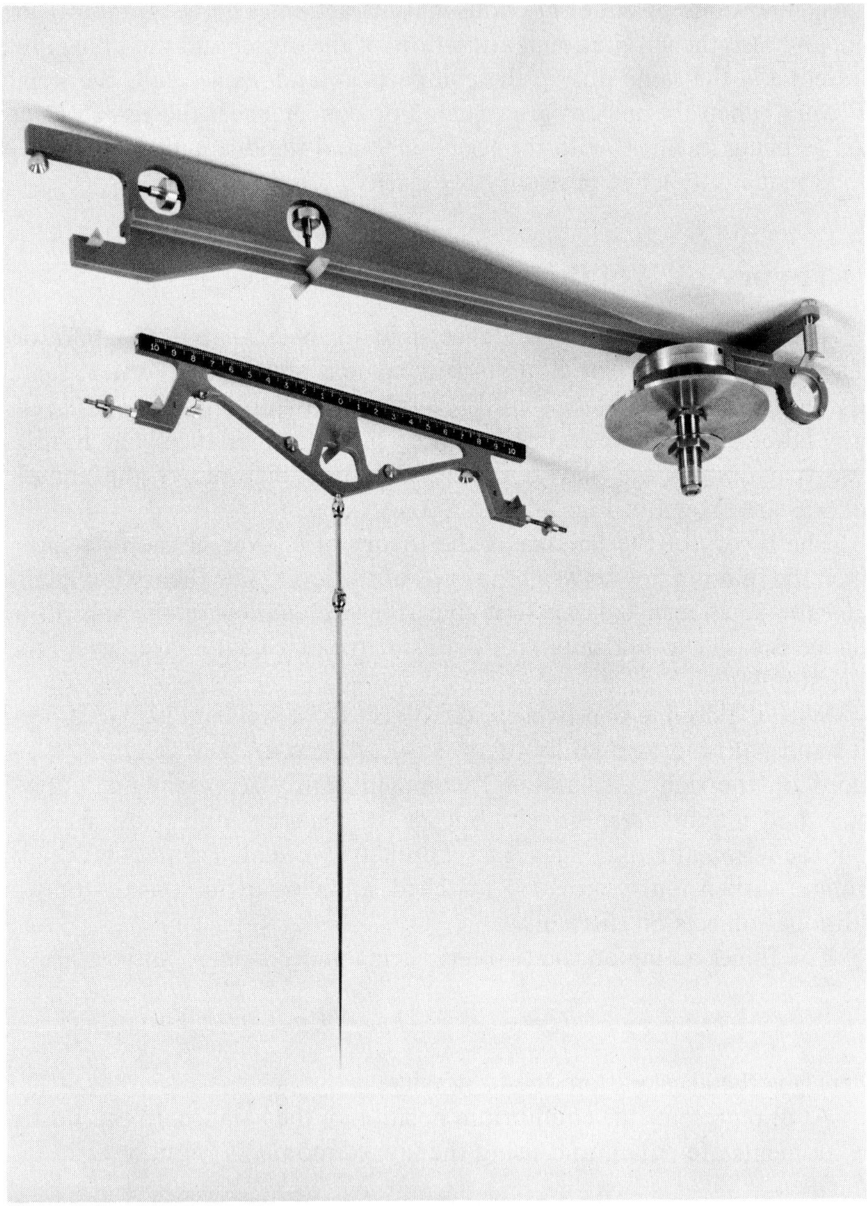

Figure 2.1. Beams for equal-arm balance and for substitution balance. (*Courtesy of Wm. Ainsworth, Inc., Denver, Col.*)

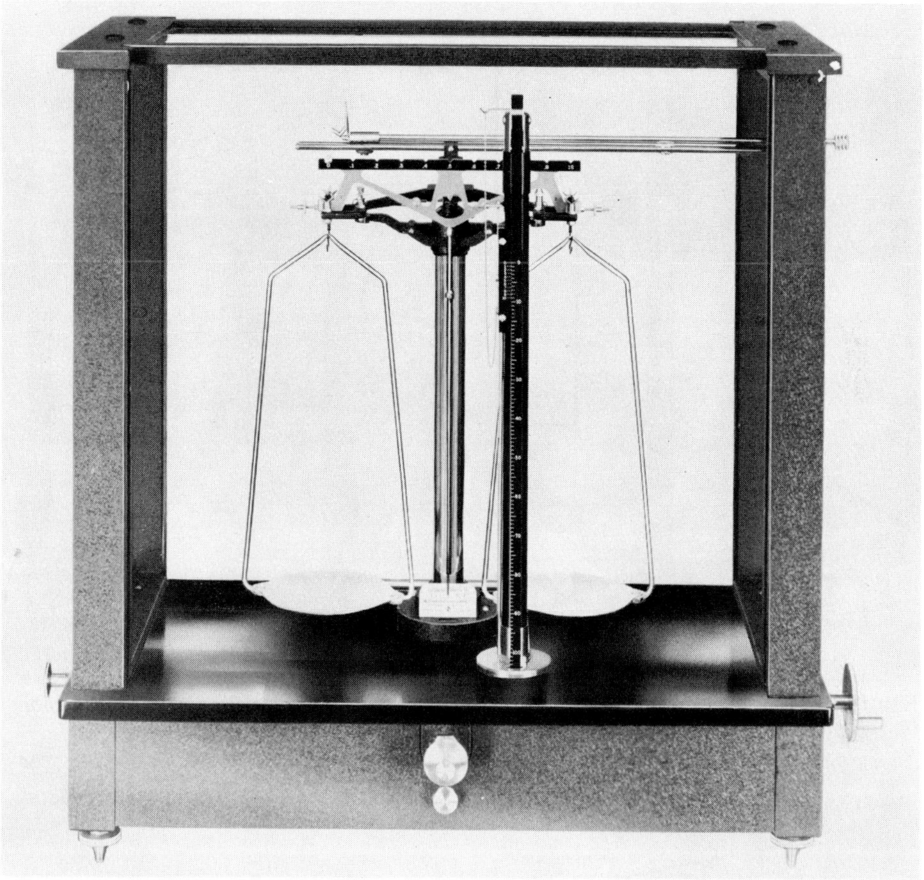

Figure 2.2. Notched beam, equal-arm balance. (*Courtesy of Wm. Ainsworth, Inc., Denver, Col.*)

When there is an excess mass on one of the pans, for example, m_0 in Fig. 2-4, the additional force $m_0 L_c$ causes a deflection of the beam in a clockwise manner about the fulcrum O. There is a restoring torque due to the mass of the beam M_B acting at its center of gravity, G_c', which causes the beam to stop and reverse its motion whenever the force of the torque exceeds the force causing the deflection. The beam would finally cease motion at the position P' so that the addition of the mass m_0 actually causes a displacement of the center of gravity through the arc $G_c G_c'$. The sensitivity of the balance is a measure of the amount of weight necessary to cause a stated displacement of the center of gravity of the beam system.

The substitution or single-pan balance is a modification of the equal-arm balance which uses a fixed weight on one arm of the balance and removable weights on the other, with the arm lengths and the total weight of the beam system adjusted so that the moments of the two arms are equal. A

Figure 2.3. Substitution (single-pan) balance. (*Courtesy of Mettler Instrument Corporation, Hightstown, N.J.*)

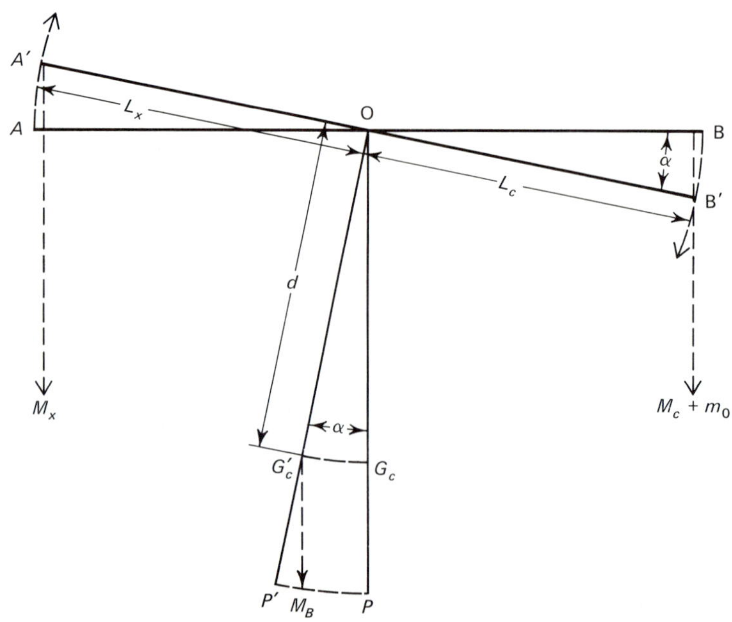

Figure 2.4. Displacement of beam by small overload.

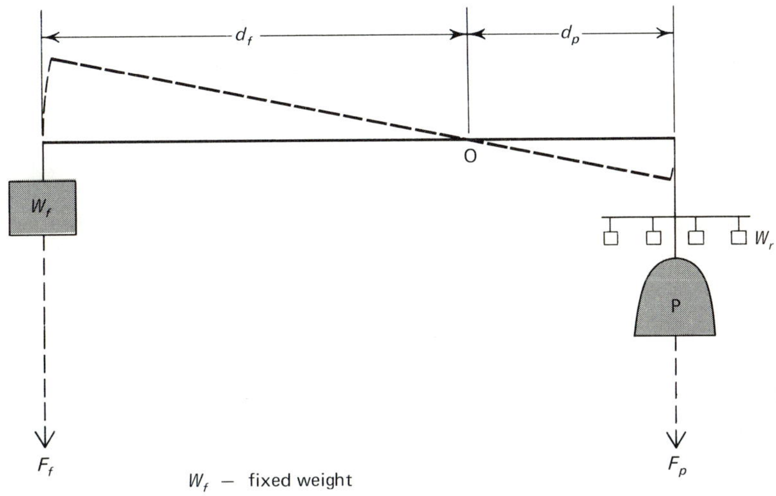

W_f — fixed weight
W_r — removable weights
d_f — length of fixed weight arm
d_p — length of pan arm
F_f — gravitational force on fixed weight arm
F_p — gravitational force on pan arm
O — fulcrum
P — pan
W_{fa} — total weight of fixed weight arm
W_{pa} — total weight of pan arm, pan, and W_r

Figure 2.5. Substitution balance beam principle.

simplified diagram is given in Fig. 2-5 and a more complete picture is shown in Fig. 2-6.

For the design in Fig. 2-5, eq. (2) above becomes

$$F_f d_f = F_p d_p \tag{4}$$

for which

$$F_p = W_{pa} g \tag{5}$$

and

$$W_{pa} = W_a + W_p + W_r \tag{6}$$

where W_a is the weight of the beam arm, W_p is the weight of the pan and suspension system, and W_r is the sum of the removable weights. Since $F_f d_f$ is constant, $F_p d_p$ must always be the same when the balance is in equilibrium.

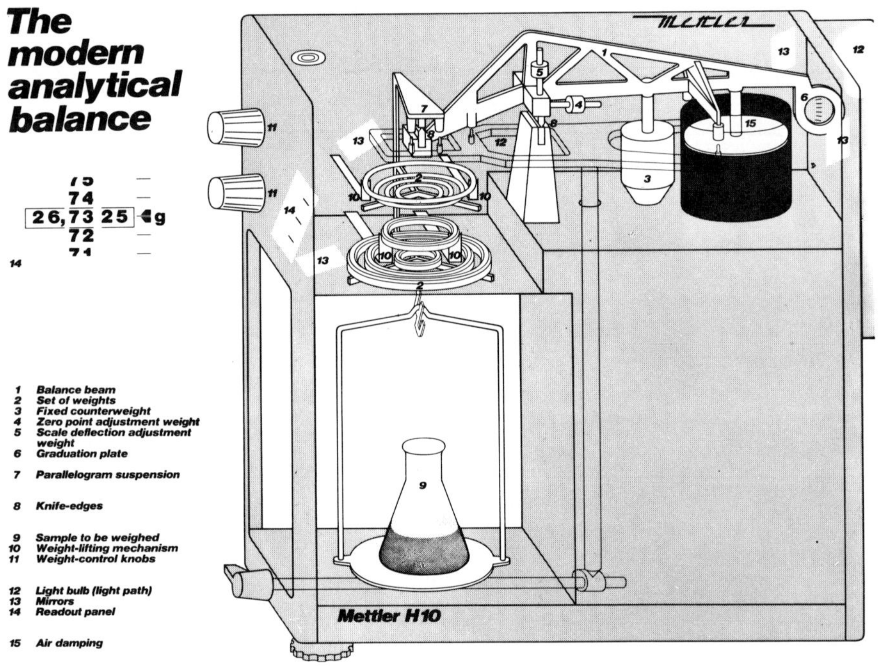

Figure 2.6. Side view of substitution balance. (*Courtesy of Mettler Instrument Corporation, Hightstown, N.J.*)

The only variable in $F_p d_p$ is W_r, which can be changed by the removal of weights. As a consequence, weights are removed in the substitution balance rather than added as in the equal-arm balance.

If an object which has the exact weight of one of the removable weights is placed on the pan and the weight is removed, the beam will remain in equilibrium since there is no effective change in the weight or moment of the beam arm. However, it is seldom that the object and a removable weight (or weights) are exactly equal since the smallest removable weight is 1 g on most laboratory balances. In these cases, the beam will be displaced from its equilibrium position due to the difference between the weight of the object and the sum of the weights removed and will assume the position of the dashed line in Fig. 2-5. An optical scale projection is used to measure this displacement.

The sensitivity of the equal-arm balance varies with the amount of load on the pans. Such variation is probably due to frictional effects which increase as the weight on the pans increases. The substitution balance has as its major feature a constant-load principle which results in the sensitivity remaining constant.

A comparison of some of the properties of equal-arm and substitution balances is given in Table 2-1.

TABLE 2.1. Comparison of Equal-Arm and Substitution Balances

Property	Equal-arm	Substitution
Sensititvity varies with load on pans	Yes	No
Buoyant effect errors can be offset by use of tare of equal volume	Yes	No
Cause of malfunction can usually be seen	Yes	No
Sensitive to addition of 0.05 mg to pan	Yes	Yes
Accuracy approximately 0.1 mg	Yes	Yes
Rapidity of weighing an object	Slower	Faster
Rapidity of weighing a finely divided solid	Equal	Equal
Operates at constant load	No	Yes
Must use separate set of weights	Yes	No
Can be calibrated using standard weights	Yes	Yes
Precision (repeatability of measurement)	Equal	Equal
Errors due to electrification	Yes	Yes

D. Substitution (Single-Pan) Balance

The substitution or single-pan balance is used in the majority of analytical laboratories. The weights are built into this type of balance, and therefore separate weights which may be dropped or may lose precision in other ways are eliminated.

The essential characteristic of the substitution balance is that it is a constant-load balance. Like the equal-arm balance, the substitution balance is a lever of the first class, with the weights and pan on one side of the fulcrum and the counterweight on the other side, as shown in Figs. 2-3, 2-5, and 2-6. Note that the weighing pan and the set of weights share a common suspension from the beam. When an object is placed on the pan, weights are removed which correspond to the weight of the object. The beam is therefore always under a constant load, and therefore the balance operates at constant sensititvity. The beam is damped, and optical scale projection is used to determine that fraction of the mass less than the smallest weights represented by the dials. The degree of deflection of the beam from its rest point position is translated on the optical scale to the mass required to restore the beam to its rest point position. By reading the weights dialed and the optical scale projection, one has a direct measure of the mass of the object.

E. Balance Properties

The balance just described must, of course, be accurate and reproducible. That is, if repeated weighings of the same object are made, the balance must reproduce the value, and the value must accurately represent the true mass of the object. For the balance to be accurate, several design features must be carefully controlled.

As mentioned earlier, on the double-pan balance the balance arms must be equal. On this balance the three knife edges must lie in the same plane and must parallel one another. This is also true for the two knife edges of the single-pan balance. The balance should be rapid or come to equilibrium in a reasonable time.

The balance must be stable. The unsupported beam oscillates about the pivotal central knife edge and returns to its original position when stable. The condition of stability is achieved when the center of gravity of the oscillating system is below the plane of support.

A balance must be sensitive; that is, it must show a reasonable response to very small weight differences. The sensitivity is measured by the ratio of the displacement to the small mass required to cause that displacement and is affected by the following:

1. The sensitivity is directly proportional to the balance arm length. Considering this factor alone, long balance arms improve sensitivity.

2. The sensitivity is inversely proportional to the mass of the oscillating system (beam, pans, and load). As this mass increases, the sensitivity diminishes. As a result, single-pan balances operate at constant but minimum sensitivity. Lightweight materials are therefore used to construct balances. Metals such as aluminum or magnesium-aluminum alloy are usually used.

3. The sensitivity is inversely porportional to the distance between the support plane and the center of gravity of the oscillating system. As this distance increases, the sensitivity decreases.

4. Finally, friction of moving parts must be held to a minimum. As friction increases, sensitivity decreases, as does the response of the balance. All regions of the balance at which friction may occur must be kept clean and free from damage.

A review of the above balance features should indicate to the student that some design features are incompatible with other design conditions. For example, the center of gravity of the oscillating system must be below the support plane in order to achieve stability. However, the greater the distance below, the less the sensitivity. Sensitivity is improved with long balance arms, but an increase of mass of the beam reduces the sensitivity.

Compromises have been made in balance construction so that a reasonable sensitivity is achieved. Beams of about 10 in. are usually encountered, and these have a truss design to provide the needed rigidity without serious loss of sensitivity due to a high mass.

F. Weighing Errors

There are quite a few errors in weighing which must be avoided or corrected for in order to obtain correct weights on an analytical balance. Such errors are grouped under four main classes: instrumental, static electrical, buoyant effect, and atmospheric effect.

1. Instrumental errors include any errors due to construction or operation of the balance or the weights used. Errors due to construction could be caused by unequal length of balance arms, chipped knife edges, corrosion on beam, distortion of beam due to excessive load, dust on beam surfaces, and contact between the vanes on the beam and the magnetic dampers. Most errors due to the weights are related to the difference between the actual weight and the labeled weight. For instance, there are usually two ten-gram weights in each weight box, and it is unusual for them both to weigh within 0.05 mg of 10 g. This can be corrected for by calibration of the weights. Such calibrations should be checked periodically since handling or dropping the weights will often cause a change in the actual weight.

2. Static electrical effects are produced on glass when it is rubbed with a cloth or some grades of paper. When such a charged piece of glass is placed on a balance pan, part of the charge is slowly dissipated into the atmosphere, but part of it is conducted by the metal portions of the balance to create two or more zones of like charge on the balance. Since like charges repel each other, force will be exerted on the pans, causing an error in the weight observed. To offset such an error, a suitable length of time must elapse between the cleaning of the glass and the weighing operation in order for the charge to be completely dissipated. A small amount of a radioactive substance placed in the balance case will dissipate the charge more rapidly due to the ionization of the air by the radioactive substance. This error is related to humidity and is more pronounced under conditions of low humidity

3. Buoyant effect errors are due to the force which buoys up any object placed in a fluid. An object will be buoyed up by the difference in the density of the object and the density of the fluid which it displaces—a cork floats on water since its density is less than that of water, while a stone sinks in water since its density is more than that of water. However, the weight of the stone in the water will be less than its weight in air, since air is less dense than water.

Such errors occur when the volume and density of the object being weighed differ from the volume and density of the weights used. These errors can be offset in some cases by using a counterpoise, or the correct weight can be determined by calculating the weight of the substance in vacuo. The weight in vacuo may be calculated by use of the equation

$$W_{vac} = W_{air} + W_{air}(0.0012/d_o - 0.0012/d_w) \tag{7}$$

where

$$W_{vac} = \text{the weight in a vacuum}$$
$$W_{air} = \text{the weight in air}$$
$$0.0012 = \text{the density of air}$$
$$d_o = \text{the density of the object}$$
$$d_w = \text{the density of the weights.}$$

An example will show how a weight in air can be corrected to the corresponding weight in vacuo.

Example. What is the weight in vacuo of a sample of aluminum (density = 2.7 g/ml) which weighs 24.3567 g in air against brass weights (density = 8.4 g/ml)?

$$W_{vac} = 24.3567 + 24.3567\left(\frac{0.0012}{2.7} - \frac{0.0012}{8.4}\right)$$
$$= 24.3567 + 24.3567(0.000444 - 0.000143)$$
$$= 24.3567 + 24.3567(0.000301)$$
$$= 24.3567 + 0.0073 = 24.3640 \text{ g}$$

It should be noted that the greater the difference in the densities of weights and object, the greater will be the correction. If the density of the object is less than the density of the weights, as in the above example, the buoyancy correction will be positive. If the density of the object is more than the density of the weights, the buoyancy correction will be negative.

If the object being weighed is a closed vessel, the air inside must be at the same pressure as atmospheric pressure when weighed. At times it is necessary to weigh a closed container which contains gases other than air. An error will be involved unless correction is made for the difference in the weight of the gas and an equal volume of air. Another buoyant effect error is caused by weighing objects which are not at the temperature of the balance. A hot object will appear to weigh less than its true weight, while a cold object will appear to weigh more.

4. Atmospheric effects. Some materials will gain weight when exposed to the atmosphere, while others will lose weight. This may be due to pickup or loss of water, oxygen, carbon dioxide, etc.

G. Problems

1. Calculate the sensitivity of the following in divisions per milligram, given the pointer deflections with the masses required to bring about those deflections.

(a) 5.6, 2.9 mg
(b) 2.5, 1.3 mg
(c) 3.7, 2.1 mg
(d) 4.0, 1.5 mg
(e) 1.8, 1.2 mg
(f) 4.5, 3.2 mg

Ans. (b) 1.9 div/mg

2. If brass weights are used to weigh the following substances in air, calculate their weight in vacuo.
(a) 5.2092 g of aluminum
(b) 14.6270 g of gold
(c) 22.0000 g of iron
(d) 37.2945 g of brass
(e) 48.7234 g of mercury
(f) 27.5216 g of platinum

Ans. (b) 14.6258 g

3. Calculate the vacuum weights of each of the substances given in problem 2 if they were taken in air against platinum weights.

Ans. (b) 14.6271 g

4. Calculate the densities of each of the following solids if the first number given is their weight in air against brass weights and the second number is their weight in vacuo.
(a) 20.0000 g, 20.0030 g
(b) 18.5000 g, 18.5010 g
(c) 9.0000 g, 8.9995 g
(d) 45.6250 g, 45.6225 g

Ans. (b) 6.1 g/ml

CHAPTER 3

Methods of Expressing Concentrations of Solutions

A. General

All the methods for expressing the concentration of a solution state in some way the amount of solute present in a given amount of the solution or the solvent. The amount may be stated as either weight or volume. In most analytical laboratories, the concentrations are expressed upon the basis of the weight of solute present in a stated volume of the solution since accurate volume measurements are performed readily. It is important for the analyst to understand the various ways in which concentration is expressed in order that he may convert from one method to another when necessary.

Concentrations may be expressed in qualitative form or in more exact terms. Thus a solution may be referred to as a concentrated solution, meaning that it has relatively large amounts of solute per unit volume, or as a dilute solution, meaning that it contains relatively smaller amounts of solute per unit of volume. These terms are used mainly with acids and bases to designate the usual article of commerce (concentrated) or dilutions of it. However, the actual concentration of concentrated acids varies widely, as is shown in Table 3-1. Other terms which have no quantitative meaning except with specifically stated systems are saturated, unsaturated, supersaturated, strong, weak, pure, contaminated, etc. In each of these, the value must be defined or stated for the system being considered and will vary from system to system.

TABLE 3.1 Composition of Concentrated Reagents

	HCl	HNO_3	H_2SO_4	$HC_2H_3O_2$	$HClO_4$	H_3PO_4	NH_4OH	NaOH
Molecular weight	36.46	63.01	98.08	60.06	100.46	98.00	35.05	40.00
Approx. specific gravity	1.19	1.42	1.84	1.05	1.67	1.69	0.90	1.53
Approx. % (w/w)	37.0	69.5	96.0	99.6	70.0	85.0	57.6	50.0
Approx. normality	12.2	15.7	36.0	17.4	11.6	44.0	14.8	19.1
Approx. milliliters per liter to prepare 1 *M* solution	83.0	64.0	56.0	57.5	86.0	69.0	67.5	52.4

A solution for which the concentration is known with the degree of accuracy required for a quantitative determination is known as a standard solution. In most quantitative work this means that the analytical concentration must be within 0.2% (2 ppt) of the stated value.

It is often necessary to distinguish between the analytical and the equilibrium concentration of a solution. The analytical concentration is that concentration which can be calculated from the volume of the solution and the weight or volume of the solute dissolved in it. It can also be considered to be a measure of the amount of solute which can be recovered by evaporation of the solvent. The equilibrium (sometimes referred to as actual) concentration is the concentration of any particular form or forms of the solute present in the solution under the existing conditions. As an example, consider a solution of succinic acid ($H_2C_4H_4O_4$) prepared by dissolving 118 g (one molecular weight) of the acid in sufficient water to make one liter of solution. The analytical concentration of such a solution is 1 *M* (one molar) (see Section B.4). After being dissolved, however, the acid ionizes stepwise to form hydrogen ions, bisuccinate ions ($HC_4H_4O_4^-$), and succinate ions ($C_4H_4O_4^{2-}$) so that the solution contains these ions as well as the non-ionized molecules of the acid. As a consequence the actual concentration of molecular succinic acid is not one molar but some smaller value. Further information (such as the ionization constants of the acid and the ionic strength) is needed to determine the exact concentration of succinic acid molecules. This exact concentration, if it is actually determined by calculation or other methods, is considered to be the equilibrium concentration of succinic acid molecules.

The methods described below are used to express analytical concentration and include essentially all the methods in use for qualitative and quantitative expression of analytical concentration. The methods used to calculate equilibrium concentrations from analytical concentrations are described in later sections.

B. Methods of Expressing Analytical Concentrations

1. Weight-Volume (w/v)

The concentration may be expressed as the weight of solute in a given volume of the solution such as grams per liter (g/liter) or milligrams per milliliter (mg/ml). The usual or clinical laboratory method of expressing the percentage concentration of a solution is in reality a weight-volume method since it represents the weight in grams of the solute present in 100 ml of the solution. Also, the concentrations of the components of blood are often expressed as mg % (mg per 100 ml of blood). Neither of these is a true percentage method but is based upon the fact that the density of many dilute solutions is approximately one. The symbol (w/v) should be shown after the % sign to ensure no misunderstanding of the concentration. Another popular weight-volume method is the expression of the concentration as parts per million (ppm) or as parts per billion (ppb). One part per million equals one milligram of the solute per liter of the solution and 1 ppb equals one microgram of solute per liter of the solution.

Problem 1. How many grams of sodium chloride are contained in (or are necessary to prepare) 500 ml of normal saline which is 0.85% (w/v)?

$$\text{g} = \% \left(\frac{\text{g}}{100\ \text{ml}}\right) \times \text{volume (ml)}$$

$$\text{g} = \frac{0.85\ \text{g}}{100\ \text{ml}} \times 500\ \text{ml} = 4.25\ \text{g} \tag{1}$$

Problem 2. (a) One liter of a 5.0% (w/v) solution of glucose contains how many grams of glucose?

$$\text{g} = \frac{5.0\ \text{g}}{100\ \text{ml}} \times 1000\ \text{ml} = 50\ \text{g}$$

(b) One gallon (3800 ml) of a 10 ppm solution of ethyl alcohol contains how many milligrams of ethyl alcohol?

$$\text{mg} = \text{ppm (mg/1000 ml)} \times \text{ml}$$

$$\text{mg} = \frac{10\ \text{mg}}{1000\ \text{ml}} \times 3800\ \text{ml} = 38\ \text{mg} \tag{1a}$$

2. Percentage Methods

The concentration of solutions may be expressed in terms of the percentage by weight or the percentage by volume in the case of a liquid solute.

The percentage by weight (w/w) is the weight in grams of solute present in 100 g of the solution, while the percentage by volume is the number of milliliters of solute present in 100 milliliters of the solution. Since percentage is parts per hundred, the weight percentage of a solution is expressed by

$$\%\,(\text{w/w}) = \frac{\text{weight of solute (g)}}{\text{weight of solution (g)}} \times 100 \tag{2}$$

and the volume percentage is

$$\%\,(\text{v/v}) = \frac{\text{volume of solute (ml)}}{\text{volume of solution (ml)}} \times 100 \tag{3}$$

Problem 3. Express the percentage by weight of a solution which weighs 200 g and which contains 25.0 g of sodium sulfate.

$$\%\,(\text{w/w}) = \frac{25.0\text{ g}}{200\text{ g}} \times 100\% = 12.5\%\,(\text{w/w})$$

Problem 4. Calculate the percentage by volume of a solution prepared by adding 50.0 ml of methanol to 200 ml of water. Assume the volumes are additive.

$$\text{Volume of solution} = 50.0 + 200 = 250\text{ ml}$$

$$\%(\text{v/v}) = \frac{50.0\text{ ml}}{250\text{ ml}} \times 100\% = 20.0\%\,(\text{v/v})$$

Problem 5. Calculate the percentage by weight of a solution prepared by dissolving 5.00 g of silver nitrate in 100 ml of water. Assume that the density of water is 1.00. (Note that this problem cannot be solved unless the density of the solvent is known.)

$$\text{wt of solvent} = 100\text{ ml} \times 1.00\text{ g/ml} = 100\text{ g}$$

$$\text{wt of solution} = 100\text{ g of solvent} + 5.00\text{ g of AgNO}_3 = 105\text{ g}$$

$$\%\,(\text{w/w}) = \frac{5.00\text{ g}}{105\text{ g}} \times 100\% = 4.76\,\%\,(\text{w/w})$$

3. Dilution Ratio

The concentration of dilute acids and bases is often stated as the dilution ratio of the concentrated acid or base with water. The ratio is expressed as two numbers separated by a colon, with the first number representing the volume of concentrated acid or base and the second representing the number of volumes of water to which the acid or base is added. The ratio is usually given in parentheses before or after the name of the solute. Examples are HCl (1:1), H_2SO_4 (1:4), and (2-3) H_3PO_4. To prepare these particular solutions, one volume of concentrated hydrochloric acid is added to an equal volume of water, one volume of concentrated sulfuric acid is added to four times its volume of water, and two volumes of concentrated phosphoric acid is added to three volumes of water. The composition of the concentrated reagents used to prepare solutions expressed as dilution ratio are shown in Table 3-1. Concentrations expressed in this way may be converted easily to volume percent if the volumes are assumed to be additive.

Problem 6. What volumes of concentrated acetic acid and of water would be used to prepare 300 ml of 1:5 solution? To solve such a problem, first obtain the total of the volumes shown in the ratio. Then divide the volume of solution to be prepared by this number to get the value of one volume.

$$1 + 5 = 6 \text{ total volumes in ratio}$$
$$300/6 = 50 \text{ ml} = \text{one volume} = \text{volume of acid used}$$
$$50 \times 5 = 250 \text{ ml} = \text{volume of water used}$$

To prepare the solution, add 50 ml of concentrated acetic acid to 250 ml of water.

4. Molar (*M*) and Formal (*F*) Concentrations

The molar concentration states the number of moles (gram molecular weight, MW) of the solute present in one liter of the solution, while the formal concentration states the number of gram formula weights present in one liter of solution. From these definitions it might appear that the two methods are identical, and for many solutions they are. The use of formal concentration in place of molar concentration was suggested originally because there is no such thing as a "molecule" of an ionic substance, since ionic substances exist as individual ions even in the pure or solid state. Also, in the case of organic compounds which do exist as molecules, association or dissociation often occurs after the molecules are dissolved. These two methods could be used to distinguish between the analytical and the equilibrium concentration, with formal representing the analytical and molar being used for the equilibrium concentration. This particular distinction, however, has never come into general use, and as a consequence some authors emphasize molar, some emphasize formal, and some use both. For our purposes in this text, we shall assume that the two methods are identical and shall use molar rather than formal concentrations. One of the main reasons for the preference of molar over formal is the general use of the term *mole* in all fields of chemistry to represent one Avogadro's number of the entity in question whether it be an atom, a molecule, an ion, an electron, or a compound formed by association of two or more molecules.

The relationship between the weight of solute and the number of moles of solute in a given volume of a solution of a given molar concentration may be developed as follows:

$$\text{moles of solute dissolved} = \text{moles of solute in solution}$$

$$\text{moles of solute dissolved} = \frac{\text{g of solute}}{\text{MW of solute (g/mole)}}$$

$$\text{moles of solute in solution} = M\ \text{(moles/liter)} \times \text{liters}$$

so that

$$\begin{aligned} M\ \text{(moles/liter)} \times \text{liters} &= \frac{\text{g of solute}}{\text{MW of solute (g/mole)}} \\ &= \text{moles of solute in solution} \end{aligned} \tag{4}$$

Equation (4) can be rearranged to give

$$\text{MW}\left(\frac{\text{g}}{\text{mole}}\right) \times M\left(\frac{\text{moles}}{\text{liter}}\right) \times \text{liters} = \text{g of solute} \tag{5}$$

which indicates how the units cancel.

In the analytical laboratory, measurements are made in milliliters more often than in liters. Substitution of milliliters per 1000 for liters in the above gives

$$\text{MW}\left(\frac{\text{g}}{\text{mole}}\right) \times M\left(\frac{\text{moles}}{\text{liter}}\right) \times \frac{\text{ml}}{1000\ \text{(ml/liter)}} = \text{g of solute} \tag{6}$$

Similarly, when needed, milligrams per 1000 can be substituted for grams, millimoles per 1000 for moles, and millimolecular weights per 1000 for molecular weight. One millimole (mmole) equals 0.001 mole and one mMW (millimolecular weight) equals 0.001 MW or molecular weights per 1000. Equations (5) and (6) can be rearranged to calculate any one of the terms as long as the other three are known. The two most common forms of the calculations used in the analytical laboratory are the calculation of the molarity of a solution prepared by dissolving a known weight in grams of solute in a given volume of solution and the calculation of the weight in milligrams of solute present in a given volume of a solution of known molar concentration. The equations for these are

$$M = \frac{\text{g of solute}}{\text{MW} \times \text{liters}} = \frac{\text{g of solute}}{\text{MW} \times \text{ml}/1000} = \frac{\text{g of solute} \times 1000}{\text{MW} \times \text{ml}} = \frac{\text{mg of solute}}{\text{MW} \times \text{ml}} \tag{7}$$

and

$$\text{MW} \times M \times \text{ml} = \text{mg of solute} \tag{8}$$

For formal concentrations, substitute formula weight (FW) for molecular weight and formal (F) for molar (M) in the above expressions.

Problem 7. A solution of sulfuric acid (MW, 98.0) contains 4.90 g of H_2SO_4 in 400 ml. Calculate the molar concentration. Using eq. (7),

$$M = \frac{\text{g of solute} \times 1000}{\text{MW} \times \text{ml}} = \frac{4.90 \text{ g} \times 1000 \text{ ml/liter}}{98.0 \text{ g/mole} \times 400 \text{ ml}} = 0.125\ M$$

Problem 8. How many grams of silver nitrate (MW, 169.9) are necessary to prepare 500 ml of 0.1250 M solution?

$$\text{g} = \text{MW} \times M \times \text{liters} = 169.9 \text{ g/mole} \times 0.1250 \text{ mole/liter} \times \frac{500}{1000} \text{ liter} = 10.62 \text{ g}$$

Problem 9. What is the molar concentration of a 0.85% (w/v) solution of sodium chloride (MW, 58.4)? Since % (w/v) represents grams per 100 ml, 1 liter contains 0.85 g × 1000 ml/100 ml or 8.5 g.

$$M = \frac{\text{g/liter}}{\text{MW}} = \frac{8.5 \text{ g}}{58.4 \text{ g}} = 0.146\ M$$

Problem 10. How many millimoles are present in 150 ml of a 0.025 M solution?

$$\text{mmoles} = \text{ml} \times M = 150 \times 0.025 = 3.75 \text{ mmoles}$$

5. Normal (N) Concentration

Since substances always react in the ratio of the equivalent weights (see Section C), it is convenient in analytical work to express the concentration as normal (N) concentration, which states the number of gram equivalent weights (equivalents, equiv.) of solute per liter of solution. The relationship between the weight of solute and the number of equivalents in a given volume of a solution of stated normality is given by

$$N\,(\text{equiv/liter}) \times \text{liter} = \text{number of equivalents} = \frac{\text{g of solute}}{\text{equiv wt of solute (g/equiv)}} \qquad (9)$$

Since volume measurements in milliliters are more common than those in liters, this is easily converted to

$$N \times \text{ml} = \text{number of meq} = \frac{\text{g} \times 1000}{\text{equiv wt}} = \frac{\text{mg}}{\text{equiv wt}} = \frac{\text{g}}{\text{meq wt}} \qquad (10)$$

One meq wt (milliequivalent weight) is usually referred to as a milliequivalent (meq). Since substances react in the ratio of their equivalents (or milliequivalents), the number of equivalents of each reactant and product must be identical for any quantitative reaction. This statement is the basis of all quantitative work and explains the reason most analytical chemists prefer to use normal rather than other methods of expressing concentration. Methods of determining the equivalent weights of substances involved in reactions are given in Section C of this chapter.

Problem 11. How many grams of sodium sulfate (MW, 142.0; equiv wt, 71.00) are necessary to prepare 200.0 ml of a 0.5000 *N* solution for precipitation reactions?

$$\text{g} = \text{equiv wt} \times N \times \text{ml}/1000 = 71.00\ \text{g/equiv} \times 0.5000\ \text{equiv/liter} \times \frac{200}{1000}\ \text{liters} = 7.100\ \text{g}$$

Problem 12. Calculate the normality for neutralization reactions of a solution of sodium hydroxide (MW, 40.00; equiv wt, 40.00) prepared by dissolving 100.0 g of sodium hydroxide in sufficient water to make 1000 ml of solution.

$$N = \frac{\text{g} \times 1000}{\text{equiv wt} \times \text{ml}} = \frac{100.0 \times 1000}{40.00 \times 1000} = 2.500\ N$$

6. Titer Methods

The titer of a solution is defined as the weight of a pure substance which corresponds to one milliliter of the solution. "Corresponds to" in this case is taken to mean (a) will react exactly with, (b) is chemically equivalent to, (c) is contained in, (d) may be obtained from, or (e) may be substituted for. This method is used primarily to state the weight of some substance *other than the solute* with which one milliliter of the solution will react exactly. The important factor here is that the concentration is expressed in terms of some substance other than the solute and that the relationship is a weight-volume relation so that volumes may be converted directly to weights by a simple multiplication. As an example, the concentration of a silver nitrate solution which is to be used to determine the amount of sodium chloride by a titration could be expressed as the number of milligrams of sodium chloride with which one milliliter of the solution will react and is chemically equivalent to. Thus one milliliter of a one normal solution of silver nitrate will react exactly with 58.44 mg of sodium chloride, so that the concentration could be expressed as being equivalent to 58.44 mg of sodium chloride per milliliter. One symbolization for titer is to use a capital *T* preceded by the formula of the substance specified (which we shall call the reactant) and followed by the formula of the solute written as a subscript. For the above silver nitrate solution this would be

$$\mathrm{NaCl}\ T_{\mathrm{AgNO_3}} = 58.44 \text{ mg}$$

and would be read as "the sodium chloride titer of the silver nitrate solution is 58.44 mg." Another method of expressing the titer is to give the name or formula of the solute followed by some statement indicating the number of milligrams of the reactant with which 1 ml of the solution will react. For the silver nitrate solution above, these could be shown as

Silver nitrate, 1 ml = 58.44 mg NaCl

or

Silver nitrate, 58.44 mg NaCl per ml

The calculation of the weight of solute present in a solution for which the concentration is expressed in the titer system involves the determination of the weight relationship of the reaction between the solute and the reactant. Since substances always react in the ratio of their equivalent weights, the normal concentration of the solution will be identical to the normal concentration of a solution of the reactant containing the titer weight in one ml. As a consequence, the first step is to calculate the normal concentration of a solution of the reactant containing the titer weight in 1 ml and then to calculate the weight of solute in a given volume of a solution of this normality. For example,

$$N_{\text{react}} = \frac{\text{titer (mg/ml)}}{\text{equiv wt (react)}} \tag{11}$$

$$g_{\text{solute}} = \text{equiv wt (solute)} \times N\text{ (reactant)} \times \text{liters (solution)} \tag{12}$$

These can be combined to give

$$g_{solute} = \frac{\text{equiv wt (solute)} \times \text{titer (mg/ml)} \times \text{ml (solution)}}{\text{equiv wt (reactant)} \times 1000} \tag{13}$$

which can be used for calculations involved in the preparation of solutions which utilize the titer system for concentration.

Problem 13. How many grams of silver nitrate (MW, 169.9) are necessary to prepare 500.0 ml of a solution which has a chloride (at wt, 35.45) titer of 0.5000 mg?

$$g(AgNO_3) = \frac{169.9 \times 0.5000 \times 500.0}{35.45 \times 1000} = 1.198 \text{ g}$$

Problem 14. Calculate the sodium carbonate (equiv wt, 53.00) titer of 0.1037 *N* hydrochloric acid.

$$\begin{aligned} \text{titer} &= N \text{ (solution)} \times \text{equiv wt (reactant)} \\ &= 0.1037 \times 53.00 = 5.496 \text{ mg/ml} \end{aligned} \tag{14}$$

Problem 15. Calculate the SO_3 (MW, 80.06) titer of a $BaCl_2$ solution which was prepared by dissolving 24.43 g of $BaCl_2 \cdot 2H_2O$ (MW, 244.3) in sufficient water to prepare 1000 ml of solution.

$$\begin{aligned} \text{titer} &= \frac{\text{g (solute)} \times \text{equiv wt (reactant)} \times 1000}{\text{equiv wt (solute)} \times \text{ml (solution)}} \\ &= \frac{24.43 \times 40.03 \times 1000}{122.15 \times 1000} \\ &= 8.006 \text{ mg/ml} \end{aligned} \tag{15}$$

Problem 16. How many milligrams of Fe_2O_3 are present in an iron ore sample if the sample required 35.57 ml of a $K_2Cr_2O_7$ solution which has a Fe_2O_3 titer of 5.000?

$$\text{mg (reactant)} = \text{titer} \times \text{ml (solution)}$$

$$\text{mg } Fe_2O_3 = \text{titer} \times \text{ml} = 5.000 \times 35.37 = 176.9 \text{ mg} \tag{16}$$

The use of the titer concept is especially favored in quality-control laboratories where the same reagent is used for the same procedure many times a day. Reading volumes and reporting weights directly in this manner short circuits the calculations otherwise needed, saves time, and avoids errors.

C. Equivalent Weights

The gram equivalent weight of a substance is a measure of the ratio by weight with which it reacts with other substances. Substances always react in the ratio of their equivalent weights; that is, one equivalent weight of a substance will react exactly with one equivalent weight of another substance to produce one equivalent weight of each product of the reaction. Another way of stating this is that the number of equivalents of one substance in a quantitative reaction is numerically equal to the number of equivalents of each other substance involved in the reaction. Since some substances react with other substances in more than one ratio, the equivalent weight will depend upon the actual reaction involved. It is very important for the student in analytical chemistry to understand the calculation and use of equivalent weights since many of the calculations encountered will involve these values. A simple method of indicating the equivalent weight without calculating the actual numerical value is to show the symbol or formula of the substance divided by the number of equivalents per unit.

The two main classes of chemical reactions are those in which there is an electron transfer (oxidation-reduction, redox) and those in which there is no electron transfer. The majority of the reactions in which there is no electron transfer may be classified as being neutralization (acid-base) or precipitation.

1. Equivalent Weights for Reactions in Which There Is No Electron Transfer

a. NEUTRALIZATION

(1) *Acids.* The equivalent weight of an acid in a neutralization reaction is the weight of the acid which contains 1 g-atom of replaceable hydrogen. For a reaction in which all the replaceable hydrogens of the acid take part in the reaction, the equivalent weight equals the molecular weight divided by the number of replaceable hydrogens in one molecule. If only a portion of the replaceable hydrogens is involved, the equivalent weight equals the molecular weight divided by the number actually involved. Examples are given below.

Complete reaction of available hydrogens:

$$\frac{HCl}{1} \quad \frac{HNO_3}{1} \quad \frac{HC_2H_3O_2}{1} \quad \frac{H_2SO_4}{2} \quad \frac{H_2CO_3}{2} \quad \frac{H_3PO_4}{3}$$

Reaction of only a portion of available hydrogens:

$$H_3PO_4 + 2NaOH \longrightarrow Na_2HPO_4 + 2HOH \qquad \frac{H_3PO_4}{2} \tag{17}$$

$$H_2SO_4 + NaOH \longrightarrow NaHSO_4 + HOH \qquad \frac{H_2SO_4}{1} \tag{18}$$

(2) *Bases.* The equivalent weight is that weight which contains one gram-ion of hydroxyl. For reactions in which all hydroxyls are not involved, divide the molecular weight by the involved number as for acids. Examples:

$$\frac{NaOH}{1} \quad \frac{Ba(OH)_2}{2} \quad \frac{Al(OH)_3}{3}$$

$$Ba(OH)_2 + HCl \longrightarrow Ba(OH)Cl + HOH \qquad \frac{Ba(OH)_2}{1} \tag{19}$$

(3) *Substances which react with acids or bases or which are produced by reaction of acids and bases.* The equivalent weight is the molecular weight divided by the number of hydrogens or hydroxyls with which 1 mole of the substance reacts or by the number of hydrogens or hydroxyls involved in the reaction which produced the substance. Examples:

$$Na_2CO_3 + HCl \longrightarrow NaHCO_3 + NaCl$$

$$\frac{Na_2CO_3}{1} \quad \frac{NaCl}{1} \quad \frac{NaHCO_3}{1} \tag{20}$$

$$Na_2CO_3 + 2HCl \longrightarrow H_2CO_3 + 2NaCl$$

$$\frac{Na_2CO_3}{2} \quad \frac{2NaCl}{2} \quad \text{or} \quad \frac{NaCl}{1} \tag{21}$$

$$2NaOH + H_2SO_4 \longrightarrow Na_2SO_4 + 2HOH \qquad \frac{Na_2SO_4}{2} \tag{22}$$

$$NaOH + H_2SO_4 \longrightarrow NaHSO_4 + HOH \qquad \frac{NaHSO_4}{1} \tag{23}$$

b. PRECIPITATION. In precipitation reactions the equivalent is that weight of the substance which contains or reacts with one gram atom of a univalent cation or anion. It is simpler, however, to define the equivalent for these reactions as the molecular weight divided by the total positive or total negative ionic charge of the cation or anion actually involved in the reaction and contained in 1 mole of the substance.

Examples:

$$AgNO_3 + NaCl \longrightarrow NaNO_3 + AgCl$$

$$\frac{AgNO_3}{1} \quad \frac{NaNO_3}{1} \quad \frac{AgCl}{1} \quad \frac{NaCl}{1} \tag{24}$$

$$BaCl_2 + K_2SO_4 \longrightarrow 2KCl + BaSO_4$$

$$\frac{K_2SO_4}{2} \quad \frac{BaSO_4}{2} \quad \frac{2KCl}{2} \quad \text{or} \quad \frac{KCl}{1} \tag{25}$$

$$3Cu(NO_3)_2 + 2Na_3PO_4 \longrightarrow Cu_3(PO_4)_2 + 6NaNO_3$$

$$\frac{Cu(NO_3)_2}{2} \quad \frac{Na_3PO_4}{3} \quad \frac{Cu_3(PO_4)_2}{6} \quad \frac{NaNO_3}{1} \qquad (26)$$

c. GENERAL METHOD. For any quantitative reaction, the total number of equivalents of one substance present in the reaction must equal the total number of equivalents of each other substance in the reaction. Consequently, if the total number of equivalents of one substance can be determined, this number may be used to calculate the equivalents of the other substances.

Thus in reaction (25), the number of equivalents of barium sulfate present is two since the ionic charge of barium is two positive and the ionic charge of sulfate is two negative. Since there are two equivalents of barium sulfate, there must also be two equivalents of barium chloride, two of potassium sulfate, and two of potassium chloride. In reaction (26), above, there are three moles of copper nitrate in the balanced reaction. Since each mole represents two equivalents (ionic charge of copper is 2), the three moles must represent six equivalents, and there must also be six equivalents of sodium phosphate, of copper phosphate, and of sodium nitrate present. Since there are two moles of sodium phosphate, each mole must contain three equivalents.

2. Equivalent Weights for Oxidation-Reduction Reactions

The equivalent weight for a substance in an electron transfer or oxidation-reduction reaction depends upon the electron change occurring in the reaction. Details of methods of balancing oxidation reactions are given in Section A of Chapter 10 and will not be discussed in this section. Balanced equations will be used to illustrate the method of determining the equivalent weights for reactions in which electron transfer occurs.

The equivalent weight is defined as the atomic, ionic, or molecular weight divided by the electron change per unit. It is important to know the exact reaction which takes place since many substances can show more than one equivalent weight depending upon the conditions under which a reaction takes place. Potassium permanganate is a good example of an oxidizing agent which can show several equivalent weights: In acid solution, the product is manganous [manganese(II) — Mn^{2+}] ion; in solutions which are nearly neutral, the product is manganese dioxide (MnO_2); and in strongly basic solutions, the product is manganate ion (MnO_4^{2-}).

To determine the total electron change and thus the equivalent weights of reactants and products, it is necessary that a balanced equation be written. The equation used can be the balanced oxidation-reduction reaction or the balanced half-cell reaction of the oxidizing agent or the reducing agent. As an example, consider the reaction of permanganate with oxalate in sulfuric acid solution. The molecular reaction is

$$2KMnO_4 + 5H_2C_2O_4 + 3H_2SO_4 \longrightarrow 2MnSO_4 + 10CO_2 + 8H_2O + K_2SO_4 \quad (27)$$

and the ionic reaction is

$$2MnO_4^- + 5HC_2O_4^- + 11H^+ \longrightarrow 2Mn^{2+} + 10CO_2 + 8H_2O \quad (28)$$

and the ionic half-cell reactions are

$$MnO_4^- + 8H^+ + 5e^- \longrightarrow Mn^{2+} + 4H_2O \quad (29)$$

$$HC_2O_4^- \longrightarrow 2CO_2 + H^+ + 2e^- \quad (30)$$

If the molecular reaction is utilized, the total electron change can be calculated from either the manganese or the oxalate. For manganese the change is from 7+ to 2+, which requires five electrons per atom, and since there are two atoms of manganese in the two molecules of potassium permanganate, the total number of electrons is ten. Similarly, the apparent charge on carbon is 3+ in oxalate and 4+ in carbon dioxide. Since ten carbons are involved, and each carbon requires one electron, a total of ten electrons are required in the reaction. The equivalent weight of the permanganate ion is one-fifth the ionic weight since five electrons are utilized, and similarly the equivalent weight of the potassium permanganate equals the molecular weight divided by five. For manganese itself, the equivalent weight is one-fifth the atomic weight so that the equivalent weight of manganous sulfate is also one-fifth the molecular weight. For the oxalate, whether it is expressed as oxalic acid, bioxalate ion, or oxalate ion, the equivalent weight will be the molecular or radical weight divided by two. For carbon and for carbon dioxide, the equivalent weight is the same as the atomic or molecular weight since one electron is involved per carbon atom. The equivalent weights of water, sulfuric acid, and potassium sulfate are seldom calculated or used for this reaction since they are not directly involved in the electron exchange. The equivalent weights, if needed, must be calculated from the total number of electrons and the number of molecules.

The identical results, except for the compounds or ions not directly involved in the electron exchange, can be calculated from the half-cell reactions. A second example is the reaction of sodium thiosulfate with iodine (actually triiodide ion) to form iodide ion and sodium tetrathionate. The molecular reaction is

$$2Na_2S_2O_3 + NaI_3 \longrightarrow Na_2S_4O_6 + 3NaI \quad (31)$$

The ionic reaction is

$$2S_2O_3^{2-} + I_3^- \longrightarrow S_4O_6^{2-} + 3I^- \qquad (32)$$

And the half-cell reactions are

$$I_3^- + 2e^- \longrightarrow 3I^- \qquad (33)$$

$$2S_2O_3^{2-} \longrightarrow S_4O_6^{2-} + 2e^- \qquad (34)$$

From the molecular reaction it is seen that the equivalent weight of the sodium thiosulfate is the same as the molecular weight since two molecules of the compound require two electrons. The equivalent weight of the sodium triiodide equals the molecular weight divided by two and that of the triiodide ion equals the ionic weight divided by two. A question arises in this case as to exactly what to use as the equivalent weight of iodine and iodide ion. Since iodine occurs in two forms (the iodide ion and the triiodide ion which is a complex formed by two atoms of iodine and one iodide ion), we must specify carefully exactly what we mean in each case. For triiodide ion, the equivalent weight is one-half the ionic weight, while the equivalent weight for elementary iodine is equal to the atomic weight. For iodide ion the equivalent weight equals 3/2 the atomic weight.

Probably the best way to keep from getting confused as to the equivalent weight of any substance in a particular reaction is to use the general method shown in Section C.l.c. for reactions in which no valence change occurs. To do this, first be sure to have a correctly balanced equation and determine the total number of electrons involved for either the oxidizing or the reducing agent. Then place this number under each reactant and product, and the equivalent weight will be the molecular weight multiplied by the ratio of the number of molecules and the number of electrons. For example, in the reaction of $K_2Cr_2O_7$ with $FeSO_4$ in the presence of sulfuric acid, the equation is

$$K_2Cr_2O_7 + 6FeSO_4 + 7H_2SO_4 \longrightarrow Cr_2(SO_4)_3 + 3Fe_2(SO_4)_3 + 7H_2O + K_2SO_4 \qquad (35)$$

and the number of electrons involved is six. The equivalent weights are

$$\frac{K_2Cr_2O_7}{6} \quad \frac{6FeSO_4}{6} \quad \text{or} \quad \frac{FeSO_4}{1} \quad \frac{Cr_2(SO_4)_3}{6}$$

$$\frac{3Fe_2(SO_4)_3}{6} \quad \text{or} \quad \frac{Fe_2(SO_4)_3}{2}$$

It should be remembered that the analyst is actually interested in the equivalent weights of only two substances in any reaction which is used as the basis of an analysis. These are the substance being used as the standard and the substance being determined. The equivalent weights of all the other reactants and of the products are not used in the calculation of the results and thus are not necessary in the analysis. In the example of iron and dichromate, which is used for the determination of iron, the equivalent weight of iron is utilized in the final calculation and the equivalent weight of dichromate must be known in order to determine the normality of the solution. These are the only two that need be known and the others are of academic interest only.

D. Conversion from One Method to Another

1. Weight-Volume to Molar or Normal

Calculate the concentration in grams per liter (g/liter) from the concentration as given and then divide by the molecular weight for molar or the equivalent weight for normal.

Problem 17. A solution of sodium iodide contains 5.00 mg/100 ml. Calculate (a) the molar concentration, (b) the normal concentration for the formation of iodate ion (IO_3^-), and (c) the normal concentration for precipitation of PbI_2.

$$\text{g/liter} = \frac{\text{mg}/1000}{\text{ml}/1000} = \text{mg/ml} \tag{36}$$

$$M = \frac{\text{g/liter}}{\text{MW}} \tag{37}$$

and

$$N = \frac{\text{g/liter}}{\text{equiv wt}} \tag{38}$$

(a).

$$\text{g/liter} = \frac{5.00\ \text{mg}}{100\ \text{ml}} = 0.0500\ \text{mg/ml}$$

$$M = \frac{0.0500}{149.9} = 0.000334\ M \qquad \text{or} \qquad 3.34 \times 10^{-4}\ M$$

(b). The half-cell equation is

$$I^- + 3H_2O \longrightarrow IO_3^- + 6H^+ + 6e^-$$

and the equivalent weight is 149.9/6 or 24.98.

$$N = \frac{0.0500}{24.98} = 0.00200\ N \qquad \text{or} \qquad 2.00 \times 10^{-3}\ N$$

(c). For precipitation of PbI_2 the equivalent weight is the same as the molecular weight since the valence of iodide ion is 1 negative.

$$N = \frac{0.0500}{149.9} = 0.000334\ N \qquad \text{or} \qquad 3.34 \times 10^{-4}\ N$$

2. Molar to Normal and Vice Versa

One mole usually contains one or more equivalents so that the normal concentration ordinarily is equal to or greater than the molar concentration. Since the equivalent weight is equal to the molecular weight divided by the number of equivalents per mole,

$$N = M \times \text{number of equiv/mole} \tag{39}$$

or

$$M = \frac{N}{\text{number of equiv/mole}} \tag{40}$$

Problem 18. What is the molar concentration of a solution of phosphoric acid which is 0.250 N for the formation of phosphate ion? The formation of phosphate from phosphoric acid requires the neutralization of all three hydrogens so that the equivalent weight is one-third the molecular weight.

$$M = N/3 = 0.250/3 = 0.0833\ M$$

Problem 19. What is the normal concentration of 0.100 M sulfuric acid for the precipitation of barium sulfate? The equivalent weight for precipitation of sulfate equals one-half the molecular weight, or $H_2SO_4/2$, so

$$N = M \times 2 = 0.100 \times 2 = 0.200\ N$$

3. Molar or Normal to Titer

To convert from molar to titer, first convert from molar to normal and then use eq. (14) to calculate the titer. To convert normal to titer, use eq. (14) directly.

Problem 20. Calculate the barium oxide titer of a 0.0500 M solution of barium chloride. Since 1 mole of barium chloride produces 1 mole of barium oxide, the equivalent weights of both barium oxide and barium chloride equal one-half the molecular weight, so

$$N = M \times 2 = 0.0500 \times 2 = 0.100\ N$$

$$\text{titer} = N \times \text{equiv wt} = 0.100 \times 153/2 = 7.65\ \text{mg/ml}$$

4. Weight Per cent (w/w) to Molar or Normal

Since the percentage by weight does not specify a volume, the density (d) (or specific gravity, sp. gr.) of the solution must also be known to convert the concentration to either molar or normal. The concentration as given is converted to grams per liter and then divided by the molecular weight or the equivalent weight. The grams of solute per liter is determined by multiplying the weight of 1 liter by the fraction of the weight due to the solute (per cent per 100). Since the weight of 1 liter equals the density times 1000, the calculation becomes

$$\text{g/liter} = d \times 1000 \times \%/100 = 10 \times d \times \% \tag{41}$$

and

$$M = \frac{10 \times d \times \%}{\text{MW}} \tag{42}$$

$$N = \frac{10 \times d \times \%}{\text{equiv wt}} \tag{43}$$

Problem 21. Concentrated sulfuric acid has a density of 1.84 g/ml and contains 86.0 per cent (w/w) of H_2SO_4. Calculate (a) the molar concentration and (b) the normal concentration.

(a) $$M = \frac{10 \times 1.84 \times 86.0}{98.0} = 16.1\ M$$

(b) $$N = \frac{10 \times 1.84 \times 86.0}{49.0} = 32.2\ N$$

E. Problems

1. A 23.7% (w/w) solution of $KHSO_4$ has a density of 1.15 g/ml. Express the concentration of this solution in the following ways: (a) molar; (b) normal for reactions in which (1) neutralization occurs, (2) $BaSO_4$ is precipitated, and (3) SO_4^{2-} is reduced to S^0; and (c) the following titers: (1) NaOH, (2) SO_3, and (3) KCl.
   Ans. (a) 2.00 M; (b) (1) 2.00 N, (2) 4.00 N, (3) 12.00 N;
   (c) (1) 80.00, (2) 160, (3) 149

2. Repeat problem 1 for a 14.0% (w/w) solution of $KHSO_4$ which has a density of 1.10 g/ml.
3. A 5.00% (w/w) solution of $NaHCO_3$ has a density of 1.04 g/ml. Calculate (a) the molar concentration; (b) the normal concentration for reactions in which (1) a base is added, (2) an acid is added, (3) $BaCO_3$ is precipitated, (4) CO_2 is produced, and (5) Na_2UO_4 is precipitated; and (c) the following titers: (1) HCl, (2) NaOH, (3) CO_2, and (4) the CO_2 titer expressed as milliliters of gas at STP instead of milligrams.
4. Calculate the equivalent weights of each substance in the following balanced reactions:
   (a) $2NaOH + H_4C_{10}H_{12}O_8N_2 \rightarrow Na_2H_2C_{10}H_{12}O_8N_2 + 2H_2O$
   (b) $4NaOH + H_4C_{10}H_{12}O_8N_2 \rightarrow Na_4C_{10}H_{12}O_8N_2 + 4HOH$
   (c) $Na_3PO_4 + 2HCl \rightarrow NaH_2PO_4 + 2NaCl$
   (d) $Fe_2(NH_4)_2(SO_4)_4 + 4BaCl_2 \rightarrow 4BaSO_4 + 2FeCl_3 + 2NH_4Cl$
   (e) $Fe(NH_4)_2(SO_4)_2 + 2BaCl_2 \rightarrow 2BaSO_4 + FeCl_2 + 2NH_4Cl$
   (f) $Fe_2(NH_4)_2(SO_4)_4 + 6NaOH \rightarrow 2Fe(OH)_3 + (NH_4)_2SO_4 + 3Na_2SO_4$
   (g) $6Fe(NH_4)_2(SO_4)_2 + K_2Cr_2O_7 + 7H_2SO_4 \rightarrow Cr_2(SO_4)_3$
   $+ 3Fe_2(SO_4)_3 + 6(NH_4)_2SO_4 + K_2SO_4 + 7H_2O$
   (h) $KBrO_3 + 9KI + 3H_2SO_4 \rightarrow KBr + 3KI_3 + 3K_2SO_4 + 3H_2O$
   (i) $2CuSO_4 + 5KI \rightarrow Cu_2I_2 + KI_3 + 2K_2SO_4$
   (j) $2KMnO_4 + 5H_2O_2 + 3H_2SO_4 \rightarrow 2MnSO_4 + 5O_2 + K_2SO_4 + 8H_2O$
5. Calculate the molar and normal concentrations for neutralization and/or precipitation reactions for each of the following solutions:
   (a) 18.194 g of $NaH_2AsO_4 \cdot HOH$ in 250 ml of solution
   (b) 4.904 g of H_2SO_4 in 200 ml of solution
   (c) 5.207 g of $BaCl_2$ in 500 ml of solution
   (d) 61.08 g of $BaCl_2 \cdot 2H_2O$ in 2000 ml of solution
   (e) 8.0423 g of $AgNO_3$ in 500 ml of solution
   (f) 1.714 g of $Ba(OH)_2$ in 333 ml of solution
   (g) 3.155 g of $Ba(OH)_2 \cdot 8H_2O$ in 333 ml of solution
   Ans. (a) 0.4000 *M*, 0.8000 *N* for neutralization
6. Calculate the molar and normal concentrations for the following solutions:
   (a) 3.167% (w/v) NaCl
   (b) 5.326% (w/v) $BaCl_2$
   (c) 10.0% (v/v) HCl
   (d) 25.0% (v/v) H_2SO_4
   (e) 25.0% (w/v) H_2SO_4
   (f) 5% (w/v) $AgNO_3$
   Ans. (a) 0.5419 *M*, 0.5419 *N*
7. Express the concentrations of the following solutions by the titer method:
   (a) The Cl^- titer of 0.500 *N* $AgNO_3$
   (b) The Cl^- titer of a solution of 8.4950 g of $AgNO_3$ in 500.0 ml water
   (c) The NaCl titer of 3.523% (w/v) $AgNO_3$
   (d) The Na_2O titer of 0.1067 *N* HCl
   (e) The $CaCO_3$ titer of 0.0200 *M* EDTA
   (f) The $CaCl_2$ titer of an EDTA solution which has a $CaCO_3$ titer of 1.000 mg/ml
   Ans. (a) 17.73
8. Calculate the normality, the molarity, the silver oxide titer, and the silver nitrate titer of a solution which contains 4.00 g of sodium chloride in 100 g of water and which has a density of 1.02 g/ml.

CHAPTER 4

Calculations of Gravimetric Analysis

A. Gravimetric (Chemical) Factors

Gravimetric analysis is based upon the law of constant composition (the proportion by weight of the constituent elements in any pure compound is always the same) and upon the law of definite proportions (the masses of substances taking part in a chemical change always show a definite and invariable ratio to each other). In gravimetric analysis, the desired constituent is separated from interfering substances and either weighed directly or else converted to a pure compound of known composition which can be weighed. Consequently, one of the most common calculations in gravimetric analysis is the conversion of a given weight of one substance (usually the precipitated compound) to the chemically equivalent weight of a second substance (usually the constituent being determined). Such a calculation is based upon the laws mentioned and upon the quantitative relationship expressed by a completely balanced equation.

The balanced equation for the reaction of sodium sulfate and barium chloride to produce barium sulfate and sodium chloride

$$Na_2SO_4 + BaCl_2 \longrightarrow BaSO_4 + 2NaCl \qquad (1)$$

states that 142.0 g of sodium sulfate (one molecular weight) reacts with 208.3 g of barium chloride (one molecular weight) to produce 233.4 g of barium sulfate (one molecular weight) and 116.9 g of sodium chloride (two molecular weights). Since these substances always react and are produced in this ratio by weight, the weight of any one of them which is produced from (or reacts with) a given weight of any one of the others may be calculated by simple ratio. Thus the weight of barium sulfate produced from 28.40 g of sodium sulfate is given by

$$\frac{X}{233.4} = \frac{28.40}{142.0} \tag{2}$$

or symbolically

$$\frac{X}{BaSO_4} = \frac{28.40}{Na_2SO_4} \tag{3}$$

Solving these gives

$$X = 28.40 \times \frac{233.4}{142.0} = 28.40 \frac{BaSO_4}{Na_2SO_4} = 28.40 \times 1.643 = 44.66 \text{ g} \tag{4}$$

Note that the ratio $BaSO_4/Na_2SO_4$ will occur in any calculation involving the weight of $BaSO_4$ produced from any given weight of Na_2SO_4. The value of this ratio may be calculated once and then used for any subsequent similar calculation. A ratio such as this which states the relationship between chemically equivalent amounts of two substances is called a gravimetric factor.

Many calculations in gravimetric analysis involve the determination of the weight of an element (constituent being determined) present in a given weight of a pure compound (substance precipitated). Chloride, for example, is often precipitated as silver chloride, which is weighed and the weight of chloride calculated from the weight of silver chloride. If a given sample produced 1.000 g of silver chloride, what weight of chloride was present?

$$\frac{\text{g of Cl}}{\text{g of AgCl}} = \frac{X}{1.000} = \frac{Cl}{AgCl} = \frac{35.45}{143.3} = 0.2474 \tag{5}$$

$$X = 1.000 \times 0.2474 = 0.2474 \text{ g}$$

Problems of this type can also be solved by analogy, as is shown below.

Since there is one atomic weight (at. wt) of Cl in each formula weight (FW) of AgCl,

$$\text{no. of at. wts of Cl} = \text{no. of FWs of AgCl}$$

$$\text{no. of FWs of AgCl} = \frac{\text{g AgCl}}{\text{FW AgCl}} = \frac{1.000}{143.3}$$

$$\text{no. of at. wts of Cl} = \frac{\text{g Cl}}{\text{at. wt Cl}} = \frac{\text{g Cl}}{35.45}$$

and

$$\frac{\text{g Cl}}{35.45} = \frac{1.000}{143.3}$$

$$\text{g Cl} = 35.45 \times \frac{1.000}{143.3} = 0.2474 \text{ g}$$

or, using symbols,

$$\text{g Cl} = \text{at. wt Cl} \times \frac{\text{g AgCl}}{\text{FW AgCl}}$$

It is often necessary to determine the weight of a compound which is chemically equivalent to a given weight of another compound. In the determination of iron, the iron present in the sample, after solution and oxidation, is precipitated as hydrous ferric oxide, which is then ignited to ferric oxide. If the compound originally present in the sample was ferrous carbonate, the weight of the carbonate can be calculated from the weight of ferric oxide produced. It should be remembered, however, that the series of equations representing the reactions involved must be completely balanced.

$$2FeCO_3 + 4H^+ \longrightarrow 2Fe^{2+} + 2H_2CO_3 \quad (6)$$

$$2Fe^{2+} + \text{oxidizing agent} \longrightarrow 2Fe^{3+} \quad (7)$$

$$2Fe^{3+} + 6OH^- \longrightarrow 2Fe(OH)_3 \quad (8)$$

$$2Fe(OH)_3 + \text{heat} \longrightarrow Fe_2O_3 + 3HOH \quad (9)$$

Consequently, 2 moles of $FeCO_3$ are necessary to produce 1 mole of Fe_2O_3 and the calculation is

$$\text{wt } FeCO_3 = \text{wt } Fe_2O_3 \frac{2FeCO_3}{Fe_2O_3} \quad (10)$$

In this case, both the original compound and the final compound contain the same element. It is possible to calculate the weight of one compound chemically equivalent to another compound which does not contain the same element provided the series of balanced reactions relating the two is known. In the analysis of arsenates, the arsenate may be quantitatively precipitated as silver arsenate, but this compound is not a weighable form. Consequently, it is dissolved in nitric acid, and the silver is precipitated as silver chloride, which is a weighable form. The weight of arsenate originally present can then be calculated from the weight of silver chloride produced. The equations are

$$Na_3AsO_4 + 3AgNO_3 \longrightarrow Ag_3AsO_4 + 3NaNO_3 \quad (11)$$

$$Ag_3AsO_4 + 3HNO_3 \longrightarrow 3AgNO_3 + H_3AsO_4 \quad (12)$$

$$3AgNO_3 + 3HCl \longrightarrow 3AgCl + 3HNO_3 \quad (13)$$

Thus, 1 mole of sodium arsenate produces 3 moles of silver chloride and the calculation is

$$\text{wt of } Na_3AsO_4 = \text{wt of AgCl} \times \left(\frac{Na_3AsO_4}{3AgCl}\right) \quad (14)$$

The important points to remember in setting up gravimetric factors are:

1. The number of molecules of each of the two substances in a gravimetric ratio must be such that they contain the same number of atoms of the particular element involved.

2. The molecular weight (or a proper multiple) of the substance weighed (precipitate) is always placed in the denominator, and the molecular weight (or a proper multiple) of the sought substance (constituent being determined) is always placed in the numerator.

3. In setting up the factor, only the substance weighed and the substance sought are used.

In following rule 1, it should be remembered that the number of oxygen atoms in the two formulas is only incidental to the analysis and need not be equal in the numerator and denominator.

B. Calculation of Percentage by Weight (General)

Since per cent represents parts per hundred, the fraction of the total weight of a material due to a particular constituent is muliplied by 100 to obtain the percentage of the constituent. The fraction due to the constituent is determined by dividing the weight due to the part by the weight of the entire sample, or

$$\% = \frac{\text{wt of constituent}}{\text{wt of sample}} \times 100 \quad (15)$$

All methods of analysis are designed to measure in some way the weight of the desired constituent present in a given sample.

C. Calculation of Per Cent of Constituent for Gravimetric Analyses

Since the per cent of the constituent is equal to 100 times the weight of the constituent divided by the weight of the sample and since the weight of the constituent is equal to the weight of the precipitate times the factor, the calculation of per cent of the constituent is

$$\%\text{ constituent} = \frac{\text{wt of ppt} \times \text{factor} \times 100}{\text{wt of sample}} \tag{16}$$

The main error made by beginning students in such a calculation is the use of an incorrect factor.

Example. Calculate the per cent SO_3 in a sample of gypsum if a 0.7560-g sample produced a precipitate of $BaSO_4$ weighing 0.9875 g.

$$\%\, SO_3 = \frac{\text{wt of ppt} \times \text{factor} \times 100}{\text{wt of sample}}$$

$$\text{factor} = \frac{SO_3}{BaSO_4} = \frac{80.06}{233.42} = 0.3430 \tag{17}$$

$$\%\, SO_3 = \frac{0.9875 \times 0.3430 \times 100}{0.7560} = 44.80\%$$

The student can check the expression to be sure that the correct factor has been used by showing the units and substances by each number to see that the units and substances cancel satisfactorily. Equation (17) would be shown in the following manner to make such a check:

$$\%\, SO_3 \left(\frac{\text{g } SO_3 \times 100}{\text{g sample}}\right) = \frac{0.9875(\cancel{\text{g } BaSO_4}) \times 0.3430 \left(\dfrac{\text{g } SO_3}{\cancel{\text{g } BaSO_4}}\right) \times 100}{0.7560 \text{ (g of sample)}}$$

$$= 44.80\% \left(\frac{\text{g } SO_3 \times 100}{\text{g sample}}\right) \tag{18}$$

If the inverse factor had been used the equation would show

$$\%\, SO_3 \left(\frac{\text{g } SO_3 \times 100}{\text{g sample}}\right) = \frac{0.9875 \text{ (g } BaSO_4) \times 2.915 \left(\dfrac{\text{g } BaSO_4}{\text{g } SO_3}\right) \times 100}{\text{g sample}}$$

$$= 380.8\% \left(\frac{(\text{g } BaSO_4)^2 \times 100}{\text{g } SO_3 \times \text{g sample}}\right) \tag{19}$$

Not only is the setup incorrect since $BaSO_4$ does not cancel, but the answer of 380.8% is obviously wrong.

In solving problems of this type which require multiplication of several numbers and division by one number or the product of several numbers, the student will find it advantageous to make the calculation using logarithms in place of longhand multiplication and division. Slide rules can be used for some calculations but are not capable of the part per thousand accuracy required in many analytical calculations.

D. Factor Weight Samples

For any particular determination, a sample weight may be selected so that the percentage of the desired constituent may be calculated from the weight of the precipitate simply by moving the decimal. Such a weight is called a factor weight and is often used when a large number of samples are to be run in order to simplify the calculation of the results. In calculation of percentages, the numerical value of the factor appears in the numerator and the weight of the sample appears in the denominator. If the sample weight taken is equal to the factor, these two values cancel each other and the calculation becomes

$$\% = \frac{\text{wt ppt} \times \cancel{\text{factor}} \times 100}{\cancel{\text{wt sample}}} = \text{wt of ppt} \times 100 \tag{20}$$

Thus, it is necessary only to weigh the precipitate and to move the decimal two places to the right to determine the per cent of the constituent being determined. In this example, each 0.01 g of the precipitate represents 1% of the constituent. If the factor weight so calculated does not approximate the optimum weight, a whole number multiple or an exact fraction of the factor weight should be used so the calculation will consist of moving the decimal and multiplying or dividing by a small number.

Example: Calculate the factor weight for the determination of the percentage of iron in an iron ore by precipitating the iron and igniting to ferric oxide.

$$\%\ \text{Fe} = \frac{\text{wt of } Fe_2O_3 \times (2Fe/Fe_2O_3) \times 100}{\text{wt of sample}} \tag{21}$$

$$\text{factor} = 0.6994$$

$$\text{Factor weight sample} = 0.6994 \text{ g}$$

If the optimum weight of sample is found to be 1.5 g, the sample to be weighed would be 2×0.6994, or 1.3988 g, and the calculation would be

$$\% \text{ Fe} = \frac{\text{wt } Fe_2O_3 \times 100}{2} \tag{22}$$

In this case, each 0.02 g of the precipitate represents 1% of iron. It should be remembered that these samples have to be weighed to the exact weight as calculated and cannot be weighed within a given range.

Any problem which states the weight of a precipitate which is equivalent to a stated percentage of constituent is considered to be a factor weight problem. To calculate the weight of sample necessary, rearrange the usual percentage calculation to calculate the weight of sample instead of the percentage.

$$\text{wt of sample} = \frac{\text{wt ppt} \times \text{factor} \times 100}{\%} \tag{23}$$

As an example, calculate the weight of a gypsum sample to be weighed if each milligram of $BaSO_4$ is to represent 0.05% SO_3. The factor $SO_3/BaSO_4$ equals 0.3430.

$$\text{g sample} = \frac{0.001 \times 0.3430 \times 100}{0.05} = 0.6860 \text{ g} \tag{24}$$

E. Quantity of Reagent Required for a Given Reaction

It is important to add the proper amount of reagent when performing a precipitation since too small an amount will cause incomplete precipitation and a large excess will often cause errors through secondary reactions. An example of a harmful secondary reaction occurs in the precipitation of chloride as silver chloride by the addition of silver nitrate to a chloride solution. In the presence of an excess of silver ion, the silver chloride forms a soluble complex by the reaction

$$AgCl + Ag^+(XS) \longrightarrow Ag_2Cl^+ \tag{25}$$

The formation of the complex ion decreases the amount of the silver chloride precipitate and thus introduces a negative error in the determination. Similarly, silver chloride also reacts with an excess of chloride ion to form the soluble $AgCl_2^-$ ion. Slight to moderate excesses will not cause appreciable error, but large excesses of the precipitant ion can cause major errors in gravimetric and other types of analysis which include the precipitation of a solid.

The amount to be added is normally the stoichiometric amount plus a small excess. The stoichiometric amount is that weight which will react exactly with the constituent being determined. In solving such a problem it should be remembered that stoichiometric quantities represent equal numbers of equivalents (or milliequivalents) of each of the reactants, and thus the number of milliequivalents of the precipitating agent must equal the number of milliequivalents of the substance with which it reacts. The number of milliequivalents present in a given volume of a solution of given normality equals the product of the normal concentration and the number of milliliters used. For solids the number of milliequivalents present is equal to the weight in grams of the substance divided by the milliequivalent weight in grams. Thus, the calculation becomes

$$\text{ml} \times N = \frac{\text{g of solid}}{\text{meq wt of solid}} \tag{26}$$

This can be solved for any one of the four quantities providing the other three are known.

Example: How many milliliters of 0.5000 N barium chloride are required to react exactly with 1.000 g of sodium sulfate? In solving such a problem, first calculate the equivalent weight (or milliequivalent weight) of the sodium sulfate (the substance for which a weight is given). The equivalent weight will be one-half the molecular weight since the valence of sulfate is two, or symbolically, $Na_2SO_4/2$. The molecular weight of sodium sulfate is 142.05, so the equivalent weight is 71.03 and the milliequivalent weight is 0.07103.

$$\text{ml} = \frac{\text{wt of solid}}{\text{meq wt (of solid)} \times N \text{ (of solution)}} = \frac{1.000}{0.07103 \times 0.5000} = 28.12 \text{ ml}$$

This type of problem can also be solved by calculating the weight of barium chloride necessary to react with 1.000 g of sodium sulfate and then calculating the volume of 0.5000 N solution which contains this weight.

F. Indirect Analysis

In actual practice it is sometimes advisable to resort to indirect methods of analysis, that is, methods which utilize arithmetic or algebraic calculations in place of actual separations and determinations for one or more constituents. There are two main methods of indirect analysis:

1. Two substances are weighed together and are then converted to some other substances which are also weighed. From these data, the per cent of each original substance can be calculated by the use of two simultaneous equations.

2. Two substances are weighed together and one of them is then determined by a separate method. From these data, the per cent of each original substance can be calculated. It should be noted that this type of problem cannot be solved by the use of simultaneous equations.

a. EXAMPLE OF METHOD 1. A sample containing only sodium chloride and potassium chloride weighs 0.2040 g. The sample is dissolved and precipitated as silver chloride, which weighs 0.4250 g. Calculate the percentage of sodium chloride in the sample.

Step 1. Let the weight of the substance for which the percentage is to be calculated equal x, let the weight of the other equal y, and set up an equation in x, y, and the weight of the original sample.

$$x + y = 0.2040 \qquad (27)$$

Step 2. Using the values x and y for the weights of the two substances, calculate the weight of the precipitate produced from each by use of the correct factor.

$$\text{wt of AgCl produced from } x \text{ g of NaCl} = x\,(\text{AgCl/NaCl}) = 2.452x \qquad (28)$$

$$\text{wt of AgCl produced from } y \text{ g of KCl} = y\,(\text{AgCl/KCl}) = 1.923\,y \qquad (29)$$

Since these two values represent the total weight of AgCl produced, the second simultaneous equation can be set up using these and the weight of AgCl.

$$2.452x + 1.923y = 0.4250 \qquad (30)$$

Step 3. Solve these two simultaneous equations for x by multiplying eq. (27) by the amount of y in eq. (30) and then subtracting the two equations to eliminate y.

$$\begin{aligned} 2.452x + 1.923y &= 0.4250 \\ 1.923x + 1.923y &= 0.3923 \\ \hline 0.529x \qquad\qquad &= 0.0327 \end{aligned}$$

$$x = 0.0327/0.529 = 0.0618 \text{ g of NaCl}$$

From eq. (27), the weight of KCl is $0.2040 - 0.0618 = 0.1422$.

Step 4. Calculate the per cent of one or both constituents as directed.

$$\% \text{ NaCl} = \frac{0.0618 \times 100}{0.2040} = 30.29\%$$

$$\% \text{KCl} = \frac{0.1422 \times 100}{0.2040} = 69.71\%$$

b. EXAMPLE OF METHOD 2. A clay sample weighing 0.7500 g produced a mixture of NaCl and KCl weighing 0.3025 g. The KCl was converted to K_2PtCl_6, which weighed 0.3874 g. Calculate the percentage of Na_2O and K_2O in the sample.

Step 1. From the data given and the proper gravimetric factor, calculate the weight of the component determined separately.

$$\text{wt of KCl} = \text{wt of } K_2PtCl_6 \left(\frac{2\text{KCl}}{K_2PtCl_6}\right) = 0.3874 \times 0.3067 = 0.1188 \text{ g}$$

Step 2. Calculate the weight of the other component by subtracting the weight of KCl from the total weight of the two compounds.

0.3025 g (wt of KCl and NaCl)
0.1188 g (wt of KCl)
0.1837 g (wt of NaCl)

Step 3. Using the proper factors, convert each of these weights to the chemically equivalent weights of the substances for which percentages are to be calculated.

$$\text{wt of } K_2O = \text{wt of KCl} \left(\frac{K_2O}{2\text{KCl}}\right) = 0.1188 \times 0.6317 = 0.0750 \text{ g}$$

$$\text{wt of } Na_2O = \text{wt of NaCl} \left(\frac{Na_2O}{2\text{NaCl}}\right) = 0.1837 \times 0.5303 = 0.0975 \text{ g}$$

Step 4. Calculate the percentage of each component.

$$\% \, K_2O = \frac{0.0750 \times 100}{0.7500} = 10.00\%$$

$$\% \, Na_2O = \frac{0.0975 \times 100}{0.7500} = 13.00\%$$

G. Problems

1. Calculate the following gravimetric factors:

	Substance sought	Substance weighed
(a)	$NaCl$	$AgCl$
(b)	Cl	$AgCl$ (*Ans.* 0.2474)
(c)	$AlCl_3$	$AgCl$
(d)	SO_3	$BaSO_4$
(e)	$Fe_2(SO_4)_3$	$BaSO_4$
(f)	$Fe_2(SO_4)_3$	Fe_2O_3
(g)	K_2O	K_2PtCl_6
(h)	Fe_3O_4	Fe_2O_3

2. Calculate the percentage of the sought substance for the following gravimetric analyses:

	Substance sought	Sample material	Sample weight (g)	Substance precipitated	Weight of precipitate (g)
(a)	Cl	Table salt	1.000	$AgCl$	2.344
(b)	SO_3	Gypsum ore	0.7560	$BaSO_4$	0.9875 (*Ans.* 44.80%)
(c)	$FeCO_3$	Iron ore	0.5000	Fe_2O_3	0.2124
(d)	CaO	Rock	0.9958	$CaCO_3$	0.2936
(e)	K_2O	Rock	1.023	K_2PtCl_6	0.1016

3. Calculate the percentage of $CaCO_3$ in a sample of limestone if a 0.7560-g sample lost 0.3301 g when ignited. (The loss in weight represents loss of CO_2.)
4. For each of the following, calculate the volume of the solution which is necessary to react exactly with the given weight of the stated solid for the reaction specified:

	Solution	Concentration (N)	Reactant	Weight of reactant (g)	Product formed
(a)	$FeCl_3$	0.1250	Na_3PO_4	0.3285	$FePO_4$
(b)	$AgNO_3$	0.2078	Na_3AsO_4	0.2678	Ag_3AsO_4 (*Ans.* 18.60)
(c)	$CaCl_2$	0.2500	Na_3PO_4	0.1964	$Ca_3(PO_4)_2$
(d)	$Na_2C_2O_4$	0.1056	$CaCl_2$	0.2931	CaC_2O_4
(e)	H_2SO_4	0.0505	$BaCl_2$	0.1310	$BaSO_4$

5. Calculate the weight of sample to be weighed if each stated unit of precipitate is to represent the stated percentage of the constituent.

	Type of sample	Weight of precipitate	Per cent of constituent	Factor
(a)	Iron ore	0.0010 g of Fe_2O_3	0.10% of Fe	0.7000
(b)	$Cu_2Cl(OH)_3$	0.0050 g of AgCl	0.20% of Cl	0.2474 (*Ans.* 0.6180)
(c)	$Cu_2Cl(OH)_3$	0.0005 g of $Cu_2(SCN)_2$	0.10% of Cu	0.5226
(d)	Gypsum	0.0020 g $BaSO_4$	0.25% of SO_3	0.3430
(e)	Lead ore	0.0010 g of $PbSO_4$	0.50% of PbO	0.7360

6. A sample containing $BaCl_2$ and $SrCl_2$ weighed 0.3705 g. The salts were converted to the carbonates, which weighed 0.3485 g. Calculate the percentage of Ba and Sr in the sample.
Ans. Ba, 39.84%
7. A rock sample weighing 0.9000 g produced a mixture of NaCl and KCl weighing 0.1025 g. This mixture was dissolved and produced a precipitate of K_2PtCl_6 weighing 0.2815 g. Calculate the percentage of Na_2O and K_2O in the rock.
8. A sample containing $CuSO_4 \cdot 5H_2O$ and $NiSO_4 \cdot 6H_2O$ weighs 1.2500 g. The sample was dissolved, and 0.2800 g of copper was plated from the solution. Calculate the per cent of each component.
9. A 1.500-g sample of a rock produced a mixture of CaO and BaO which weighed 0.5193 g. The oxides were converted to the sulfates, which weighed 0.9072 g. Calculate the percentage of Ca and Ba in the rock.
10. A mixture of Fe_2O_3 and Al_2O_3 weighs 0.7100 g. On ignition in H_2 only the Fe_2O_3 is affected and is reduced to Fe. The mixture then weighs 0.6318 g. Calculate the per cent of Al in the original mixture.
11. A mixture of $BaCl_2 \cdot 2H_2O$ and LiCl weighs 0.6000 g and produces a precipitate of AgCl weighing 1.440 g. Calculate the percentage of Ba and of Li in the sample.

CHAPTER 5

Calculations of Titrimetric Analysis

A. General

In a titrimetric analysis, a solution (the titrant) is added from a burette into a titration vessel containing the substance being titrated. The purpose of the titration is to determine the volume of the titrant which will react quantitatively (exactly, stoichiometrically) with the substance being titrated. Consequently, titration may be defined as the determination of the volume of a solution which will react exactly with a given volume or weight of another substance.

The stoichiometric (equivalence) point in a titration is attained when exactly equivalent quantities of the titrant and substance being titrated have been brought into reaction. In many titrations this condition is accompanied by some detectable change produced by a reactant or product. The detectable change may be visual (such as a change of color or the formation or disappearance of a precipitate) or it may be a change in physical or chemical properties (such as a change in potential, light absorption, or density), which usually must be detected by a special instrument.

Often the detectable change does not appear at the exact equivalence point since it may be caused by a slight excess or deficiency of the titrant. For this reason, the appearance of the detectable change is known as the end point and not as the equivalence point. The difference in the volume of titrant necessary to reach the equivalence point and that necessary to reach the end point is known as the end-point error.

In some titrations a detectable change is not furnished by the reactants or products. In this case another substance must be added which will show a detectable change at or near the equivalence point. Such a substance is called an indicator and usually furnishes a visual change.

At the stoichiometric point the number of equivalents (or milliequivalents) of the titrant added to the titration vessel must equal the number of equivalents (or milliequivalents) of the titrated substance originally placed in the vessel. This is the basis of all calculations involved in titrimetric analysis; that is, at the equivalence point

$$\text{meqs of titrant} = \text{meqs of titrated substance} \qquad (1)$$

The titration data must furnish a means of calculating the number of milliequivalents of either the titrant or the titrated substance in order that the number of milliequivalents of the other may be determined. For solutions the number of milliequivalents of the solute equals the product of the volume (milliliters) and normality, and for solids the number of milliequivalents equals the weight of the solid divided by the milliequivalent weight. Consequently, the titration data must include an accurately known volume of a standard solution or an accurately known weight of a pure substance.

B. Standards

Standards are substances or solutions of known composition and are classified as primary and secondary. Usually the word *standard* by itself is applied only to solids, and the word *solution* is added when referring to a solution of known composition.

A primary standard is a solid of such purity that it can be used to prepare a standard solution by dissolving an accurately known weight in sufficient solvent to make an accurately known volume of solution. A secondary standard is a substance whose composition has been determined against a primary standard by some method of analysis. A suitable primary standard will meet the following requirements:

1. It should react with the substance being standardized or determined by a single, definite, quantitative chemical reaction. A reaction is considered to be quantitative if only a negligible amount of at least one reactant is left unreacted at the equivalence point.
2. The purity must be known definitely, and the types and amounts of impurities, if present, should be determinable by relatively simple procedures.
3. It should not change weight or composition upon exposure to the atmosphere. A solid which changes weight upon exposure is difficult to weigh accurately and consequently is not satisfactory as a primary standard.

4. It should be capable of being dried at normal temperatures, usually 105–110°C, without change of composition. This characteristic rules out most hydrates since hydrates tend to lose all or part of their water of hydration at or below these temperatures.
5. It should have a high equivalent weight so that relatively large samples may be used in standardization procedures.

Typical primary standards include potassium hydrogen phthalate ($KHC_8H_4O_4$) (also known as potassium acid phthalate and potassium biphthalate and often abbreviated as KHP), sulfamic acid (HSO_3NH_2), and sodium carbonate (Na_2CO_3) for acids and bases; potassium dichromate ($K_2Cr_2O_7$), arsenious oxide (As_2O_3), potassium iodate (KIO_3), and sodium oxalate ($Na_2C_2O_4$) for oxidation-reduction reactions; and sodium chloride (NaCl) and silver nitrate ($AgNO_3$) for precipitation reactions. Primary standard-grade chemicals may be purchased from the usual suppliers, and some may be obtained from the National Bureau of Standards. For drugs, the U.S.P. reference standards are usually used.

Secondary standards may be preferable in some cases, especially for procedures for which a suitable primary standard is not readily available. As an example, the standard samples of rocks, alloys, metals, and other materials furnished by the National Bureau of Standards are considered to be the equal of primary standards even though the composition is determined by analysis.

C. Standard Solutions

The concentration of a standard solution must be known within the limits of accuracy necessary in the actual analysis for which it is being used. In the beginning course in quantitative analysis, a solution will be considered a standard solution if its concentration is known accurately to within two parts per thousand of the stated value. Standard solutions are also classified as primary or secondary. The concentration of a secondary standard solution is determined by titration against a known weight of a standard solid or a known volume of a standard solution.

There are two methods of preparing a primary standard solution. The simplest is to dissolve a known weight of a primary standard solid in sufficient solvent to make a known volume of the solution. The second method involves determining the concentration by a gravimetric procedure of an accurately known volume of the solution. It should be noted in both of these methods that the two requirements are an accurately known weight of a pure solid and an accurately known volume of the solution.

An example of the first method is the preparation of standard silver nitrate in the laboratory determination of chloride. A known weight of silver nitrate is dissolved in water, transferred to a volumetric flask, and made to volume in the flask. The number of milliequivalents of silver nitrate (as determined by the weight dissolved) equals the number of milliequivalents of silver nitrate in the solution. The calculation is

$$N = \frac{\text{g of solute}}{\text{ml of solution} \times \text{meq wt of solute}} \qquad (2)$$

If 8.5938 g of silver nitrate (equiv wt, 169.9) is dissolved and diluted to 500.0 ml, the N may be calculated as follows:

$$N = \frac{8.5938\ \text{g}}{500\ \text{ml} \times 0.1699\ (\text{g/meq})} = 0.1012\ N\ (\text{meq/ml})$$

Often it is impossible to prepare a solution by weighing a primary standard solid since none is available. In this case the solution is prepared by dissolving an approximate weight of the solid (or volume of liquid) in enough solvent to make an approximate volume of the solution. An aliquot of this solution is then pipetted and the weight of solute determined by precipitation of a compound from which the weight of the solute may be calculated. Thus a solution of potassium chloride may be standardized by precipitating the chloride in an aliquot as silver chloride and weighing the silver chloride. A solution so standardized would be considered to be a primary standard. The calculation of the normality is

$$N = \frac{\text{wt of precipitate}}{\text{volume (ml) of aliquot} \times \text{meq wt of precipitate}} \qquad (3)$$

Note that the milliequivalent weight of the precipitate, not the solute, is used in this calculation. If a 25.00-ml aliquot of a solution of potassium chloride produces a precipitate of silver chloride (equiv wt, 143.34) weighing 0.3583 g, the normality is calculated by

$$N = \frac{0.3583\ \text{g}}{25.00\ \text{ml} \times 0.14334\ (\text{g/meq})} = 0.1000\ N$$

The concentration of a secondary standard solution is usually determined by titration against a standard solid or a standard solution. An example of titration against a primary standard solid is the standardization of sodium hydroxide using potassium acid phthalate (KHP). In this procedure, samples of KHP (equiv wt, 204.22) are weighed, dissolved, and then titrated with the sodium hydroxide solution using phenolphthalein as the indicator. If a 0.7546-g sample of biphthalate required 34.79 ml of the sodium hydroxide, the normality of the sodium hydroxide may be calculated by

$$N = \frac{\text{wt of solid}}{\text{ml of solution} \times \text{meq wt of solid}} \tag{4}$$

or

$$N = \frac{0.7546 \text{ g}}{34.79 \text{ ml} \times 0.20422 \text{ (g/meq)}} = 0.1062\ N$$

An example of a titration against a standard solution is the determination of the concentration of a thiocyanate solution against standard silver nitrate solution using ferric alum indicator. Usually an aliquot of the silver nitrate is pipetted and titrated with the thiocyanate from the burette. However, both solutions could be measured by burette. In this case both reactants are solutions and

$$\text{ml} \times N \text{ (of solution 1)} = \text{ml} \times N \text{ (of solution 2)} \tag{5}$$

If 10.00 ml of 0.1015 N silver nitrate required 20.30 ml of the thiocyanate, the N of thiocyanate may be calculated by

$$N = \frac{\text{ml} \times N \text{ (of } AgNO_3)}{\text{ml of thiocyanate}} \tag{6}$$

or

$$N = \frac{10.00 \times 0.1015}{20.30} = 0.0500\ N$$

D. Calculation of Per Cent of Constituent from Titrimetic Analysis

The two main types of titrimetric analysis are direct titration and indirect or back titration. In a direct titration the volume of a standard solution required to react with the constituent is determined by adding the titrant to the sample until the end point is reached. In the indirect method, an excess of a standard solution is added to the sample, and the amount of excess is determined by titration with a second standard solution. The sample being determined may be a weighed sample of the solid material dissolved in water or may be an aliquot of an accurately prepared solution of the unknown. If the end point and stoichiometric point do not coincide, a blank should be run and the volume of titrant required by the blank should be subtracted from the volume of titrant used in the titration. A blank is a determination which is performed without an actual sample being present. The blank solution for a titration is a volume of water equal to that used to dissolve the sample. The blank solution is then carried through the analysis, and the amount of titrant necessary to produce the end point is determined.

1. Direct Titration

The basis of the calculation of direct titration is the fact that at the equivalence point the milliequivalents of titrant added (milliliters $\times$ N) must equal the number of milliequivalents of the constituent present in the sample. Since the number of milliequivalents of the titrant is known, the weight of the constituent may be found by multiplying by the milliequivalent weight. Thus the calculation of percentage becomes

$$\%\text{ constituent} = \frac{\text{wt of constituent} \times 100}{\text{wt of sample}} \tag{7}$$

$$\text{wt of constituent} = \text{number of meqs} \times \text{meq wt of constituent} \tag{8}$$

$$\text{number of meqs} = \text{ml} \times N \text{ of titrant} \tag{9}$$

or

$$\%\text{ constituent} = \frac{\text{ml} \times N\text{ (of titrant)} \times \text{meq wt (of constituent)} \times 100}{\text{wt of sample}} \tag{10}$$

As an example, calculate the per cent of chloride in an unknown if a 0.4179-g sample required 34.67 ml of 0.1012 N silver nitrate by a direct titration method. The atomic weight of Cl is 35.453.

$$\%\text{ Cl}^- = \frac{34.67 \times 0.1012 \times 0.035453 \times 100}{0.4179} = 29.77\%$$

If an aliquot of an accurately prepared solution is used in place of the weighed sample, the weight of the sample in the aliquot may be calculated and used in the calculation. For example, suppose that 8.3580 g of the unknown was dissolved and diluted to 500.0 ml and a 25.00-ml aliquot of this solution was used as the sample. The calculation of the weight of the sample used is

$$\text{wt of sample} = 8.3580 \times \frac{25.00}{500.0} = 0.4179\text{ g} \tag{11}$$

As an example of the calculation of a determination in which a blank is necessary, calculate the percentage of chloride in a sample if a 25.00-ml aliquot of a solution of 5.5920 g in 500.0 ml required 33.47 ml of 0.1098 *N* $AgNO_3$ by the Mohr method. In a blank determination, 0.12 ml of the $AgNO_3$ was required.

$$\% \text{ Cl}^- = \frac{(33.47 - 0.12) \times 0.1098 \times 0.035453 \times 100}{5.592(25.00/500.0)}$$

$$= \frac{33.35 \times 0.1098 \times 0.035453 \times 100}{5.592 \times 0.0500} = 46.43$$

Calculations for replicate determinations which use the same size aliquots of a given solution so that the sample weights are identical may be simplified by calculating the value of all the terms except milliliters. This value, which is often represented by Q, may then be multiplied by the volume of titrant to obtain the percentage. For example, assume that three titrations of 25.00-ml aliquots of the unknown in the last example required 33.47, 33.52, and 33.39 ml, respectively. Q may be calculated for all the samples as

$$Q = \frac{0.1098 \times 0.035453 \times 100}{5.592 \times 0.05} = 1.392 \qquad (12)$$

The calculation for the percentage for the three replicates then becomes

$$\% \text{ Cl}^- = 33.35 \times 1.392 = 46.43$$

$$\% \text{ Cl}^- = 33.40 \times 1.392 = 46.50$$

$$\% \text{ Cl}^- = 33.27 \times 1.392 = 46.32$$

2. Indirect Titration

In indirect methods, an excess of one standard solution is added to a solution of the sample and the amount of excess is determined by titration with a second standard. An example is the Volhard method for chloride, in which an excess of standard silver nitrate is added to the sample, and the amount of excess silver nitrate is determined by titration with a standard solution of thiocyanate using ferric alum as the indicator. Since some of the silver nitrate reacts with all the chloride present and the remainder reacts with the thiocyanate, the total number of milliequivalents of the silver nitrate must be equal to the sum of the milliequivalents of chloride plus the milliequivalents of thiocyanate. Since the milliequivalents of silver nitrate is given by the volume and the normality of the standard solution used and the milliequivalents of thiocyanate is determined by the volume and normality of the solution used in the back titration, the milliequivalents of chloride may be found by subtraction of the milliequivalents of thiocyanate from the milliequivalents of silver nitrate.

$$\text{meqs } AgNO_3 = \text{meqs } Cl^- + \text{meqs KSCN} \tag{13}$$

or

$$\text{meqs } Cl^- = \text{meqs } AgNO_3 - \text{meqs KSCN} \tag{14}$$

or

$$\text{meqs } Cl^- = (\text{ml} \times N \text{ of } AgNO_3) - (\text{ml} \times N \text{ of KSCN}) \tag{15}$$

and

$$\% \, Cl^- = \frac{\text{meqs } Cl^- \times \text{meq wt } Cl^- \times 100}{\text{wt of sample}} \tag{16}$$

or

$$\% \, Cl^- = \frac{[(\text{ml} \times N \text{ of } AgNO_3) - (\text{ml} \times N \text{ of KSCN})] \times \text{meq wt } Cl^- \times 100}{\text{wt of sample}} \tag{17}$$

As an example, calculate the per cent of Cl^- in an unknown if a 0.3469-g sample required 15.56 ml of 0.0509 *N* KSCN to titrate the excess $AgNO_3$ remaining from 25.00 ml of 0.1018 *N* solution added to the sample.

$$\% \, Cl^- = \frac{[(25.00 \times 0.1018) - (15.56 \times 0.0509)] \times 0.035453 \times 100}{0.3469}$$

$$\% \, Cl^- = \frac{(2.545 - 0.792) \times 0.035453 \times 100}{0.3469}$$

$$\% \, Cl^- = \frac{1.753 \times 0.035453 \times 100}{0.3469} = 17.92\%$$

E. Optimum Weight of Sample for Standardizations

The optimum weight of standard to be used in the standardization of solutions which have been prepared so that only the approximate concentration is known is the weight which will cause the use of the largest feasible volume of titrant but which will not require a volume of titrant which exceeds the total readable volume (capacity) of the burette. The use of the largest feasible volume will decrease the relative error due to misestimation of the positions of the mensiscus before and after delivery of titrant, but the use of more than one burette-full increases the meniscus error since the position must be read four instead of two times. Since the maximum meniscus error in two estimations of the meniscus is normally 0.04 ml, the error in four estimations will be 0.08 ml, which will double the relative error unless a large portion of the burette volume is used both times. If a volume close to two burette volumes is used, the titration will require an excessively long time to perform. For a 50-ml burette, volumes between 20 and 50 ml are considered satisfactory.

The error made in preparation of solutions of approximate normalities can be either positive or negative. It is often large but is seldom greater than 30 per cent. That is, the concentration of an approximately 0.1 *N* solution will seldom be less than 0.0700 *N* or greater than 0.1300 *N*.

If the solution is prepared so that the concentration is approximately 0.1 *N* and the weight of standard used is that weight which would require exactly 50.00 ml of exactly 0.1000 *N* solution, more than one burette-full will be needed if the concentration turns out to be less than 0.1000 *N*. Consequently the weight of standard to be used is that which will require no more than one burette-full if the maximum negative error was made in the preparation. This is approximately 70 per cent of the weight required to react with exactly one burette-full of a solution of the estimated normality. The calculation of the potimum weight then becomes

$$\text{optimum wt of standard} = \text{volume of burette} \times \text{estimated } N \times \text{meq wt of standard} \times 0.7 \quad (18)$$

As an example, solutions of sodium hydroxide are prepared to an approximate normality by preparing an approximately 50 per cent (w/w) solution and diluting a calculated volume of this solution to the volume required with carbon dioxide-free water. The solution is then standardized against potassium biphthalate (KHP). If a solution which is approximately 0.1 *N* is prepared in this way, the optimum weight of KHP for the standardization using a 50-ml burette is

$$\text{wt KHP} = 50 \times 0.1 \times 0.20422 \times 0.7 = 0.7148 \text{ g}$$

If the maximum negative error had been made in the solution and 0.7148 g of KHP had been used, the volume of titrant required would be

$$\text{ml} = \frac{\text{wt of KHP}}{N \times \text{meq wt}} = \frac{0.7148}{0.0700 \times 0.20422} = 49.99 \text{ ml}$$

If the maximum positive error had been made in the solution and 0.7148 g of KHP had been used, the volume of titrant required would be

$$\text{ml} = \frac{0.7148}{0.1300 \times 0.20422} = 26.92 \text{ ml}$$

Thus the use of the 0.7 safety factor in the calculation of the weight of standard to be weighed will ensure that a satisfactory volume of titrant will be used in the titration provided the error in preparation of the solution does not exceed 30 per cent.

F. Maximum Weight of Unknown Sample To Be Weighed

Since no titration should require more titrant than the capacity of the burette, the maximum weight of an unknown sample is that weight which requires no more than one burette-full of the titrant. However, it is also necessary that the sample weight be large enough to require a substantial portion of the capacity of the burette in order that the relative meniscus error be as small as possible. Since standard solutions are used in the titrations of unknowns, the problem becomes one of estimating the actual percentage of constituent in the sample. While it is obvious that the maximum possible percentage of the constituent in the sample is 100%, it should be remembered that the actual percentage due to the constituent may be very small, and that a sample weight calculated on the basis that the constituent is 100% of the sample could cause the use of volumes of titrant which are too small for satisfactory accuracy in the determination. That is, the relative error in estimating the volume would be greater than can be allowed to obtain the requisite accuracy needed in the analysis.

The weight to be used can be estimated readily only if the analyst has prior information concerning the composition of the unknown material. Often the maximum and minimum percentages of the constituent in such samples are known from previous analyses on similar materials. Sometimes the actual compound containing the constituent is known even though the fraction of the weight of the unknown material due to this compound is not known.

If the maximum percentage of the constituent is known from previous analysis, the calculation of the maximum weight can be made in the following manner. The usual calculation of percentage from a direct titration is given by eq. (10):

$$\%\ \text{constituent} = \frac{\text{ml} \times N \times \text{meq wt (of constituent)} \times 100}{\text{wt of sample}} \qquad (10)$$

which, upon rearranging and using general terms, becomes

$$\text{max wt sample} = \frac{\text{volume of burette} \times N \text{ of titrant} \times \text{meq wt of constituent} \times 100}{\text{max \% of constituent in sample}} \tag{19}$$

As an example, an analyst is required to determine the percentage of iron in an iron ore, and his previous experience indicates that this type of sample will seldom contain more than 30 per cent iron. What is the maximum weight of sample which he should weigh for the determination if a 50-ml burette and 0.1000 *N* standard solution is to be used and the equivalent weight of iron in the titration selected is 55.85?

$$\text{max wt of sample} = \frac{50.00 \times 0.1000 \times 0.05585 \times 100}{30}$$

$$= 0.9308 \text{ g}$$

If the compound containing the constituent is known, the maximum weight can be calculated assuming that this compound represents 100 per cent of the unknown material by the following method:

$$\text{max wt sample} = \frac{\text{volume of burette} \times N \text{ of titrant} \times \text{meq wt of compound} \times 100}{100} \tag{20}$$

As an example, a chloride unknown containing potassium chloride (equiv wt, 74.56) and inert material is to be analyzed by a direct method using a 25.00-ml burette and 0.2000 *N* silver nitrate. Calculate the maximum weight of sample to be weighed.

$$\text{max wt sample} = \frac{25.00 \times 0.2000 \times 0.07456 \times 100}{100}$$

$$= 0.3728 \text{ g}$$

If no satisfactory method of calculating the maximum weight of sample is available, the analyst will usually take an approximate weight of sample, dissolve and titrate rapidly, neglecting interferences, to an approximate end point. From the approximate data thus determined, he can calculate the maximum weight of sample by simple ratio. That is,

$$\frac{\text{approx wt of sample used}}{\text{approx vol of titrant used}} = \frac{\text{wt of sample to be weighed}}{\text{volume of burette to be used}} \tag{21}$$

G. Problems

1. Calculate the normality of a solution prepared by dissolving 8.5938 g of $AgNO_3$ and diluting to 500.0 ml. *Ans.* 0.1012 *N*
2. Repeat problem 1 for (a) 2.5439 g of NaCl in 250.0 ml, (b) 1.3216 g of $BaCl_2$ in 100 ml, (c) 3.9274 g of NaOH in 1000 ml, and (d) 2.1696 g of $K_2Cr_2O_7$ in 500.0 ml.
3. Calculate the normality of a NaCl solution if a 25.00-ml aliquot produced a precipitate of AgCl weighing 0.4015 g.
   Ans. 0.1120 *N*
4. Repeat problem 3 for the following solutions:

Solute	Aliquot	ppt	wt of ppt (g)
(a) $BaCl_2$	20.00	$BaSO_4$	0.4783
(b) $CaCl_2$	15.00	$CaCO_3$	0.2222
(c) KOH	10.00	K_2PtCl_6	0.4275
(d) HCl	25.00	AgCl	0.6539

5. A 0.7546-g sample of KHP required 34.79 ml of a NaOH solution. Calculate the normality of the solution.
   Ans. 0.1062 *N*
6. Repeat problem 5 for the following weights of standards and volume in solutions:
   (a) 0.4263 g of $AgNO_3$ required 25.37 ml of NaCl solution
   (b) 0.1065 g of Na_2CO_3 required 38.83 ml of H_2SO_4 solution
   (c) 0.1021 g of NaCl required 43.69 ml of $AgNO_3$ solution
7. Calculate the percentage of chloride in an unknown if a 0.4179-g sample required 34.67 ml of 0.1012 *N* $AgNO_3$ by the Fajans method.
   Ans. 29.77%
8. Calculate the per cent of chloride in an unknown if a 3.6380-g sample was dissolved and diluted to 500.0 ml. A 25.00-ml aliquot of this solution required 23.92 ml of 0.1069 *N* $AgNO_3$ by the Mohr method. A blank of 0.12 ml was necessary.
   Ans. 49.59%
9. For the solution prepared in problem 8, the $AgNO_3$ remaining after addition of 25.00 ml of 0.1258 *N* $AgNO_3$ to a 25.00-ml aliquot required 9.54 ml of 0.0629 *N* KSCN. Calculate the per cent Cl.
10. For the solution prepared in problem 8, a 20.00-ml aliquot required 21.62 ml of 0.0943 *N* $AgNO_3$ by the Fajans method. Calculate the per cent of Cl.
11. Calculate the optimum weight of standard for the following standardizations:

Standard	Solution	Normality	Volume of burette
(a) KHP	KOH	0.2	25.00 (*Ans.* 0.7147)
(b) KHP	NaOH	0.5	50.00
(c) $AgNO_3$	NaCl	0.05	10.00
(d) Na_2CO_3	HCl	0.1	50.00
(e) $C_4H_{10}O_2NOH$	H_2SO_4	0.25	25.00

12. Calculate the maximum weight of sample to be weighed for an unknown atacamite ore [$Cu_2Cl(OH)_3$] in which the per cent Cl is to be determined by the Fajans method using 0.1250 *N* $AgNO_3$ and a 25.00-ml burette.
Ans. 0.6677
13. Calculate the maximum weight of sample for the unknown in problem 12 if the copper was to be determined iodimetrically by titration with 0.0500 *N* $Na_2S_2O_3$ using a 50.00-ml burette. In this titration the equivalent weight of copper is the same as the atomic weight.

CHAPTER 6

Equilibrium

A. The Activity Concept

When Arrhenius first proposed his theory of ionization it was believed that all substances existed as neutral molecules in the pure state and that these molecules dissociated or broke apart after dissolving to form charged particles called ions. Strong electrolytes were those compounds which dissociated to a great extent, and weak electrolytes were those which were ionized only to a slight extent after being dissolved. One of the main arguments for the dissociation into ions was the fact that solutions of electrolytes conducted the electric current and that solutions of strong electrolytes were better conductors than weak electrolytes. Another factor was the fact that solutions of strong electrolytes lowered the freezing point of water more than solutions of equal concentrations of nonelectrolytes. Since the lowering was attributed to the number of particles in solution, strong electrolytes evidently existed as ions in solution. On this basis it was predicted that the freezing points of solutions of strong electrolytes should be lower by a factor of the number of ions produced; that is, the freezing point of a sodium chloride (NaCl) solution should be two times lower than that of a solution of sugar of the same concentration, and the freezing point of a solution of aluminum chloride ($AlCl_3$) should be four times lower than that of the sugar solution. Since the freezing-point depression of such solutions turned out to be a little less than predicted, Arrhenius postulated that strong electrolytes were not completely dissociated in water solution.

We know now that most strong electrolytes are completely ionic even in the pure or solid state so that the Arrhenius theory must be modified to take this into account. Evidently the ions in solutions of strong eleclrolytes are not as completely effective or efficient in taking part in physical or chemical reactions as they are in very dilute solutions. The concept of active concentration (activity, active mass) has been introduced to explain the behavior of solutions of strong electrolytes. Strong electrolytes are assumed to be completely ionic in solution but are not completely effective (efficient) so that the active concentration is less than the actual concentration. The two concentrations are related by the expression

$$A = fM \tag{1}$$

where A is the active concentration, M is the molar concentration, and f is a dimensionless number known as the activity coefficient. For simplicity in designating concentrations it is customary to enclose the formula of the solute ion or molecule in parentheses to indicate activity or active concentration and in brackets to indicate molar concentration. Using the general formula for an acid as the solute, eq. (1) can be written as

$$(\mathrm{HA}) = f_{\mathrm{HA}}[\mathrm{HA}] \tag{2}$$

A complete explanation of the activity concept is beyond the scope of this text, but a simplified explanation will be attempted to give the student some concept of effective concentration. The effective concentration of an ion is a function of the number of like ions present (the analytical molar concentration of the ion) and also of the number of other ions present in the solution. As an ion moves about in solution, it is attracted by ions of opposite charge and repelled by ions of like charge. These attractive and repulsive effects keep the ion from moving freely and thus decrease the efficiency of the ion. The effect can be compared to the plight of a student who is to meet a friend at the classroom door after class is dismissed. If there are only a small number of students in the room, his progress toward the door is not hindered to a great extent. If, however, the room is filled with students and he sits in the front of the room, his progress toward a door at the back of the room is seriously impeded by the other students, and, as a consequence, he cannot be completely effective, timewise, in reaching the door. The larger the number of students, the greater the interference with his motion and the less effective he can be. The same situation exists for an ion in solution—the greater the number of ions, the less effective any single ion can be.

Debye and Hueckel related the activity coefficient to the number and charge of all the ions in the solution by using the concept of ionic strength. The ionic strength (μ, Greek letter mu) is defined by the equation

$$\mu = \tfrac{1}{2}\Sigma\,(M_i Z_i^2) \tag{3}$$

where M_i is the analytical molar concentration of an ion, Z_i is the charge on the ion, and the Σ (Greek letter sigma) is read as "the sum of." In words, the ionic strength equals one-half the sum of the products of the molar concentration and the ionic charge (valence) squared for all the ions in the solution. Since the ions are placed in solution in the form of compounds, the ionic strength due to any one solute compound may be calculated. Using the general formula $A_x^{m+}B_y^{n-}$ to represent a solute compound, the ionic strength will be

$$\mu = \frac{M(xm^2 + yn^2)}{2} \tag{4}$$

The total ionic strength of the solution is the sum of the ionic strengths due to the individual compounds. Equation (4) shows that the ionic strength due to a compound is either equal to the molar concentration or is a multiple of the molar concentration. The multiplier is dependent upon the ionic charge of the ions present in the compound: for 1 : 1 types (NaCl), the multiplier is 1; for 2 : 1 or 1 : 2 [Na_2SO_4 or $Ca(NO_3)_2$], the multiplier is 3; for 2 : 2 ($CaSO_4$), the multiplier is 4; for 1 : 3 or 3 : 1 (K_3PO_4 or $FeCl_3$), the multiplier is 6; for 3 : 2 (Fe_2S_3), the multiplier is 15.

As an example of the calculation of ionic strengths, determine the ionic strength of a solution which is 0.08 *M* in NaCl, 0.04 *M* in K_2SO_4, and 0.10 *M* in $HC_2H_3O_2$. In the case of acetic acid, remember that only ions affect the ionic strength and that since acetic acid is a weak electrolyte which exists mainly as molecules in the solution it will not contribute appreciably to the ionic strength of the solution.

Compound	Calculation	μ
NaCl	$\frac{0.08(1 \times 1^2 + 1 \times 1^2)}{2}$	0.08 or $M \times 1$
K_2SO_4	$\frac{0.04(2 \times 1^2 + 1 \times 2^2)}{2}$	0.12 or $M \times 3$
$HC_2H_3O_2$		Negligible
Total		0.20

Debye and Hueckel also showed that the activity coefficient was related to the ionic strength by the equation

$$\log f_i = \frac{-0.5Z^2\sqrt{\mu}}{1 + \sqrt{\mu}} \tag{5}$$

Values calculated by this equation agree well with the experimental values up to an ionic strength of approximately 0.10 and can be used for approximations up to an ionic strength of about 0.25. Accurate activity coefficient values for ionic strengths above 0.1 must be determined experimentally. As an example of the use of eq. (5), calculate the activity coefficients of the sodium, potassium, chloride, and sulfate ions in the solution given above in which the ionic strength is 0.20.

For sodium, potassium, and chloride,

$$\log f_i = \frac{-0.5 \times 1^2 \times \sqrt{0.20}}{1 + \sqrt{0.20}}$$

$$\sqrt{0.20} = 0.45$$

$$\log f_i = \frac{-0.5 \times 1^2 \times 0.45}{1.45} = -0.16$$

$$f_i = 10^{-0.16} = 10^{0.84} \times 10^{-1} = 6.9 \times 10^{-1}$$

$$= 0.69$$

For sulfate the calculation is

$$\log f_i = \frac{-0.5 \times 2^2 \times \sqrt{0.20}}{1 + \sqrt{0.20}}$$

$$= \frac{-0.5 \times 4 \times 0.45}{1.45} = -0.62$$

$$f_i = 10^{-0.62} = 10^{0.38} \times 10^{-1} = 2.4 \times 10^{-1}$$

$$= 0.24$$

B. Equilibrium Constants

If substances A and B react to form products C and D, and if C and D when mixed under proper conditions will react to form A and B, the reaction is said to be reversible; that is, it can take place in either direction depending upon the conditions. Most reactions are reversible to some extent, even though for some reactions the reverse reaction is so slow that it would take eons of time for detectable quantities to react.

All reversible reactions will attain equilibrium if left in a closed system without change of conditions for a long enough period of time, and the concentrations of the reactants and products at equilibrium are consequently a measure of the extent of the reaction. Some reactions attain equilibrium almost instantaneously, while others may require extremely long periods of time.

The state of equilibrium occurs in a chemical reaction when the rate of the forward reaction equals the rate of the reverse reaction. That is, in the general reaction

$$A + B \rightleftharpoons C + D \tag{6}$$

equilibrium is attained when the rate at which A and B react to form C and D is exactly equal to the rate at which C and D react to form A and B. Since the rates of formation and decomposition are equal, the concentrations will remain unchanged.

The equilibrium constant expression shows the relationship between the concentrations of reactants and products at equilibrium and may be derived by use of the law of mass action, which states that the rate of a chemical reaction at constant temperature is directly proportional to the product of the active masses of the reacting substances.

Applying the law of mass action to the general reversible reaction of A and B to form C and D, which is assumed to be at equilibrium, the rates for the forward and reverse reactions become

$$R_f \propto (A) \times (B) \qquad \text{and} \qquad R_r \propto (C) \times (D) \tag{7}$$

in which the sign $\propto$ means "is proportional to." Any proportionality may be converted to an equality by multiplying one of the factors by a number which makes the two factors equal. Thus

$$R_f = k_1 \times (A) \times (B) \qquad \text{and} \qquad R_r = k_2 \times (C) \times (D) \tag{8}$$

Since the system is at equilibrium, $R_f = R_r$, and

$$k_1 \times (A) \times (B) = k_2 \times (C) \times (D) \tag{9}$$

which upon rearrangement becomes

$$\frac{k_1}{k_2} = \frac{(C) \times (D)}{(A) \times (B)} = K_{eq} \tag{10}$$

This expression is known as the thermodynamic equilibrium constant and states that the product of the activities of the products divided by the product of the activities of the reactants is a constant at a constant temperature. By convention, the products are shown in the numerator for the reaction as written. K_{eq} for the reverse reaction is the inverse of K_{eq}, or $1/K_{eq}$. If the concentrations are expressed in terms of molar concentrations, this expression becomes

$$K_{eq} = \frac{f_c \times [C] \times f_d \times [D]}{f_a \times [A] \times f_b \times [B]} = \frac{f_c \times f_d}{f_a \times f_b} \times \frac{[C] \times [D]}{[A] \times [B]}$$

or

$$\frac{f_a \times f_b}{f_c \times f_d} \times K_{eq} = \frac{[C] \times [D]}{[A] \times [B]} = K_{prac} = K_M \qquad (11)$$

The product of the activity coefficient term and the thermodynamic equilibrium constant is known as the practical equilibrium constant. Since the activity coefficients vary with concentration, the value at any concentration can be calculated by use of eq. (5) and used when necessary.

By convention, pure liquids, pure solids, and ideal gases at 1 atm partial pressure are assumed to have unit activity and need not be shown in equilibrium constant expressions.

In the dilute solutions normally used in quantitative analysis, the activity coefficients are often almost equal to one and usually are almost equal to each other. Assuming this to be true, the activity coefficient term becomes unity, and the equilibrium constant expression may be stated as

$$K_{eq} = \frac{[C] \times [D]}{[A] \times [B]} \qquad (12)$$

This form is used in the majority of the beginning courses in quantitative analysis. The error actually involved in disregarding activity coefficients is appreciable but is assumed to be negligible in order to simplify calculations.

Since reaction occurs only when the proper number of units of each reactant are either in contact or in close proximity to each other, the concentration of each product or reactant must be raised to the power of its multiplier in the balanced equation of the reaction. Thus K_{eq} for the reaction

$$2A + 3B \rightleftharpoons C + 4D$$

is

$$K_{eq} = \frac{[C] \times [D]^4}{[A]^2 \times [B]^3} \qquad (13)$$

The equilibrium constant expression may be applied to any equilibrium whether it be homogeneous (all reactants and products in one phase) or heterogeneous (reactants and products in more than one phase). Most homogeneous equilibria involved in analysis are those in which all reactants and products occur in a solution, and most heterogeneous equilibria apply to a precipitate in contact with a solution.

The value of the equilibrium constant and thus the concentrations of reactants remaining when equilibrium is established is a measure of the completeness of the reaction. A reaction is said to be *complete* if only negligible quantities of at least one of the reactants remain at equilibrium. Since *negligible* is a relative term which expresses the relationship of one number to another larger number, it must be defined quantitatively for any particular system. In analytical work, a negligible amount is defined as an amount which is difficult to detect and to determine and is taken to be 0.1 mg or less. This amount in 100 ml represents a concentration of approximately 10^{-6} or 10^{-7} *M*. A reaction which involves equal numbers of moles of each reactant and product is assumed to be quantitatively complete if K_{eq} has a value of 10^6 or more. The K_{eq} value for quantitative completeness must be higher than this for equilibria which involve differing numbers of moles of reactants and/or products.

C. Factors Which Affect an Established Equilibrium

After equilibrium has been established in a reaction under a given set of conditions, the equilibrium may be upset or shifted by variations in these conditions. According to LeChatelier's law, the system will react in such a way as to minimize the effect of the change; that is, the system will shift in the direction which tends to relieve the stress placed upon it because of the change.

Any variation of concentration, temperature, or pressure (if a gas is involved) will cause a change in the equilibrium. Addition of a catalyst has no measurable effect upon an established equilibrium since it either has no effect upon the rates at equilibrium or else it affects both the forward and reverse reactions equally. A catalyst is used primarily to speed the attainment of equilibrium.

Of the above factors, variation in temperature will affect the numerical value of the equilibrium constant as well as causing a shift in the equilibrium. Variation of the other factors will cause a shift in the equilibrium but will not affect the numerical value of the equilibrium constant.

The effect of variation of these factors can be evaluated if one considers the fact that concentrations of reactants and products remain constant until conditions are changed. If the change causes an increase in one of these concentrations, the equilibrium is shifted due to the resulting increase in the rate of the reaction which uses up the added substance, thus causing the momentarily increased concentration to be decreased. Consider the

general reaction given above ($A + B \rightleftharpoons C + D$) at equilibrium and assume that K_{eq} has a value of 1 and that the concentration of each substance present is 0.5 *M*. Thus

$$\frac{[C][D]}{[A][B]} = 1 = \frac{0.5 \times 0.5}{0.5 \times 0.5} \tag{14}$$

Assume that enough of substance C is added to the system to momentarily increase its concentration to 1 *M*. Under these conditions the rate of the reaction of C with D is increased due to the law of mass action. This tends to decrease the concentration of both C and D and to increase the concentration of both A and B. A new equilibrium will be established when the rates again become equal. The value of the equilibrium constant will remain the same, but the concentrations of the substances will differ in the new equilibrium. To show this mathematically, assume that the concentration of C is reduced to $1 - x$ at the new equilibrium. The concentration of D will be $0.5 - x$ and the concentration of A and B will both be $0.5 + x$. Substituting these values in K_{eq}, we have

$$\frac{(1 - x)(0.5 - x)}{(0.5 + x)(0.5 + x)} = 1$$

$$(1 - x)(0.5 - x) = (0.5 + x)(0.5 + x)$$

$$0.5 - 1.5x + x^2 = 0.25 + x + x^2$$

$$-2.5x = -0.25$$

$$x = 0.1$$

The concentrations at equilibrium are

$$A = 0.5 + 0.1 = 0.6$$

$$B = 0.5 + 0.1 = 0.6$$

$$C = 1.0 - 0.1 = 0.9$$

$$D = 0.5 - 0.1 = 0.4$$

Substitute these in K_{eq} and

$$\frac{0.9 \times 0.4}{0.6 \times 0.6} = \frac{0.36}{0.36} = 1$$

Note that the equilibrium has been shifted in the direction of the reactants A and B; that is, the concentrations of these substances are larger in the new equilibrium than in the original. Note also that the concentration of D is decreased, while the concentration of C is larger than the original concentration due to the addition of more C but is less than the momentary concentration attained upon addition of more C.

D. Applications of the Equilibrium Constant

1. Ionization Constants

a. INTRODUCTION. Electrolytes are substances which conduct the electric current in water solution and are classified as strong, moderate, and weak. Since the conduction of current through a solution depends upon the number and type of ions present, those substances which furnish relatively large numbers of ions are considered to be strong and those which furnish relatively small numbers are weak. Those in between are classified as moderate or moderately strong.

Most strong electrolytes are soluble compounds which are completely ionic even in the solid or pure state. The act of solution simply separates these ions and allows them to move freely in the solution. A few strong electrolytes, such as HCl, are molecular in the pure state but react with water upon dissolving to become essentially completely ionic. Since there will be relatively large numbers, the ions are not completely effective in reactions because of interference with each other, and as a consequence their activity coefficients will be less than one. Soluble weak electrolytes exist as uncharged molecules in the pure or solid state, and after solution a relatively small fraction decomposes to form ions. This reaction to form ions from molecules is called ionization and is a reversible reaction.

Strong electrolytes include the strong acids (HCl, $HClO_3$, $HClO_4$, H_2SO_4, HNO_3, etc.), the strong bases (NaOH, KOH, etc.), and practically all soluble salts. Weak electrolytes include the slightly soluble ionic substances (barium sulfate, silver chloride, ferric hydroxide, etc.), the weak acids (organic, boric, carbonic, nitrous, hypochlorous, etc.), the weak bases (NH_4OH, organic amines, etc.), and a very few soluble salts (mercuric chloride, lead acetate, complex ions, etc.). The moderate electrolytes include the secondary ions of some diprotic acids (bisulfate), some acids (phosphoric), and the hydroxides of the alkaline earth metals, which are completely ionic but only moderately soluble.

Since the ionization of a weak electrolyte is a reversible reaction which will attain equilibrium, the equilibrium constant can be applied and, in this case, is called the ionization constant (K_i). The ionization constants of acids are often symbolized as K_a and the ionization constants of bases as K_b. The ionization constants of weak electrolytes by convention relate to the reaction which produces ions; thus for the weak acid HA, the reaction is written

$$HA \rightleftharpoons H^+ + A^- \tag{15}$$

and the ionization constant is

$$K_a = \frac{[H^+][A^-]}{[HA]} \tag{16}$$

b. CALCULATION OF IONIZATION CONSTANTS FROM PER CENT OF IONIZATION AND MOLAR CONCENTRATION. If the very small concentration of hydrogen ion produced by the ionization of water is neglected, the concentration of the hydrogen ion in a solution of a weak acid HA will equal the molar concentration (M_a) of the compound multiplied by the fraction which has been ionized, or

$$[H^+] = M_a \times \text{fraction ionized} \tag{17}$$

The fraction ionized equals the per cent of ionization divided by 100 and

$$[H^+] = \frac{M_a \times \%}{100} \tag{18}$$

Since the hydrogen ion and the anion are produced in equal numbers, the anion concentration will equal the hydrogen ion concentration if the hydrogen ion concentration from the ionization of water is negligible compared to that from the ionization of the acid. The concentration of the nonionized molecules of the acid will be equal to the molar concentration minus the fraction which has ionized, or

$$[HA] = M_a - \frac{M_a \times \%}{100} = M_a - [H]^+ \tag{19}$$

Substituting these values in the ionization constant,

$$K_a = \frac{(M_a \times \%/100) \times (M_a \times \%/100)}{M_a - (M_a \times \%/100)} \tag{20}$$

As an example, calculate the ionization constant of benzoic acid, which is 2.50% ionized in 0.1 *M* solution. The student should review the material in Appendix 10 concerning the use of logarithms and exponents in calculations before trying to follow the steps in this example.

$$[H^+] = [A^-] = \frac{M_a \times \%}{100} = \frac{0.1 \times 2.50}{100} = 0.0025$$

$$= 2.5 \times 10^{-3} = 10^{0.398} \times 10^{-3} = 10^{-2.602}$$

$$[HA] = M_a - \frac{M_a \times \%}{100} = 0.1 - 0.0025 = 0.0975$$

$$= 9.75 \times 10^{-2} = 10^{0.989} \times 10^{-2} = 10^{-1.011}$$

$$K_a = \frac{10^{-2.602} \times 10^{-2.602}}{10^{-1.011}} = \frac{10^{-5.204}}{10^{-1.011}} = 10^{-4.193}$$

$$= 10^{0.807} \times 10^{-5} = 6.41 \times 10^{-5}$$

or

$$K_a = \frac{(2.5 \times 10^{-3})^2}{9.75 \times 10^{-2}} = \frac{6.25 \times 10^{-6}}{9.75 \times 10^{-2}} = 6.41 \times 10^{-5}$$

In calculations involving equilibrium constants in analytical chemistry, it is seldom necessary to express exponents to more than three places, and in many instances two places are satisfactory. At present three places will be used in all such claculations.

c. CALCULATION OF IONIC CONCENTRATIONS FROM K AND M. As indicated in the previous section, the concentrations of the hydrogen ion and the anion in a solution of a weak acid are essentially equal, and the concentrations of the nonionized molecules equal the molar concentration of the acid (M_a) minus the concentration of the hydrogen ion since each dissociation of the acid produces one such ion. Thus

$$[H^+] = [A^-]$$

$$[HA] = M_a - [H^+]$$

and

$$K_a = \frac{[H^+]^2}{M_a - [H^+]} \tag{21}$$

or

$$[H^+]^2 + K_a[H^+] - K_a M_a = 0 \tag{22}$$

Equation (22) is in the general quadratic form of $x^2 + bx + c = 0$, for which

$$x = \frac{-b \pm \sqrt{b^2 - 4c}}{2} \tag{23}$$

in which $b = K_a$ and $c = -K_a M_a$. The usual quadratic form for the general equation $ax^2 + bx + c = 0$ is

$$x = \frac{-b \pm \sqrt{b^2 - 4ac}}{2a} \tag{24}$$

Equation (24) reduces to eq. (23) when $a = 1$, which is the general case for ionization calculations.

If the ionization constant is 10^{-4} or smaller, that is, if the per cent of ionization in a 1 M solution is less than 1 %, the number of molecules which have dissociated is negligible compared to the number which have not so that $M_a - [H^+]$ is approximately equal to M_a. In such cases, the use of M_a is permissible and the expression for $[H^+]$ becomes

$$[H^+] = \sqrt{K_a M_a} \tag{25}$$

For example, calculate the molar hydrogen ion concentration in 0.100 M benzoic acid for which K_a is 6.61×10^{-5}.

$$6.61 = 10^{0.820}$$

and substitution in eq. (25) gives

$$[H^+] = \sqrt{10^{0.820} \times 10^{-5} \times 10^{-1}} = \sqrt{10^{0.820} \times 10^{-6}}$$

and, since dividing an exponent by 2 is equivalent to extracting the square root,

$$\begin{aligned} [H^+] &= 10^{0.820/2} \times 10^{-(6/2)} \\ &= 10^{0.410} \times 10^{-3} \\ &= 2.57 \times 10^{-3} M \qquad (\text{since } 10^{0.410} = 2.57) \end{aligned}$$

However, to calculate the hydrogen ion concentration in a 0.100 M solution of chloroacetic acid for which K_a equals 1.41×10^{-3}, it is necessary to solve the quadratic equation since in this case M_a is not approximately equal to $M_a - [H^+]$. Thus

$$b = K_a = 1.41 \times 10^{-3} = 0.00141 = 10^{0.149} \times 10^{-3} = 10^{-2.851}$$

$$b^2 = (1.41 \times 10^{-3})^2 = (10^{-2.851})^2 = 10^{-5.702} = 10^{0.298} \times 10^{-6}$$

$$= 1.99 \times 10^{-6} = 0.00000199$$

$$c = -K_a M_a = -1.41 \times 10^{-3} \times 0.1 = -1.41 \times 10^{-4} = -0.000141$$

$$-4c = -(4 \times -1.41 \times 10^{-4}) = 5.64 \times 10^{-4} = 0.000564$$

Substituting these values in eq. (23),

$$[H^+] = \frac{-b \pm \sqrt{b^2 - 4c}}{2} = \frac{-0.00141 \pm \sqrt{0.00000199 + 0.000564}}{2}$$

$$= \frac{-0.00141 \pm \sqrt{0.000566}}{2} = \frac{-0.00141 \pm \sqrt{10^{-3.247}}}{2}$$

$$= \frac{-0.00141 \pm 10^{-1.624}}{2} = \frac{-0.00141 \pm 0.0238}{2}$$

$$= \frac{0.0224}{2} = 0.0112 = 1.12 \times 10^{-2}\ M$$

If the simplified formula (25) is used, the calculated value for $[H^+]$ is 1.16×10^{-2} M, which represents an error of approximately 3.6% or 36 ppt. Such errors increase as the value of K increases so that a larger error will be involved for K_a values in the range of 10^{-2} and 10^{-1}.

Equations (23) and (25) can also be used to calculate the $[H^+]$ in solutions of many polybasic (polyprotic) weak acids using K_1 for K_a since the numbers of hydrogen ions produced by the second and subsequent ionization steps will be negligible in most cases compared to those produced by the primary ionization provided K_1 and K_2 differ by at least 10^4.

2. Solubility Product

a. GENERAL. In any saturated solution of an ionic compound in contact with undissolved solute, an equilibrium exists between the undissolved solute and the dissolved ions since the rate of dissolving equals the rate of crystallization.

$$MA_{(solid)} \rightleftharpoons M^+ + A^- \qquad (26)$$

Since pure solids are considered to be at unit activity and thus need not be shown in equilbrium constant expressions, the equilibrium constant for the heterogeneous equilibrium between the solid MA and its ions would be

$$K_{eq} = [M^+] \times [A^-] = K_{sp} \tag{27}$$

Thus the product of the concentrations of the ions in a saturated solution, each raised to the appropriate power, is a constant which is known as the solubility product and given the symbol K_{sp}. In analytical work, the solubility product is important since it is related to the maximum concentrations which may remain unprecipitated in solution. Whenever the momentary concentrations upon mixing exceed the solubility product, precipitation will occur and will continue until the product of the ion concentrations again equals the solubility product. (See Chapter 7 for a complete discussion of the theory of precipitate formation.)

b. CALCULATION OF K_{sp} FROM SOLUBILITY. The steps involved in this calculation consist of converting the solubility to the molar solubility (M_{sol}), calculating the ionic concentrations from M_{sol}, and substituting these values in the solubility product expression. As an example, calculate the solubility product of nickel(II) hydroxide (MW, 92.7) if the solubility is 1.82×10^{-2} mg/100 ml. First, convert the solubility to moles per liter (M_{sol}) by expressing the solubility as grams per liter and dividing by the molecular weight.

$$\begin{aligned} M_{sol} &= \frac{1.82 \times 10^{-2}\ (\text{mg/100 ml}) \times 1000\ \text{ml/100 ml} \times 1\ \text{g/1000 mg}}{92.7\ \text{g/mole}} \\ &= \frac{1.82 \times 10^{-4}\ (\text{g/liter})}{92.7\ (\text{g/mole})} = 1.97 \times 10^{-6}\ M\ (\text{moles/liter}) \end{aligned} \tag{28}$$

The concentration of an ion is equal to M_{sol} multiplied by the number of those ions in the formula of the compound, or

$$[\text{ion}] = M_{sol} \times \text{number of these ions in the formula} \tag{29}$$

In our case

$$[Ni^{2+}] = M_{sol} \times 1 = 1.97 \times 10^{-6} = 10^{0.294} \times 10^{-6} = 10^{-5.706}$$

$$\begin{aligned} [OH^-] &= M_{sol} \times 2 = 2 \times 1.97 \times 10^{-6} = 3.94 \times 10^{-6} \\ &= 10^{0.596} \times 10^{-6} = 10^{-5.404} \end{aligned}$$

Substituting these values in K_{sp}, we obtain

$$\begin{aligned} K_{sp} &= [Ni^{2+}] \times [OH^-]^2 = 10^{-5.706} \times (10^{-5.404})^2 \\ &= 10^{-5.706} \times 10^{-10.808} = 10^{-16.514} = 10^{0.486} \times 10^{-17} \\ &= 3.06 \times 10^{-17} \end{aligned}$$

c. CALCULATION OF SOLUBILITY FROM K_{sp}. This is the reverse calculation from that above and involves calculating M_{sol} from K_{sp} and converting this to the solubility desired. For compounds in which the ions have the same valence, M_{sol} is simply the square root of K_{sp} since the ion concentrations are equal to each other. For example, to calculate the solubility in milligrams per 100 ml of lead carbonate for which K_{sp} is 2.50×10^{-11}, express the ion concentrations in terms of M_{sol},

$$[Pb^{2+}] = [CO_3^{2-}] = M_{sol} \times 1$$

and substitute in the K_{sp} expression to solve for M_{sol},

$$K_{sp} = [Pb^{2+}] \times [CO_3^{2-}] = (M_{sol})^2 = 2.50 \times 10^{-11}$$
$$= 25.0 \times 10^{-12} = 10^{1.398} \times 10^{-12} = 10^{-10.602}$$
$$M_{sol} = \sqrt{25.0 \times 10^{-12}} = 5.00 \times 10^{-6}\ M$$

or

$$M_{sol} = \sqrt{10^{-10.602}} = 10^{-5.301} = 10^{0.699} \times 10^{-6} = 5.00 \times 10^{-6}$$

Then convert M_{sol} to the solubility in milligrams per 100 ml,

$$\text{MW} \times M_{sol} \times \text{ml} = \text{mg}$$

or

$$267 \times 5.00 \times 10^{-6} \times 10^2 = 1335 \times 10^{-4}\ \text{mg} = 1.34 \times 10^{-1}\ \text{mg/100 ml}$$

For compounds in which the ions do not have the same ionic charge, the calculation of M_{sol} is more complicated since the ionic concentrations have to be raised to the power of the number of ions involved. Thus, to calculate solubility in grams per 100 ml of cupric hydroxide for which K_{sp} is 8.52×10^{-20},

$$[Cu^{2+}] = M_{sol} \times 1$$
$$[OH^-] = M_{sol} \times 2$$
$$K_{sp} = [Cu^{2+}] \times [OH^-]^2 = M_{sol} \times (2\ M_{sol})^2 = 4(M_{sol})^3$$
$$= 8.52 \times 10^{-20}$$
$$(M_{sol})^3 = \frac{8.52 \times 10^{-20}}{4} = 2.13 \times 10^{-20} = 10^{0.328} \times 10^{-20}$$
$$= 10^{-19.672}$$

$$M_{sol} = 10^{-19.672}/3 = 10^{-6.557} = 10^{0.443} \times 10^{-7}$$
$$= 2.77 \times 10^{-7}\ M$$
$$\text{MW} \times M_{sol} \times \text{liter} = 97.6 \times 2.76 \times 10^{-7} \times 10^{-1}$$
$$= 2.69 \times 10^{-6}\ \text{g/100 ml}$$

d. CALCULATION OF AMOUNT OF ION OR COMPOUND LEFT UNPRECIPITATED. One way of estimating the effectiveness of a precipitation is to calculate the weight of the original ion left unprecipitated under the conditions of precipitation. Most precipitations are performed by adding a definite volume of the precipitant to a known volume of solution containing the sample being analyzed. The majority of precipitations in any method of analysis are planned in conjunction with the sample weight and with the usual range of percentage of the substance being determined to give a slight to moderate excess of the precipitating agent. If the final volume of the solution and the final concentration of the precipitating ion is known, the concentration and thus the weight of the sought ion or substance which is left unprecipitated can be calculated.

As an example, calculate the milligrams of chromate ion left unprecipitated if sufficient lead nitrate is added to a solution of sodium chromate to make the final lead ion concentration equal to 0.01 M and the final volume equal to 100 ml.

$$K_{sp} = [Pb^{2+}] \times [CrO_4^{2-}] = 1.82 \times 10^{-14}$$

or, rearranging and solving for CrO_4^{2-},

$$[CrO_4^{2-}] = \frac{K_{sp}}{[Pb^{2+}]} = \frac{1.82 \times 10^{-14}}{0.01} = 1.82 \times 10^{-12}\ M$$

and the milligrams of chromate ion left unprecipitated is

$$116 \times 1.82 \times 10^{-12} \times 10^{2} = 2.11 \times 10^{-8}\ \text{mg of } CrO_4^{2-}$$

In working these problems, it should be noted that the K_{sp} of the precipitate and the concentration of the precipitant is used to calculate the concentration of the sought ion left unprecipitated and that the atomic, radical, or molecular weight of the sought substance or ion is used to calculate the actual weight left unprecipitated. Thus in the above example, 116 is the radical weight of CrO_4^{2-}, not the molecular weight of sodium chromate.

In a saturated solution of lead chromate made by dissolving lead chromate in water, the concentration of the lead ion and also of the chromate ion would be $(K_{sp})^{1/2}$ or 1.35×10^{-7} M, and 100 ml of the solution would contain 1.57×10^{-3} mg of chromate ion. Thus the addition of the excess of lead ion decreased the amount of chromate ion left in the solution to a value which is approximately 1.3×10^{-4} times as large as the amount left in a saturated solution. This is an example of the common ion effect, which states that the solubility of a slightly soluble compound can be decreased by the addition of a common ion in excess. However, too large an excess may cause complex formation with resultant increase in solubility. It is important to remember that the common ion effect actually works properly only with slight to moderate excesses of the precipitant ion.

3. Factors Which Affect Solubility at a Given Temperature

a. EFFECT OF DIVERSE IONS ON SOLUBILITY. In general, an increase in the concentration of ions not common to the precipitate increases the solubility of a precipitate since an increase in the ionic strength increases the value of the practical ionization constant and thus the molar concentrations of the ions of the precipitate in the saturated solution. As an example, the solubility of barium chromate is 4.8×10^{-5} *M* in a solution with ionic strength of 0.25 and is 1.6×10^{-5} *M* in pure water.

b. EFFECT OF HYDROGEN ION CONCENTRATION ON SOLUBILITY. While the solubility of salts of strong acids such as silver chloride is not greatly affected by moderate concentrations of strong acids such as nitric acid, the salts of weak acids may be completely dissolved by small to moderate concentrations of strong acids. As an example, calculate the solubility of calcium oxalate in a 0.05 *M* sodium oxalate solution to which enough hydrochloric acid has been added to make the hydrogen ion concentration equal to 0.1 *M*. We must first consider the equilibrium existing among the various forms of oxalate ($H_2C_2O_4$, $HC_2O_4^-$, and $C_2O_4^{2-}$), since only $C_2O_4^{2-}$ appears in the solubility product of calcium oxalate and affects the solubility. The ionization constants for oxalic acid are

$$H_2C_2O_4 \rightleftharpoons H^+ + HC_2O_4^-$$

for which

$$K_1 = \frac{[H^+][HC_2O_4^-]}{[H_2C_2O_4]} = 6.5 \times 10^{-2} \tag{30}$$

$$HC_2O_4^- \rightleftharpoons H^+ + C_2O_4^{2-}$$

for which

$$K_2 = \frac{[H^+][C_2O_4^{2-}]}{[HC_2O_4^-]} = 6.1 \times 10^{-5} \tag{31}$$

To calculate the ratio of $[H_2C_2O_4] : [HC_2O_4^-] : [C_2O_4^{2-}]$ in a solution with a total oxalate concentration of 0.05 *M*, let $[H_2C_2O_4]$ be 1, rearrange eq. (30) and (31) to calculate $[HC_2O_4^-]$ and $[C_2O_4^{2-}]$, and calculate $[HC_2O_4^-]$ and $[C_2O_4^{2-}]$ at the hydrogen ion concentration specified. Thus for the 0.1 *M* hydrogen ion

$$[HC_2O_4^-] = \frac{[H_2C_2O_4] \times K_1}{[H^+]} = \frac{1 \times 6.5 \times 10^{-2}}{10^{-1}} = 0.65 \tag{32}$$

$$[C_2O_4^{2-}] = \frac{[HC_2O_4^-] \times K_2}{[H^+]} = \frac{0.65 \times 6.1 \times 10^{-5}}{10^{-1}} = 0.0004 \tag{33}$$

The ratio of $[H_2C_2O_4] : [HC_2O_4^-] : [C_2O_4^{2-}]$ thus is 1 : 0.65 : 0.0004 and the fraction due to oxalate is 0.0004/1.6504 or 0.00024. Since the total concentration is 0.05 *M*, the actual $[C_2O_4^{2-}]$ is 1.2×10^{-5} *M*. Using the K_{sp} of calcium oxalate as 2.0×10^{-9}, the concentration of calcium ions and thus the solubility of CaC_2O_4 is

$$[Ca^{2+}] = \frac{2.0 \times 10^{-9}}{1.2 \times 10^{-5}} = 1.7 \times 10^{-4}\ M$$

By a similar calculation the solubility at a hydrogen ion concentration of 1.0 *M* is 1×10^{-2} *M*, at a hydrogen ion concentration of 0.001 *M* it is 7.0×10^{-7} *M*, and at a hydrogen ion concentration of 10^{-5} *M* it is 4.7×10^{-8} *M*. It should be noted that the 100-fold increase in hydrogen ion from 0.001 to 0.1 *M* caused more than a 200-fold increase in the solubility, and the 10-fold increase from 0.1 *M* to 1.0 *M* caused a 60-fold increase. Another way of showing this effect is to calculate the milligrams of calcium left unprecipitated when the *p*H is raised slowly by addition of ammonia to 100 ml of a strong acid solution which contains 2.0 mmoles of calcium and 5 mmoles of oxalate. The calcium does not start to precipitate until the *p*H reaches a value between zero and 1, and 0.68 mg of calcium ($40 \times 1.7 \times 10^{-4} \times 10^2$) is left unprecipitated at a *p*H of 1. At a *p*H of 3, only 0.0028 mg ($40 \times 7 \times 10^{-7} \times 10^2$) of calcium is left in solution, and the precipitation is quantitatively complete. These values have been obtained neglecting the effect of ionic strength on the ionization constants and solubility product. By utilizing the correct values of the practical constants at an ionic strength of 0.16, the solubility of calcium oxalate at *p*H 1 turns out to be 5.0×10^{-4} *M*, or approximately three times as large as the value obtained neglecting the effect of diverse ions.

c. CALCULATION OF CONCENTRATIONS OF SPECIES PRESENT AT A GIVEN HYDROGEN ION CONCENTRATION. It is often necessary to calculate the equilibrium concentration of the molecular acid or one of the ions formed by stepwise ionization of a polyprotic acid as in the previous problem concerning the solubility of calcium oxalate in an acid solution. Calculations of this type are also necessary in finding the relative concentrations of the two forms of an indicator at a definite hydrogen ion concentration. A general formula for calculating the concentration of any species is derived below, in which the total concentration of all the related species in the solution is symbolized as M_t and the fraction due to any one species is symbolized by α (Greek letter alpha), with subscripts denoting the particular species. Thus the molecular form is indicated by a subscript of zero and the ion produced by the primary ionization by a subscript of 1. The actual or equilibrium concentration of any species is then obtained as the product of M_t and α. Using the general formula for a triprotic acid, H_3A, the ionization steps with corresponding constants are:

$$H_3A \rightleftharpoons H^+ + H_2A^- \qquad K_1 = \frac{[H^+] \times [H_2A^-]}{[H_3A]} \tag{34}$$

$$H_2A^- \rightleftharpoons H^+ + HA^{2-} \qquad K_2 = \frac{[H^+] \times [HA^{2-}]}{[H_2A^-]} \tag{35}$$

$$HA^{2-} \rightleftharpoons H^+ + A^{3-} \qquad K_3 = \frac{[H^+] \times [A^{3-}]}{[HA^{2-}]} \tag{36}$$

and the total concentration is given by

$$M_t = [H_3A] + [H_2A^-] + [HA^{2-}] + [A^{3-}] \tag{37}$$

Equations (34), (35), and (36) may be rearranged to solve for each equilibrium concentration in terms of the equilibrium concentration of H_3A as follows.

$$[H_2A^-] = \frac{K_1[H_3A]}{[H^+]} \tag{38}$$

$$[HA^{2-}] = \frac{K_2 \times [H_2A^-]}{[H^+]} = \frac{K_2 \times (K_1 \times [H_3A]/[H^+])}{[H^+]} = \frac{K_1K_2[H_3A]}{[H^+]^2} \tag{39}$$

$$[A^{3-}] = \frac{K_3[HA^{2-}]}{[H^+]} = \frac{K_3 \times (K_1K_2[H_3A]/[H^+]^2)}{[H^+]} = \frac{K_1K_2K_3[H_3A]}{[H^+]^3} \tag{40}$$

Substitution of these values in eq. (37) gives

$$M_t = [H_3A] + \frac{K_1[H_3A]}{[H^+]} + \frac{K_1K_2[H_3A]}{[H^+]^2} + \frac{K_1K_2K_3[H_3A]}{[H^+]^3} \tag{41}$$

The values for α can now be determined as

$$\alpha_0 = \frac{[H_3A]}{M_t} = \frac{[H^+]^3}{[H^+]^3 + K_1[H^+]^2 + K_1K_2[H^+] + K_1K_2K_3} \tag{42}$$

$$\alpha_1 = \frac{[H_2A^-]}{M_t} = \frac{K_1[H^+]^2}{[H^+]^3 + K_1[H^+]^2 + K_1K_2[H^+] + K_1K_2K_3} \tag{43}$$

$$\alpha_2 = \frac{[HA^{2-}]}{M_t} = \frac{K_1K_2[H^+]}{[H^+]^3 + K_1[H^+]^2 + K_1K_2[H^+] + K_1K_2K_3} \tag{44}$$

$$\alpha_3 = \frac{[A^{3-}]}{M_t} = \frac{K_1K_2K_3}{[H^+]^3 + K_1[H^+]^2 + K_1K_2[H^+] + K_1K_2K_3} \tag{45}$$

The equilibrium concentration of each species may now be calculated by multiplying M_t by the proper α, or

$$[\text{Species}] = M_t\alpha_{species} \tag{46}$$

It should be noted in eqs. (42) through (45) that the denominator is the same in all calculations for α and that the numerator is identical to the term in the denominator which has a number one higher than the subscript on α. Thus in eq. (44) for α_2, the numerator is the third term in the denominator. Utilizing this fact and the more general formula for a polyprotic acid H_nA, the denominator for calculation of α will be

$$[H^+]^n + [H^+]^{n-1}K_1 + [H^+]^{n-2}K_1K_2 + \cdots + [H^+]K_1K_2\cdots K_{n-1} + K_1\cdots K_n \tag{47}$$

and the value for a particular α will use this denominator and the proper term of the denominator as the numerator.

An example is the calculation of the molar concentration of HPO_4^{2-} in a 0.100 *M* solution of phosphate in which the hydrogen ion concentration is 10^{-7} *M*.

$$K_1 = 10^{-2.12}, \qquad K_2 = 10^{-7.21}, \qquad K_3 = 10^{-12.32}$$

At $[H^+] = 10^{-7}$, the denominator will be

$$\text{Denominator} = (10^{-7})^3 + (10^{-7})^2 \times 10^{-2.12} + 10^{-7} \times 10^{-2.12} \times 10^{-7.21} + 10^{-2.12} \times 10^{-7.21} \times 10^{-12.32}$$

or

$$\text{Denominator} = 10^{-21} + 10^{-16.12} + 10^{-16.33} + 10^{-21.65}$$

Disregarding the first and last term,

$$\text{Denominator} = 7.6 \times 10^{-17} + 4.7 \times 10^{-17} = 12.3 \times 10^{-17}$$

so that

$$\alpha_2 = \frac{4.7 \times 10^{-17}}{12.3 \times 10^{-17}} = 0.382$$

and

$$[HPO_4^{2-}] = M_t\alpha_2 = 0.100 \times 0.382 = 0.0382\ M$$

d. EFFECT OF COMPLEX FORMATION ON SOLUBILITY. As mentioned previously, most precipitation procedures are designed to add only a slight to moderate excess of the precipitant in order to offset secondary reactions. The majority of the harmful secondary reactions are caused by the reaction of the precipitate with the large excess of precipitant to form a soluble complex. This causes some of the solid precipitate to dissolve or prevents its formation. A good example of such a compound is AgCl, which can react with excess chloride to form $AgCl_2^-$ and with excess silver ion to form Ag_2Cl^+. Such reactions can cause appreciable errors in gravimetric procedures and separations. As an illustration, calculate the silver ion concentration (which is a measure of the solubility) in a 0.1 *M* solution of NaCl in contact with solid AgCl (1) assuming no complex formation and (2) assuming complex formation. The K_{sp} of AgCl is 1.56×10^{-10} and the stability (formation) constant of $AgCl_2^-$ is 1.91×10^5.

$$(1)\quad [Ag^+] = \frac{K_{sp}}{[Cl^-]} = \frac{1.56 \times 10^{-10}}{10^{-1}} = 1.56 \times 10^{-9}\ M$$

For (2), the $[Ag^+]$ must satisfy both K_{sp} and K_{stab}. The equilibrium constants for complexes are designated as stability constants (K_{stab}) for the formation of the complex and as instability constants (K_{instab}) for the ionization of the complex. For the dichloroargentate(I) ion in this example, the formation equation and constant would be

$$Ag^+ + 2Cl^- \rightleftharpoons AgCl_2^- \qquad K_{stab} = \frac{[AgCl_2^-]}{[Ag^+] \times [Cl^-]^2} \tag{48}$$

and the ionization or instability equation and constant would be

$$AgCl_2^- \rightleftharpoons Ag^+ + 2Cl^- \qquad K_{instab} = \frac{[Ag^+] \times [Cl^-]^2}{[AgCl_2^-]} \tag{49}$$

The numerical values of K_{stab} and K_{instab} are the inverse of each other. Since many tables list stability constants and many list instability constants, the student must be careful to use the correct value. One simple method of distinquishing the two is that the stability constant will ordinarily be a very large number, while the instability constant will be a small number. In terms of exponents, the exponents for stability constants are positive, while the exponents for instability constants are negative. Solving eq. (48) for $[Ag^+]$ gives

$$[Ag^+] = \frac{[AgCl_2^-]}{K_{stab}[Cl^-]^2} \tag{50}$$

and since $[Ag^+]$ must satisfy both K_{sp} and K_{stab},

$$[Ag^+] = \frac{K_{sp}}{[Cl^-]} = \frac{[AgCl_2^-]}{K_{stab}[Cl^-]^2} \tag{51}$$

and solving this for $[AgCl_2^-]$, we obtain

$$\begin{aligned}[AgCl_2^-] &= K_{sp} K_{stab}[Cl^-] \\ &= 1.56 \times 10^{-10} \times 1.91 \times 10^5 \times 10^{-1} = 3.00 \times 10^{-6}\ M\end{aligned}$$

This value represents more than a 1000-fold increase in the solubility of silver chloride over that calculated from K_{sp} only and points out the danger of adding too large an excess of the precipitant.

Another factor to consider with complex formation is the presence of another complexing agent in the solution which will prevent quantitative precipitation of the desired solid. Silver ion forms a soluble complex with thiosulfate ions, and this fact is used in photography to remove unactivated silver halides from the exposed film in the development and fixing process. To illustrate the effect of the presence of such agents in the precipitation medium, we can calculate the solubility of AgCl in 0.01 *M* sodium thiosulfate. The instability or dissociation constant for $Ag(S_2O_3)_2^{3-}$ is 6.0×10^{-14}.

The equation for the reaction involved is

$$AgCl_{(s)} + 2S_2O_3^{2-} \rightleftharpoons Ag(S_2O_3)_2^{3-} + Cl^- \tag{52}$$

and the final chloride ion concentration is a measure of the solubility. Since the silver ion concentration in the solution must satisfy both K_{sp} and K_{instab},

$$[Ag^+] = \frac{K_{sp}}{[Cl^-]} = \frac{K_{instab}[Ag(S_2O_3)_2^{3-}]}{[S_2O_3^{2-}]^2} \tag{53}$$

Neglecting the small free silver ion concentration in the solution, the concentration of the complex equals the concentration of chloride, and

$$[Cl^-]^2 = [Ag(S_2O_3)_2^{3-}]^2 = \frac{K_{sp}[S_2O_3^{2-}]^2}{K_{instab}}$$

$$= \frac{1.56 \times 10^{-10} \times 10^{-4}}{6.0 \times 10^{-14}} = 0.26$$

$$[Cl^-] = [Ag(S_2O_3)_2^{3-}] = \sqrt{0.26} = 0.51\ M$$

Thus the solubility of AgCl in 0.01 *M* thiosulfate is approximately 0.51 *M*, and AgCl could not be precipitated quantitatively from this solution. This aspect of complex formation is often used to offset the interference of an ion in an anlytical procedure by addition of a complex-forming ion to tie up the interfering ion in a slightly ionized complex. This removal of interference is called masking.

4. Calculation of Separations by Fractional Precipitation

If two ions in a solution form slightly soluble compounds with the same ion and the two compounds have sufficiently different solubilities, one compound may be precipitated quantitatively before the other starts to precipitate. The effectiveness of such a separation depends on the solubility products and the ionic concentrations in saturated solutions of the two compounds. If a solution contains A^- and B^- ions which form slightly soluble compounds with M^+ ions, slow addition of M^+ will cause the solubility product of the less soluble compound to be exceeded and that compound to precipitate before the solubility product of the more soluble compound is attained. To determine which compound is precipitated first, it is necessary to calculate the minimum concentration of M^+ which will cause precipitation of each. Since the lowest of these two values will be exceeded first as the concentration of M^+ is increased from zero, the compound which requires the lowest minimum concentration of M^+ will precipitate first. A quantitative separation is theoretically possible if the concentration of this ion remaining when the second starts to precipitate has been reduced to a negligible value.

The Mohr method for chloride is a good example of the application of this principle to an analytical method. In this method, chloride ion is titrated with silver nitrate using potassium chromate as the indicator. Since silver chloride is white and silver chromate is a reddish brown, the formation of the first silver chromate produces a change of color which is utilized as the end point. In the usual procedure the chromate ion concentration is about 0.002 *M* and the chloride ion concentration is usually assumed to be approximately 0.02 *M*. The solubility product of silver chloride is 1.56×10^{-10} and of silver chromate it is 9.0×10^{-12}. Thus the minimum concentrations of Ag^+ for precipitation of AgCl and Ag_2CrO_4 are

$$[Ag^+]_{min} \text{ for AgCl} = \frac{K_{sp}}{[Cl^-]} = \frac{1.56 \times 10^{-10}}{2 \times 10^{-2}} = 7.8 \times 10^{-9}\ M$$

$$[Ag^+]_{min} \text{ for } Ag_2CrO_4 = \sqrt{\frac{K_{sp}}{[CrO_4^{2-}]}} = \sqrt{\frac{9 \times 10^{-12}}{2 \times 10^{-3}}}$$

$$= \sqrt{10^{0.653} \times 10^{-9}} = 10^{-(8.347/2)}$$

$$= 10^{-4.174} = 10^{0.826} \times 10^{-5}$$

$$= 6.70 \times 10^{-5}\ M$$

and silver chloride will precipitate first. The concentration of chloride left unprecipitated when silver chromate starts to precipitate will be that which is present in a solution having the silver ion concentration necessary to cause precipitation of silver chromate and may be calculated as

$$[Cl^-] = \frac{1.56 \times 10^{-10}}{6.70 \times 10^{-5}} = 2.33 \times 10^{-6}\ M$$

This represents a chloride ion concentration of approximately 0.0083 mg/100 ml left unprecipitated, and so the separation is theoretically quantitative. It should be remembered that such calculations indicate only that the separation is theoretically possible and do not prove that the actual separation is possible or even probable under laboratory conditions.

As a second example, calculate whether or not the separation of Pb^{2+} and Ba^{2+} by slow addition of K_2CrO_4 is theoretically possible if the concentration of each ion is approximately 0.10 *M*. The solubility product for $PbCrO_4$ is 1.8×10^{-14} and for $BaCrO_4$ it is 2.4×10^{-10}.

$$[CrO_4^{2-}]_{min} \text{ for } PbCrO_4 = \frac{K_{sp}}{[Pb^{2+}]} = \frac{1.8 \times 10^{-14}}{10^{-1}}$$
$$= 1.8 \times 10^{-13}\ M$$

$$[CrO_4^{2-}]_{min} \text{ for } BaCrO_4 = \frac{K_{sp}}{[Ba^{2+}]} = \frac{2.4 \times 10^{-10}}{10^{-1}}$$
$$= 2.4 \times 10^{-9}\ M$$

Lead chromate will be precipitated first, and the concentration left unprecipitated when barium chromate starts to precipitate will be

$$[Pb^{2+}] = \frac{K_{sp}}{[CrO_4^{2-}]} = \frac{1.8 \times 10^{-14}}{2.4 \times 10^{-9}} = 7.5 \times 10^{-6}\ M$$

and the milligrams of Pb^{2+} left in 100 ml would be

$$207 \times 7.5 \times 10^{-6} \times 100 = 0.16 \text{ mg}$$

and so the separation is theoretically possible.

E. Problems

1. Calculate the ionic strength of the following solutions:
   (a) 0.100 M NaCl
   (b) 0.050 M K_2SO_4
   (c) 0.150 M $Al(C_2H_3O_2)_3$
   (d) 0.250 M $Ca_3(C_6H_5O_7)_2$ (citrate)
   (e) A solution which is 0.010 M in K_3PO_4, 0.020 M in K_2HPO_4, 0.030 M in $KHSO_4$, and 0.40 M in $Al_2(SO_4)_3$ (assume that no precipitate forms)
   Ans. (b) 0.150
2. Calculate the activity coefficient for the stated ion in a solution of the stated ionic strength.

	Ion	Ionic strength	
(a)	Na^+	0.025	
(b)	Ca^{2+}	0.250	(*Ans.* 0.21)
(c)	H^+	0.100	
(d)	Al^{3+}	0.010	
(e)	Cl^-	0.225	

3. Calculate the ionization constants for the following monobasic acids:
   (a) An acid which is 0.900% ionized in 0.100 M solution
   (b) An acid which is 1.05% ionized in 0.0100 M solution

(c) An acid which is 2.50% ionized in 1.00 *M* solution
(d) An acid which is 0.0250% ionized in 0.0100 *M* solution
(e) An acid which is 1.23% ionized in 10.00 *M* solution
Ans. (b) 1.11×10^{-6}

4. Calculate the molar hydrogen ion concentration in each of the following solutions:
(a) 0.125 *M* acetic acid ($pK_a = 4.74$)
(b) 0.100 *M* arsenic acid (H_3AsO_4) ($pK_1 = 2.22$)
(c) 0.250 *M* hypochlorous acid ($pK_a = 7.53$)
(d) 0.500 *M* phosphoric acid ($pK_1 = 2.12$)
(e) 1.00 *M* benzoic acid ($pK_a = 4.18$)
Ans. (b) 0.0217

5. Calculate the K_{sp} of each of the following compounds from the solubility data given:

Compound	Solubility
(a) $Ag_2C_2O_4$	3.30×10^{-2} g/liter
(b) Ag_2S	5.70×10^{-16} mg/ml (*Ans.* 4.9×10^{-53})
(c) $Cd(OH)_2$	2.60×10^{-2} mg/100 ml
(d) CdS	1.00×10^{-9} mg/liter
(e) Bi_2S_3	3.10×10^{-18} g/100 ml

6. Calculate the solubility of the following compounds:

Compound	Express solubility as	pK_{sp}
(a) $Cu(OH)_2$	Milligrams per liter	19.07
(b) $Co(OH)_2$	Milligrams per 100 ml	15.60 (*Ans.* 0.037)
(c) $Ca_3(PO_4)_2$	Grams per liter	28.70
(d) AgCl	Milligrams per ml	9.81
(e) $Cr(OH)_3$	Molar	30.22

7. Calculate the number of milligrams of Mn^{2+} left unprecipitated in 100 ml of a 0.100 *M* solution of $MnSO_4$ to which enough sodium sulfide has been added to make the final S^{2-} ion concentration equal 0.001 *M*. Assume no change in volume due to addition of the sodium sulfide.

Ans. 6.04×10^{-9}

8. Repeat problem 7 for the following:

Ion left	Compound present	Compound added	Final ion concentration (*M*)
(a) SO_4^{2-}	Na_2SO_4	$BaCl_2$	$Ba^{2+} = 0.005$
(b) CrO_4^{2-}	K_2CrO_4	$PbCl_2$	$Pb^{2+} = 0.010$
(c) Fe^{3+}	$FeCl_3$	NaOH	$OH^- = 1.0 \times 10^{-4}$
(d) Ag^+	$AgNO_3$	KI	$I^- = 0.033$

9. Calculate the molar solubility of magnesium fluoride in a solution of 0.100 *M* sodium fluoride to which enough strong acid has been added to make the molar hydrogen ion concentration equal to 0.100.

Ans. 0.111 *M*

10. Repeat problem 9 for the following solubilities:

M_{sol} of	In	Final molar hydrogen ion concentratron
(a) PbF_2	0.050 *M* NaF	0.010
(b) CaC_2O_4	0.250 *M* $Na_2C_2O_4$	2.00×10^{-4}
(c) $BaCrO_4$	0.100 *M* K_2CrO_4	5.00×10^{-5}
(d) $CrPO_4$	0.500 *M* Na_3PO_4	1.00×10^{-9}

11. Calculate the silver ion concentration which is a measure of the molar solubility of AgCN in a 0.050 *M* solution of NaCN in contact with solid AgCN (a) assuming no complex formation and (b) assuming complex formation.

Ans. (a) 3.2×10^{-13}, (b) 1.0×10^{6}

12. Repeat problem 11 for the molar solubility of the compound listed in the solution specified:

M_{sol} of	In
(a) AgCl	0.010 *M* $Na_2S_2O_3$
(b) AgCl	0.500 *M* NH_3
(c) AgI	1.00 *M* NH_3
(d) $BaSO_4$	0.010 *M* Na_4EDTA

13. Calculate the molar concentration of $H_2PO_4^-$ in a 0.050 *M* solution of Na_3PO_4 to which enough strong acid has been added to make the final hydrogen ion concentration equal 1.00×10^{-8}.

Ans. 0.043 *M*

14. Repeat problem 13 for the molar concentrations of the following ions:

Ion	Compound (0.050 *M*)	Final molar hydrogen ion concentration (*M*)
(a) $HC_2O_4^-$	$Na_2C_2O_4$	1.00×10^{-3}
(b) $H_2AsO_4^-$	Na_3AsO_4	2.50×10^{-6}
(c) $HC_8H_4O_4^-$ (phthalate)	$K_2C_8H_4O_4$	1.00×10^{-4}
(d) $HC_6H_5O_7^{2-}$ (citrate)	$Na_3C_6H_5O_7$	5.00×10^{-5}

15. If hydrogen sulfide is bubbled into 100 ml of an acidified solution which is 0.100 *M* in lead nitrate and also 0.100 *M* in manganous nitrate, which sulfide will precipitate first? Calculate the number of milligrams of this ion remaining unprecipitated in the solution when the second ion starts to precipitate.

Ans. 1.34×10^{-10} mg of Pb^{2+}

16. Repeat problem 15 for the following solutions:
    (a) 0.100 *M* lead nitrate and 0.100 *M* nickel nitrate
    (b) 0.050 *M* cupric sulfate and 0.050 *M* manganous sulfate
17. If 15 *M* NaOH is added dropwise to 100 ml of a solution which is 0.050 *M* in $FeSO_4$ and 0.050 *M* in $Fe_2(SO_4)$, which ion will precipitate first? Calculate the molar concentration of this ion remaining when the second starts to precipitate, assuming no change in volume due to addition of the NaOH.
18. Repeat problem 17 for the following solutions:
    (a) 0.005 *M* $CuSO_4$ and 0.100 *M* $Ce_2(SO_4)_3$
    (b) 0.250 *M* $CoCl_2$ and 0.005 *M* $CrCl_3$

CHAPTER 7

Theory of Precipitation

A. Introduction

In Chapter 6, the factors which affect the solubility of a solid in a liquid were studied without consideration of the factors which affect the actual formation and purity of a precipitate in a solution. In this section, the steps involved in the formation of a precipitate, the factors which affect particle size, and the types of impurities which contaminate the solid will be discussed.

It will be beneficial to review the terms *saturated* and *supersaturated* as applied to solutions of solids before beginning the discussion. If an ionic solid is placed in a solution, the solid begins to dissolve from the surface, and thus the concentrations of the ions in the solution continually increase. However, as the ions move about in the solution, they may approach the surface close enough to be reincorporated into the crystal. The rate of dissolving is a function of the temperature and the surface area but can be assumed to be constant for a given solid and solution. The rate of reincorporation into the solid, however, is a function of the concentration of the ions in the solution and consequently increases as more solid dissolves. Under these circumstances, an equilibrium must be attained at which the rate of dissolving is just offset by the rate of reincorporation into the solid. This solution is said to be a saturated solution. Thus a saturated solution contains all the dissolved solid which it could dissolve at that temperature in contact with undissolved solid and may be defined as a solution in which the dissolved solute will remain in equilibrium with solid solute. It should be pointed out that the solution would still be saturated if it were decanted or filtered from the solid as long as the temperature remained constant. The product of the concentrations of the ions (raised to the appropriate powers) in a saturated solution equals the solubility product.

The solubility of most solids increases with increase in temperature so that if a given amount of a solvent is saturated with a solid at an elevated temperature, any excess solid removed, and the solution cooled, it is possible that no solid would separate from the solution. Such a solution would be termed to be supersaturated since it contains more dissolved solid than it could dissolve at that temperature in contact with undissolved solute. Supersaturation is a metastable condition, and the supersaturation often may be broken by agitation or by addition of crystals of the solute or other small particles which can act as nucleation centers.

It should also be beneficial to review the types of dispersions which are encountered in analysis, especially in relation to particle size. Dispersions are classified as being homogeneous or heterogeneous. Solutions are examples of homogeneous dispersions in which the ions or molecules of the solute are the dispersed phase and the molecules of the solvent are the dispersion medium. The dispersed phase in a solution is distributed homogeneously throughout the mixture, does not settle out, and cannot be removed by filtration. Since the maximum dimension of many ions or molecules is less than 10 Å (angstroms), the dispersed phase in solutions is usually considered to be one nm or less in size. Suspensions of solids in liquids, such as sand in water, are examples of hetreogeneous dispersions. In suspensions, the particles are large enough to settle out by gravity alone if left undisturbed a sufficient length of time. The minimum particle size which will settle out by gravity alone is usually considered to be in the range of 100 nm. The dispersed phase in suspensions are particles composed of many atoms or molecules which are dispersed heterogeneously among the molecules of the dispersion medium. They will settle out and can be removed by ordinary filtration.

Dispersions in which the size of the dispersed phase is between those of solutions and suspensions are known as colloids or as colloidal dispersions. They are heterogeneous since the dispersed phase consists of particles which are composed of a large number of atoms or molecules. Colloids do not settle out and the dispersed phase cannot be separated by ordinary filtration since the particles are small enough to pass through the finest filters. Colloids are stable and exhibit the Tyndall effect and Brownian movement. The Tyndall effect refers to the fact that the path of a beam of light through a colloid can be seen since the particles are large enough to reflect light even though they are too small to be seen by the human eye. The unordered or random motion of the colloidal particles is known as Brownian movement. The particles appear to move in straight lines with sudden changes of direction and velocity. The stability of colloids results, at least in part, from the fact that colloidal particles are charged and thus repel each other due to the electrical repulsion between like charges. Colloidal particles also exhibit a very large surface area or have a large surface to mass ratio.

B. Formation of Precipitates

If two soluble compounds containing the ions which form an insoluble solid are mixed in solution, the insoluble compound may or may not separate from the solution depending upon the concentrations of the ions. A precipitate will form only if the instantaneous concentrations of the ions upon mixing exceed the solubility product of the insoluble substance. The actual steps necessary for the formation of the precipitate can be very involved, but for a compound formed from two ions, the ions first come together to form ion pairs, which are held together loosely by electrostatic attraction. Such ion pairs can grow by addition of ions or ion pairs to form ion clusters, which can be separated easily until some minimum size cluster is attained. This minimum size is a critical size above which the cluster will continue to grow and form a particle with a definite crystalline pattern. Below this critical size, the clusters will separate more rapidly than they grow. The critical size varies for individual substances but is often between six and ten ions. The cluster is called a nucleus after attaining the critical size for growth.

Thus the steps involved in the formation of the solid particles are first the formation of ion pairs, then the formation of ion clusters, followed by the formation of nuclei which are capable of growing in the presence of concentrations which exceed the solubility product. The ion pairs and clusters can form in solutions in which the concentrations are not great enough to form a precipitate, but nuclei will not form until the solubility product is exceeded. The period of time involved in the formation of nuclei of a size sufficient to continue growth is called the induction period. It varies in length according to the precipitate and the conditions, being short for silver chloride, moderate for barium sulfate, and long for magnesium ammonium phosphate.

C. Particle Size of Precipitates

The final particle size is an important property of the precipitate which governs, at least in part, both the purity and the filterability of the precipitate. One of the main factors which affect particle size is the speed of precipitation. Von Weimarn studied the effect of rate of formation of the precipitate upon the particle size and found that the faster the precipitation, the smaller the particle size. He also found that the rate was dependent upon the extent of supersaturation and defined relative supersaturation as $Q - S$, where Q is the molar concentration of the mixed reagents before precipitation occurs (the instantaneous concentrations upon mixing) and S is the molar solubility of the precipitate when the system has come to equilibrium. According to von Weimarn, the rate of formation is governed by the ratio of the relative supersaturation to the molar solubility, or

$$\text{rate} = \frac{Q - S}{S} \tag{1}$$

Actually the rate of formation of nuclei differs from the rate of crystal growth, and it has been found that final particle size is a function of both rates. These rates depend upon the extent of supersaturation, with the rate of formation of nuclei being governed by $(Q - S)^n$ (where n may be as large as 4) and the rate of crystal growth being affected by the surface area of the particles also, or $A(Q - S)$ (where A is the surface area). The values of these two rates are such that after the first group of nuclei are formed, crystal growth predominates over nuclei formation provided $Q - S$ is small. If $Q - S$ is large, nuclei formation predominates over crystal growth and many small particles will be formed.

The variation in particle size can also be explained by use of supersolubility curves (temperature-concentration curves which plot the concentrations above which supersaturation may not exist) compared to ordinary solubility-temperature curves. Theoretically, a supersaturated solution may remain indefinitely at a constant temperature without precipitation provided no crystals or nucleation sites are added. However, at any given temperature there is an instantaneous concentration upon mixing for any solid above which nucleation will occur whether or not nucleation sites are present. A typical graph is shown in Fig. 7-1. The upper or dotted curve represents the supersolubility curve and the lower the regular solubility curve. The area above the dotted curve is called the labile region. Whenever the instantaneous concentrations are in this region, nucleation occurs rapidly. The area between the two curves is known as the metastable region. In this region, no new nuclei will be formed, and the particles will grow until the concentrations equal those in a saturated solution. The area below the lower curve is the unsaturated region.

If the concentration of a precipitant at temperature T is raised slowly and uniformly throughout the solution, no precipitation will occur until the concentration represented by the point P on the upper curve is attained. At this point particles will form, and the particles will grow until the concentration has been lowered to the value at point S on the lower curve. If

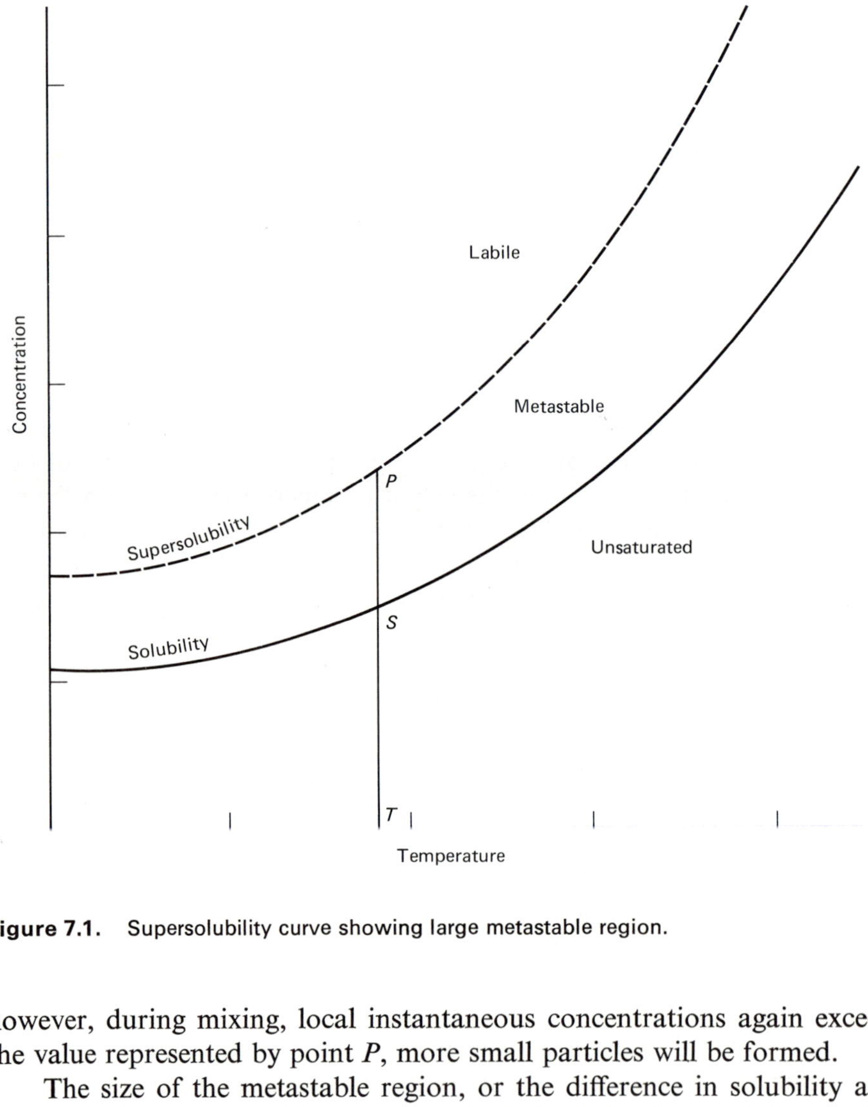

Figure 7.1. Supersolubility curve showing large metastable region.

however, during mixing, local instantaneous concentrations again exceed the value represented by point *P*, more small particles will be formed.

The size of the metastable region, or the difference in solubility and supersolubility curves, affects the length of the induction period and thus has a great effect upon the size of the precipitate particles. As shown in Fig. 7-2, the metastable area for barium sulfate is much larger than the one for silver chloride, and, as a result, barium sulfate particles are larger than silver chloride particles. The metastable area is even smaller for ferric hydroxide and other compounds which tend to form gelatinous or flocculant-type precipitates.

From the above, it is apparent that any factor which affects the solubility of the precipitate will affect the particle size and that slow and uniform mixing is important if a small number of large crystals (rather than a large number of small crystals) is desired.

Figure 7.2. Diagram of relative sizes of metastable regions for various type precipitates. **A**, solubility curve for AgCl and $BaSO_4$; **B**, supersolubility curve for AgCl; **C**, supersolubility curve for $BaSO_4$; **D**, solubility curve for $Fe(OH)_3$; and **E**, supersolubility curve for $Fe(OH)_3$. Particle size for gelatinous precipitates is usually less than 0.02 μm (micrometers, formerly known as micron, μ), for curdy precipitates from 0.02 to 0.1 μm, and for crystalline precipitates 0.1 μm or more.

D. Growth of an Individual Particle of a Precipitate

To understand the incorporation of impurities with a precipitate, it is necessary to understand the growth of individual particles in the solution. Most precipitates are formed by addition of a solution of the precipitant ion to a solution of the ion to be precipitated so that particles actually form and grow in the presence of an excess of one of the ions of the precipitate. Let us now consider this using calcium oxalate as an example when it is formed by the slow addition of dilute sodium oxalate to a solution of calcium chloride. As we have seen, the ions form ion pairs, then ion clusters, and finally nuclei which can continue growth. Calcium oxalate forms in

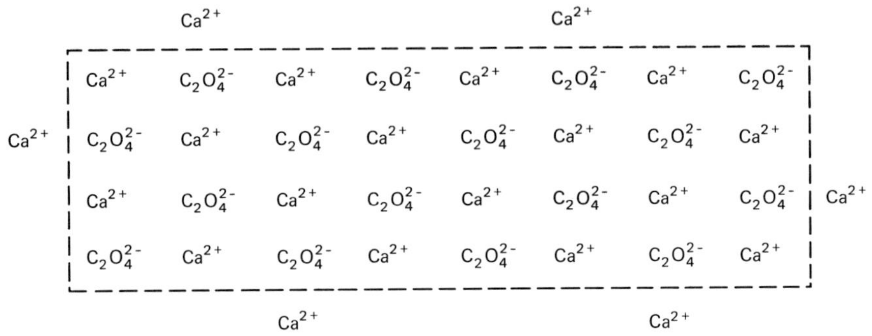

Figure 7.3. One plane of a calcium oxalate crystal in the presence of excess Ca^{2+} ions showing the adsorption of Ca^{2+} ions at negative sites on the surface.

cubic arrangement so that after a short while a particular crystal contains many ions arranged in the general pattern shown in Fig. 7-3. The crystal will grow by addition of both calcium and oxalate ions at particular points on the surface. Thus an oxalate ion is attracted to a calcium ion on the surface, and a calcium ion is attracted to an oxalate ion on the surface. This is caused by the fact that each ion on the surface has a slight charge since it is not completely surrounded by other ions of opposite charge. A calcium ion in the interior of the crystal is surrounded by six oxalate ions (four in the plane of the ion and one each in the two adjacent planes). As a consequence, its charge is equally distributed among six ions or one-sixth of its charge to each ion. A calcium ion on the surface, however, is surrounded by only five oxalates and as a consequence has a slight residual positive charge equal approximately to one-sixth of its charge. An oxalate ion is attracted to this positive area on the surface and takes its position in the crystal lattice. Similarly calcium ions are attracted to the area of oxalate ions and cause continued growth.

Since the oxalate is being added to the calcium, the calcium will be in excess until equivalent quantities are added. Consider a calcium oxalate particle in the solution if the addition is stopped before an equivalent quantity has been added. It will still try to grow by addition of both calcium and oxalate, but the oxalate will soon give out, while calcium ions will still be present. Calcium ions continue to be attracted to and become attached to the surface so that the particle assumes a positive charge. Since the system as a whole must be electrically neutral, this charge will be neutralized by an excess of chloride ions in the immediate vicinity of the particle. These, however, will not be as firmly attached to the crystal as the calcium ions. The calcium ions which are attached to the crystal in excess of the stoichiometric amount are said to be adsorbed on the crystal. The depth of the solution around the crystal in which the charge is neutralized is the counter-ion region.

E. Impurities in and on Precipitates

Precipitate impurities may occur on the surface of the precipitate particle or inside the precipitate particle. Those on the surface may be adsorbed or postprecipitated, while those inside the particle may be occluded or coprecipitated. Each of these will be discussed in detail.

1. Adsorption

The calcium ions discussed in the previous section are said to be adsorbed on the crystal so that adsorption is a surface or boundary phenomenon. The concentration of the material adsorbed on the surface will be different from its concentration in the body of either of the two phases which constitute the boundary.

Adsorption on ionic solids always occurs in two layers, called the primary layer and the secondary layer, as shown in Fig. 7-4. In the case of the calcium oxalate particle discussed previously, the calcium ion is considered to be in the primary layer since it is held to the particle by some force of attraction, and the chloride ions in the counterion region would be considered to be in the secondary layer. The ions in the secondary layer are free to move about in the solution, while those in the primary layer are not.

One of the main questions concerning adsorption of interest to the analyst is the type of ions that tend to be adsorbed. Considering first the primary layer, ions which are common to the precipitate and are in excess in the solution will be adsorbed more strongly than any other ions. If there are no common ions present in excess, then any ion which forms an insoluble compound with either of the ions of the precipitate may be adsorbed. In case there is more than one such ion present, the ion which forms the most insoluble compound will tend to be adsorbed the most strongly. Considering calcium oxalate, a carbonate ion would be adsorbed in preference to a nitrate ion, and a barium ion would be adsorbed rather than a potassium ion.

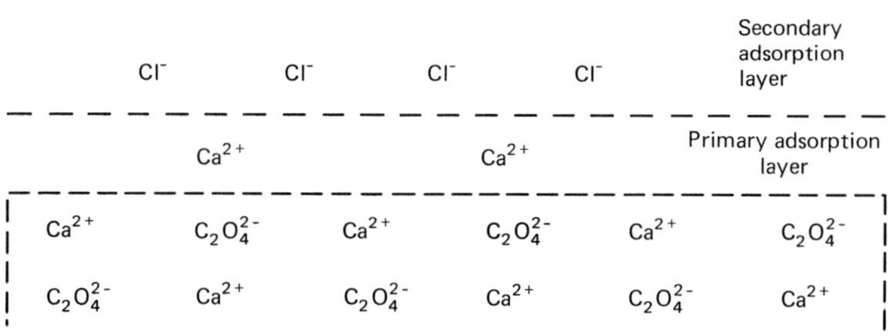

Figure 7.4. Surface of a calcium oxalate particle in the presence of excess Ca^{2+} ions showing primary adsorption of Ca^{2+} ions and secondary adsorption of Cl^- ions.

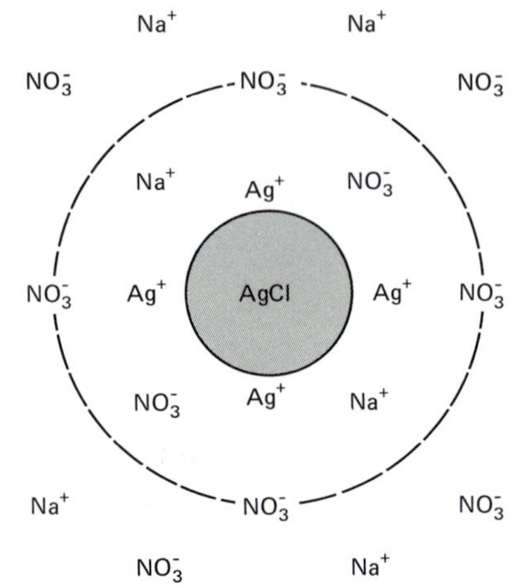

Figure 7.5. Colloidal particle of AgCl in the presence of excess Ag^+ ion showing the primary adsorption of Ag^+ ions and the counterion region in which the charge is neutralized.

One result of adsorption is the fact that the particles become charged and thus repel each other since they all have the same charge. The force of repulsion is a function of the size of the charge and the distance of separation. For any given size of charge the repelling force increases as the particles come closer together. This effect does not extend beyond the counterion region since the charge is neutralized by the excess of oppositely charged ions in this region, as shown in Fig. 7-5. There are also cohesive forces (often called van der Waals' forces) between the particles which tend to attract the particles to each other so that they may form larger aggregates. This cohesive force is similar to gravity in that it increases as the mass increases and as the distance of separation decreases. For two charged particles to be able to combine to form a larger particle or aggregate, the cohesive force must be greater than the repelling force due to like charges as the particles come close together. The combination of two or more charged small particles to form a larger particle is called agglomeration. It is also known as flocculation and as coagulation. Factors which favor agglomeration are those which tend to decrease the depth of the counterion region and include both the concentration and charge of the counterions. An increase in concentration of the counterion will decrease the depth of the counterion region, and the larger the charge of the counterion, the greater the effect of any given concentration.

The agglomeration of small particles differs from crystal growth since the agglomerated particles retain their original form and do not actually increase in size, as is the case when particles grow by addition of ions to the crystal lattice. The difference between crystalline and agglomerated particles is shown in Fig. 7-6. The agglomerated particles carry both the primary

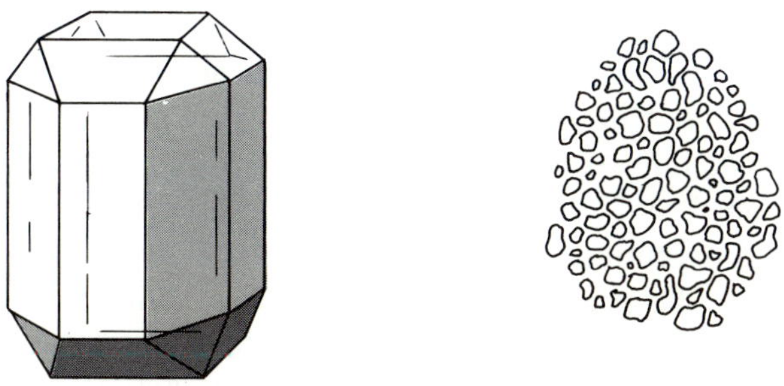

Figure 7.6. Contrast between crystalline and curdy precipitate particles.

layer and the counterlayer of adsorbed ions into the agglomerated particle since they must remain neutral to be held together by the cohesive forces. Removal of the counterions by washing will cause the particles to redisperse or peptize. As a consequence, the wash solution for an agglomerated precipitate must contain an electrolyte to prevent peptization of the agglomerates.

2. Postprecipitation

The second type of impurity often found on the surface of a precipitate is the postprecipitated impurities which form on the surface of a previously formed precipitate particle. The postprecipitation of zinc sulfide on copper (II) sulfide is an excellent example of this type of impurity. In slightly acid solution, cupric sulfide will be precipitated quantitatively by the addition of excess hydrogen sulfide, but zinc sulfide will not form under these conditions. If both copper and zinc ions are present, the copper sulfide will precipitate and, if filtered immediately, will be almost free of zinc ion. If allowed to remain in contact with the solution for an appreciable time, the copper sulfide particles will become coated with precipitated zinc sulfide and thus become contaminated by zinc. The extent of the contamination is governed at least in part by time because the amount of postprecipitated impurity will increase with time until a maximum is reached. The effect is the same whether the zinc ion is present before precipitation or added after the precipitate is formed.

Postprecipitation is probably an adsorption phenomenon since the copper sulfide is precipitated in the presence of an excess of sulfide ion, and thus sulfide ion is adsorbed in the primary layer of the particles. This attracts zinc ion as the secondary ion so that the solubility product of the zinc sulfide is probably exceeded in the counterion region. As a result, zinc sulfide precipitates on the surface of the copper sulfide particles.

Another example of postprecipitation occurs in the oxalate separation of calcium and magnesium in limestone analysis. In this procedure, an excess of oxalate is used in the precipitation step. Calcium oxalate precipitates, but magnesium oxalate forms a complex and remains in solution. If the calcium is filtered immediately, little or no magnesium will be carried down with the calcium oxalate. If the solution is allowed to stand for some time (more than 1 or 2 hr) magnesium oxalate will postprecipitate and contaminate the precipitate.

Even though the primary adsorption of the common ion in excess is probably the main factor in postprecipitation, all cases cannot be explained by adsorption alone. One possible explanation is that a supersaturated solution of the second material is formed on the surface of the particle and that the surface offers nucleation sites to break the supersaturation and cause the secondary precipitate to form. Other factors which might be involved include active sites on the precipitate surface which cause deformation of ions and exert a catalytic effect on the secondary precipitation.

3. Occlusion

Occlusion is the physical entrapment of impurities in the crystal as the particle grows rapidly. The impurities may be trapped as mother liquor or as individual ions or molecules. Essentially, the particle is growing so fast that ions in the immediate vicinity are simply surrounded by the growing particle and cannot escape back into the mother liquor before being completely enclosed by the crystal. Such impurities may not be removed by washing, but they are decreased in some cases during aging of the precipitate.

4. Coprecipitation

The process by which soluble impurities become incorporated into a precipitate particle during its formation is known as coprecipitation. Coprecipitation may occur either by isomorphous replacement (sometimes referred to as mixed crystals or as solid solution formation) or by adsorption followed by entrapment or occlusion.

Coprecipitation due to solid solution formation in which the impurity actually enters the crystal lattice of the precipitate may be illustrated by the coprecipitation of barium chromate with barium sulfate. The addition of barium chloride to a solution which is 0.02 *M* in chromate ion and 0.1 *M* in hydrogen ion causes no precipitation of barium chromate. Addition of barium chloride to a solution which is 0.02 *M* in sulfate ion and 0.1 *M* in hydrogen ion causes the quantitative precipitation of barium sulfate. Addition of barium chloride to a solution which is 0.02 *M* in chromate, 0.02 *M* in sulfate, and 0.1 *M* in hydrogen ion causes the precipitation of barium sulfate contaminated with approximately 5 per cent by weight of barium chromate. Even though barium chromate alone would not precipitate under these conditions, the presence of the barium sulfate lattice allows the chromate ion to be incorporated into the precipitate.

There are relatively few cases of solid solution formation as the cause of impurities, since the two compounds must not only form in the same crystal pattern, but the radii of the ions of like charge must be within ten per cent of each other.

Coprecipitation caused by adsorption and entrapment is much more common in analytical procedures. As an example, barium sulfate may contain either anions or cations which are coprecipitated in this way. The cations tend to predominate if barium is added to sulfate, and the anions predominate if the sulfate is added to the barium solution. This can be explained by consideration of the adsorption of the common ion which is in excess and the counterions which would be attracted to the particle. Assume that less than the stoichiometric amount of barium is added to a sulfate solution. The particles will adsorb sulfate ions in the primary layer and cations in the secondary layer. If more barium is now added, the rapid growth of the particles in the immediate vicinity of mixing can entrap the secondary layer of cations in the precipitate particle. Similarly, addition of the sulfate to the barium solution will cause entrapment of the anions in the secondary layer. The extent of occlusion by adsorption and entrapment is affected by the solubility of the entrapped substance. Thus calcium is entrapped more than sodium with barium sulfate, and nitrate is entrapped more than chloride. The extent of coprecipitation is also a function of the ratio of concentrations of the impurity ion and the ion being precipitated. The coprecipitation increases as the concentration of the ion being precipitated decreases, that is, as the ratio of the impurity to precipitated ion becomes large.

In some procedures, coprecipitation is desirable and actually the basis of the separation. As an example, manganese coprecipitates with the hydrous oxide of iron (ferric hydroxide) when precipitated at high *p*H with sodium hydroxide. Even though manganese is present in too small a concentration to precipitate itself, the high *p*H causes the hydrous oxide particles to become negatively charged and attract the manganese in the secondary layer. This particular process is known as gathering and is used in several analytical procedures.

F. Purification of Precipitates

Precipitate impurities must often be removed or reduced in order to obtain a satisfactory result in a determination. Some impurities may be removed by simple washing, but most require more sophisticated treatment.

Adsorbed impurities may be removed by simple aqueous washing of crystalline precipitates, but the wash solution must contain an electrolyte to offset peptization for curdy or flocculent precipitates. In this case, the electrolyte ion actually replaces one of the adsorbed ions, either in the primary or the secondary layer. If the precipitate is to be dried, the compound formed and left on the precipitate must be volatile. In the washing of silver chloride precipitates which contain adsorbed silver ions in the primary layer and nitrate ions in the secondary layer, the wash solution is usually dilute nitric acid since the hydrogen ion will displace the silver ions from the primary layer to form nitric acid which will volatilize when heated. The hydrous oxide precipitates cannot be washed with an acid solution, and so ammonium compounds are usually used in the wash solutions since most ammonium compounds volatilize completely when heated.

Postprecipitated impurities cannot be removed by simple washing either with or without an electrolyte. Postprecipitated impurities may be reduced by dissolving the precipitate and reprecipitating, being careful to filter as promptly as possible after the precipitate is formed.

Occluded and coprecipitated impurities are not affected by washing since they occur in the body of the precipitate particles. In some cases, they can be removed only by solution and reprecipitation. The amount of impurities occluded or coprecipitated will be less in the second precipitation since their concentrations will be lower.

Some precipitates may be partially purified by aging or digestion. Aging is allowing the precipitate to remain in contact with the mother liquor for extended periods of time, and digestion is the same process except at elevated temperatures for shorter lengths of time.

Very small particles of a precipitate are more soluble than larger particles. Since the mother liquor constitutes a saturated solution, small particles will tend to dissolve during aging or digestion and deposit on the larger crystals. Thus the small particles will disappear and the larger particles will grow larger. The impurities trapped during the growth of the smaller particles will be returned to the solution and probably will not be trapped during the slow growth of the larger particles. Even though the particles are not small enough to have increased solubility, the smaller particles will dissolve during aging or digestion and reprecipitate on the larger particles since the smaller surface area represents a more stable energy relationship.

G. Types of Precipitates

Precipitates are classified as crystalline, curdy, and gelatinous or flocculent. Crystalline precipitates are characterized by easily filterable particles which are relatively pure. As a consequence, all analysts prefer to work with crystalline precipitates. Curdy precipitates are those which are formed as colloidal particles but which agglomerate to form larger aggregates which are of filterable size. Curdy precipitates are usually readily filterable, somewhat more impure than crystalline precipitates, and must be washed with an electrolyte solution. Gelatinous precipitates are also flocculated colloids, but the particle size is smaller than the particles of a curdy precipitate. Gelatinous precipitates are difficult to filter, contain more impurities than curdy precipitates, and also must be washed with an electrolyte solution to offset peptization.

H. Rules for Precipitation

Some of the general rules which should be followed in the formation of a precipitate which is relatively pure and easy to filter include the following.

1. The precipitate should be formed under conditions which tend to increase the solubility. If necessary the conditions can be changed to decrease the solubility after the precipitate is formed. To do this, the precipitate should be formed at elevated temperatures and in the presence of compounds which will increase the solubility. Thus barium sulfate is precipitated from solutions which contain small amounts of hydrochloric acid since barium sulfate is more soluble in acid than in water, due to the formation of some bisulfate ions.

2. The solutions should be as dilute as possible and should be mixed slowly with vigorous agitation.

3. A slight to moderate excess of the precipitant ion should be added to take advantage of the common ion effect. Large excesses should be avoided since they may cause complex formation.

4. The precipitate should be aged or digested under conditions of increased solubility if possible. However, precipitates which are subject to postprecipitation should be filtered as soon as possible after formation.

5. If necessary, the precipitate should be dissolved and reprecipitated in order to decrease impurities.

I. Precipitation from Homogeneous Solution

1. Introduction and General Considerations

In ordinary precipitation, two solutions are mixed together and, as precipitation occurs, there are great changes in the concentrations of the precipitant ion and the ion being precipitated in the immediate vicinity of mixing. As a result, temporary supersaturation in excess of the supersolubility curve will occur often in the immediate vicinity of mixing before agitation causes the solution to become uniform. These are the conditions which lead to rapid formation of a large number of small particles and the consequent adsorption and entrapment of impurities in the precipitate. To understand what happens, let us consider a drop of barium chloride which has been placed in a solution of sodium sulfate without mixing. The barium ions will migrate outward into the sulfate solution, and the sulfate ions will migrate into the barium solution. At the juncture of the two solutions, barium sulfate will be precipitated and deplete the concentration of both ions. The barium sulfate particles closest to the barium solution will be in a region of excess barium ion and will adsorb barium ions in the primary layer to become positively charged. Similarly, the barium sulfate particles closest to the sulfate solution will adsorb sulfate ions and become negatively charged. Thus some of the particles will tend to entrap cations and some will entrap anions so that both types of impurities will be present in the crystals being formed. These conditions continue as the ions migrate toward each other and form more particles. As a result, precipitates formed by the mixing of two solutions tend to be impure and to consist largely of small particles.

If the precipitant ion could be generated slowly and uniformly throughout the solution, the localized concentration effects which occur when solutions are mixed would be eliminated. In such cases there should be very few areas in which the instantaneous concentrations exceeded the supersolubility curve, so that a relatively small number of particles should be formed, and these should be able to grow slowly without adsorption and entrapment of counterions. In other words, excessive supersaturation with consequent small particles and occluded impurities should not occur since the concentration of the precipitant ion is raised slowly and uniformly throughout the solution by a chemical reaction or a physical change.

This process of generating the precipitating ion or condition uniformly throughout the solution is known as precipitation from homogeneous solution. The essential idea in precipitation from homogeneous solution is the gradual increase in the avialable concentration of the precipitant ion so that the solubility product is exceeded uniformly throughout the solution at a slow or moderate rate which allows crystal growth to predominate over nucleation. Precipitation from homogeneous solution is used to improve separations, to study and reduce coprecipitation, to form larger crystalline particles, and to produce precipitates which are purer and easier to filter.

This type of precipitation can be applied to any system in which the necessary reagent can be generated slowly and uniformly throughout the solution by a chemical reaction or a physical change. Physical changes which may be used include the volatilization of an organic solvent or of ammonia. These changes either decrease the solubility or lower the *p*H. Chemical reactions can be used which will generate the actual ion or compound needed or which will produce hydrogen or hydroxyl ions to raise or lower the *p*H.

For any of these, the rate must be satisfactory. A reaction which is too slow will cause too much time to be spent in the analysis, while a reaction which is essentially instantaneous will cause undesired local concentration excesses. There are many examples of reactions which can be used for control of *p*H or generation of individual ions or molecules. A few typical ones will be discussed.

2. Types of Reactions

a. CONTROL OF *p*H. In most cases, slow rise of *p*H is desired rather than slow increase in acidity. The procedure used most often is the hydrolysis of urea at elevated temperatures. Urea hydrolyzes to form ammonia and carbon dioxide so slowly at room temperature that the necessary rise in *p*H would require an excessively long time. The rate at 100°C is satisfactory, being neither too fast nor too slow. The hydrolysis reactions are often written as

$$CO(NH_2)_2 + H_2O \longrightarrow 2NH_3 + CO_2 \tag{2}$$

$$2NH_3 + 2H^+ \rightleftharpoons 2NH_4^+ \tag{3}$$

but it has been shown that the hydrolysis is actually a rearrangement to form ammonium cyanate followed by hydrolysis and reaction with hydrogen ion as follows:

$$CO(NH_2)_2 \longrightarrow NH_4^+ + OCN^- \tag{4}$$

$$OCN^- + 2H^+ + H_2O \longrightarrow NH_4^+ + CO_2 \tag{5}$$

The ammonium cyanate does not accumulate below *p*H 5 and the rate of hydrolysis is almost independent of *p*H between *p*H 1 and 7. If the elevated temperature is not desired, cyanates can be used at room temperature to raise the *p*H, essentially by reaction (5). Other methods include the hydrolysis at elevated temperature of acetamide, hexamethylenetetramine, mixtures of halide and halate, and thiosulfate. The hydrolysis of 2-hydroxyethyl acetate can be used to lower *p*H slowly at elevated temperature since the acetic acid formed by hydrolysis reacts with hydroxyl ions and since the ethylene glycol (which is the other product) is neutral. For ammoniacal solutions the *p*H can be lowered by simple volatilization of the ammonia.

b. GENERATION OF OTHER REAGENTS. Various substances can be used at room or at elevated temperatures for slow generation of particular ions. Sulfamic acid and methyl or ethyl sulfate are utilized for sulfate, ethylene or propylene chlorohydrin for chloride, methyl or ethyl oxalate for oxalate, methyl phosphate or mataphosphoric acid for phosphate, trichloroacetic acid for carbonate, oxidation of chromic ion by bromate to form chromate, and oxidation of iodine by chlorate to form iodate.

In the case of precipitation using organic reagents, the reagent may be formed by some suitable reaction in the solution. Examples are the slow formation of dimethylglyoxime by the reaction of biacetyl and hydroxylamine and the hydrolysis of 8-acetoxyquinoline to form 8-hydroxyquinoline. Since the chelate or metal salt of the organic compound is usually much more soluble in organic solvents than in water, this type of compound may be precipitated by slow volatilization of an organic solvent such as acetone.

3. Uses of Precipitation from Homogeneous Solution

a. PRECIPITATION OF HYDROUS OXIDES. Gelatinous precipitates such as the hydrous oxides are very difficult to filter and are usually rather impure. Precipitation from homogeneous solution produces precipitates which are denser, purer, and much easier to filter. The presence of an organic anion often improves the characteristics of the precipitate. The precipitation is usually effected by the adjustment of *p*H to a value just below incipient precipitation, urea added, and the solution maintained at or near the boiling point for one or two hours or until precipitation is complete. In the presence of some organic acids (such as formic acid), the iron precipitate contains basic salts [such as basic ferric formate—$FeOCHO_2$ or $Fe(OH)_2$ CHO_2] and a ferric oxide monohydrate (FeOOH). These precipitates are much purer than those obtained by heterogeneous precipitation, and the purity can be improved even further by use of a two-step procedure in which the solution is cooled rapidly and filtered when the precipitation is approximately 90 to 95 per cent complete. After filtration, the solution is again heated to complete the precipitation and the two precipitates are combined. These precipitates are much easier to filter and to wash than those produced by addition of ammonia or sodium hydroxide.

b. PRODUCTION OF RELATIVELY LARGE CRYSTALS. Substances such as silver chloride which form as colloidal particles and then agglomerate to form curdy precipitates may be precipitated in the form of comparatively large crystalline particles by precipitation from homogeneous solution. Crystals with dimensions of as much as 0.2 mm have been produced by destruction of the silver amine complex [$Ag(NH_3)_2^+$] by the hydrolysis of 2-hydroxyethyl acetate and by the volatilization of ammonia from solutions which contain an excess of chloride ion. Nickel dimethylglyoximate, aluminum oxinate, thorium iodate, cadmium and other sulfides, and many other compounds which tend to form in small particles have been precipitated in crystalline form in much larger particles by precipitation from homogeneous solution.

C. STUDY OF COPRECIPITATION. Precipitation from homogeneous solution has been used to study coprecipitation, and it has been shown that coprecipitation often occurs during the later stages of the precipitation when the concentration of the ion being precipitated becomes small compared to the concentration of the coprecipitated impurity. In the coprecipitation of yttrium with ferric periodate formed by liberation of periodate from paraperiodic acid by hydrolysis of acetamide, it was found that coprecipitation was essentially negligible until approximately 99.5 per cent of the ion had been precipitated. Coprecipitation increased rapidly in the final stages of precipitation after 99.9 per cent of the iron had been precipitated. This explains the increased purity of the precipitate in the two-step method.

CHAPTER 8

Gravimetry and Precipitation Titrimetry

A. Gravimetry

A large number of gravimetric methods have been used in the analysis of materials since the beginning of careful analytical determinations. As satisfactory titrimetric methods and, more recently, instrumental methods are developed for many constituents, fewer and fewer gravimetric procedures are used. However, many gravimetric methods are still the method of choice for the analysis of a relatively large number of constituents determined in analytical laboratories. Often, if the results of the analysis are to be used as legal evidence in a court, a gravimetric procedure will be used because of its greater chance of being accepted and understood in a court of law.

Four representative gravimetric procedures which illustrate the three types of precipitates encountered in analytical methods together with one example of precipitation from homogeneous solution will be presented in this chapter. The basic theory and calculations of gravimetry dealing with the formation and solubility of precipitates, the completeness of precipitation, and the impurities found on or in precipitates have been discussed in previous chapters. The three types of precipitates encountered in gravimetric procedures are the crystalline precipitates exemplified by $BaSO_4$, the curdy precipitates of which AgCl is the most common example, and the gelatinous precipitates such as the hydrous oxides of iron and aluminum usually called ferric hydroxide and aluminum hydroxide.

1. Determination of Sulfate as Barium Sulfate

The precipitation of barium sulfate in the gravimetric determination of sulfate is performed by the addition of a barium chloride solution to a hot solution of sulfate which has been acidified with hydrochloric acid.

$$Ba^{2+} + SO_4^{2-} \longrightarrow BaSO_4 \qquad (1)$$

The hydrochloric acid prevents the precipitation of slightly soluble barium salts such as barium carbonate and barium phosphate. Since barium sulfate is more soluble in acid solutions than in water due to the formation of bisulfate (HSO_4^-) ions, the hydrochloric acid also favors the formation of larger particles according to the von Weimarn rule.

The barium sulfate precipitate tends to be contaminated with occluded and coprecipitated impurities, and quantitatively exact determinations can be accomplished only on sulfuric acid solutions and sodium sulfate solutions under well-defined conditions. The extent of the contamination and the type of contamination depends in part upon which reagent is added to the other. If the barium chloride solution is added to the sulfate solution, coprecipitation of cations predominates, but if the sulfate is added to the barium, coprecipitation of anions predominates. The extent of coprecipitation also depends upon the relative concentrations of the impurities in the solution and upon the solubility of the barium or sulfate compounds formed with the impurities.

Nitrate and chlorate are coprecipitated to a greater extent than other common anions such as the halides. Chloride is coprecipitated less than most other anions and is the anion of choice whenever possible. Almost all the heavy metals are coprecipitated with barium sulfate, with the ion of higher valence for any particular metal being coprecipitated more than ions of lower valence. Thus iron(III) (ferric) ions are coprecipitated much more than iron(II) (ferrous) ions. Ferric iron is present more often than other heavy metal ions and consequently causes major interference. The interference of ferric ion can be offset to some extent by reduction to ferrous ion but cannot be completely removed except by prior removal of the ferric ion by precipitation or complexation. Coprecipitation can cause either positive or negative errors depending upon the relationship of the equivalent weights of the metals or anions in question and upon which component of the precipitate is present in excess. Thus ferric ion causes a negative error for sulfate since iron has a lower equivalent weight than barium. In the analysis of barium, coprecipitation of ferric ion would cause a positive error since it would increase the total weight of the precipitate in the presence of excess sulfate. Similarly, nitrate causes a positive error in the determination of sulfate but would cause a negative error in the determination of barium since the nitrate is converted to the oxide upon ignition.

Cations such as lead and strontium which form insoluble sulfates in acid solution and anions such as chromate which form insoluble barium salts in acid solution must be absent since coprecipitation is extensive for such ions.

The precipitate is usually filtered through fine-porosity quantitative paper, washed with hot water until free from chloride, and ignited after the paper has been charred carefully and the carbon burned off. Oxidizing conditions must exist during the charring, burning, and ignition procedures to prevent reduction of the precipitate to the sulfide.

2. Determination of Chloride as Silver Chloride

Chloride is often determined by precipitation as silver chloride by addition of a solution of silver nitrate to a chloride solution which has been acidified with nitric acid. After a suitable aging or digestion period (preferably in the dark), the precipitate is filtered, washed with very dilute nitric acid, dried, and weighed. The precipitate must be protected from the light since silver chloride is reduced by light. When the proper procedure is followed, practically all sources of error can be eliminated, and the method is one of the most accurate procedures among analytical determinations. In fact, much of the early work on the determination of accurate atomic weights involved the conversion of the element to the chloride followed by the analysis of the chloride by precipitation as silver chloride.

The solution of chloride is acidified with nitric acid to prevent the precipitation of insoluble silver salts of weak acids such as carbonates and phosphates. The nitric acid also speeds the coagulation of the precipitate which forms as colloidal particles upon addition of the silver nitrate. The nitric acid further aids in the purification of the precipitate by displacing adsorbed silver ions to form adsorbed nitric acid which is volatilized in the drying process.

A large excess of silver nitrate must be avoided since silver chloride forms a soluble complex ion Ag_2Cl^+ in the presence of excess silver ion. Formation of such ions causes incomplete precipitation with a consequent negative error. A slight to moderate excess is desirable, however, to decrease the solubility of the precipitate by the common ion effect and thus decrease the amount of chloride ion left unprecipitated.

The precipitate forms immediately upon addition of the silver nitrate solution in colloidal particles which adsorb silver ions in the primary layer and nitrate ions in the secondary layer. The resulting particles are positively charged due to the adsorbed silver ions and tend to remain dispersed as colloidal particles which are too small to be of filterable size. Upon aging overnight in the presence of nitric acid, the particles coagulate to form filterable-sized particles called curds. The actual particles are still the same size but have clumped together to form a relatively large conglomerate which can be filtered. The adsorbed silver ions are removed in part by displacement by hydrogen ions during the aging process. If sufficient time is not available to allow overnight aging, the precipitate can be coagulated by digestion for approximately one hour near the boiling point of water.

The precipitate is usually filtered through a Gooch crucible or through a sintered glass crucible of medium porosity and must be washed with very dilute nitric acid to prevent peptization of the precipitate. If water is used as the wash liquid, the ions in the secondary layer are more easily removed than those in the primary layer. As a result the particles again become charged, repel each other, and redisperse as colloidal particles. The process of redispersion is known as peptization. The precipitate must be washed until there is no test for Ag^+ ions in the filtrate. The precipitate is then dried at 110–130°C, at which temperature essentially all the water is removed from the precipitate.

The precipitate must be protected from the light, especially while wet, since it is photosensitive and is easily reduced by light. This is the same reaction which makes photography possible since photographic film is essentially a dispersion of colloidal silver halides in a polymeric base. When light strikes the film, the silver halides are sensitized or partially reduced to the metal. The development process is simply the further reduction of all the sensitized halides to the metal by a weak reducing agent which is not strong enough to reduce unsensitized silver halides. Unreacted halides are then removed by dissolving in a sodium thiosulfate solution.

The reactions involved in the reduction of the silver chloride precipitate in the solution before filtration may be shown as

$$2AgCl + \text{light} \longrightarrow 2Ag^0 + Cl_2^0 \qquad (2)$$

$$3Cl_2^0 + 3H_2O \longrightarrow 6H^+ + ClO_3^- + 5Cl^- \qquad (3)$$

$$5Cl^- + 5Ag^+ \longrightarrow 5AgCl \qquad (4)$$

The formation of chloride ions occurs in a solution which contains excess silver ion and more silver chloride is precipitated. Since the metallic silver formed in the decomposition remains with the precipitate, a positive error occurs. If, however, the decomposition takes place after filtration, the error will be negative due to loss of elementary chlorine.

3. Determination of Iron as Ferric Oxide

Iron is often determined by the addition of ammonia to a hot solution containing iron(III) (ferric) ions to form the hydrous oxide. The precipitate is a gelatinous mass of coagulated colloidal particles which are often described as microcrystalline since they do not show a definite crystal pattern until after being aged for some time. The precipitate is a hydrous oxide of indefinite composition which may be represented as $Fe_2O_3 \cdot xH_2O$. For simplicity it is usually designated as $Fe(OH)_3$ and called ferric hydroxide even though the ratio of Fe_2O_3 to H_2O is seldom 1 : 3. Ferric hydroxide is very insoluble with K_{sp} in the range of 10^{-37} so that it may be quantitatively precipitated even at a *p*H of 4. The gelatinous precipitate rapidly clogs the pores of any filter paper so that a paper of course porosity must be used and the precipitate washed as much as possible by decantation before transfer to the filter. After filtration and washing with an electrolyte solution to prevent peptization, the paper is charred, the carbon removed by heating over a Bunsen flame, and the precipitate ignited at temperatures near 1000°C before weighing. Oxidizing conditions must be maintained during the charring, carbon removal, and ignition steps to prevent reduction of the precipitate to Fe_3O_4.

Any cation (such as aluminium and chromic) which forms insoluble hydroxides with ammonia will contaminate the precipitate. Many cations which will not form hydroxide precipitates with ammonia are coprecipitated with the iron, especially at high *p*H. At low *p*H (below approximately 8) the particles are positively charged and repel cations, but at higher *p*H the particles become negatively charged due to adsorption of hydroxide ions and thus attract cations into the secondary layer. As a result, most procedures call for slow addition of ammonia to the hot solution until a faint odor of ammonia persists after ammonia fumes have been dispelled by blowing across the surface. If the amount of iron is small, it is difficult to determine when precipitation starts upon addition of the ammonia. The best indication is a slight deepening of the light yellow or tan color of the solution due to the formation of colloidal particles of the hydrous oxide. If the solution is then digested for a short period of time, the precipitate will coagulate into the usual gelatinous mass. Many procedures call for coagulation by gentle boiling for a few minutes after the formation of the precipitate.

The fact that cations are attracted to the precipitate at high *p*H values is used in the "gathering" process by which trace amounts of certain metals are coprecipitated with the hydrous oxide of iron. The procedure calls for the addition of a concentrated solution of NaOH which raises the *p*H quickly to values in the range of 12 to 14, followed by coagulation at elevated temperatures. Trace amounts of manganese can be precipitated in this way for determination by satisfactory methods.

Fluoride and pyrophosphate ions must be absent since these two anions form very stable complexes with ferric ion, which prevent the precipitation of the hydrous oxide.

The precipitate is usually washed with warm solutions of an ammonium salt, either ammonium nitrate or ammonium chloride, to prevent peptization of the precipitate and to form a volatile compound in the primary and secondary layers which will be removed during the ignition step.

4. Precipitation of Nickel Dimethylglyoximate from Homogeneous Solution

Nickel(II) ion is often precipitated as dimethylglyoximate by the addition of ammonium hydroxide to an acidic solution containing nickel(II) ion and dimethylglyoxime (DMG). The precipitate thus formed is flocculent, fluffy, and difficult to filter. Formation of the precipitate by precipitation from homogeneous solution by the hydrolysis of urea in the presence of nickel and dimethylglyoxime produces a precipitate which is formed in relatively large particles and which is easy to filter. The basic theory and advantages of precipitation from homogeneous solution are covered in Chapter 7.

Dimethylglyoxime is essentially specific for nickel since it reacts only with nickel in solutions which are nearly neutral (*p*H 5 and above). It will also precipitate palladium from dilute hydrochloric acid solutions and bismuth from solutions of high *p*H (11 to 11.5). No other single element interferes, but cobalt and ferric iron together will hinder the nickel from precipitating in the usual bright red form. Dimethylglyoxime forms a chelate with nickel which can be represented by the equation in Fig. 8-1.

Chelates are derivatives of organic compounds which have two functional groups so arranged that a ring structure can be formed with a metal cation. The stable ring structure may be contained in a neutral compound or may exist as an ion. Chelating precipitation agents are usually much more specific than inorganic precipitants. Insoluble chelate compounds are also less contaminated by coprecipitation or adsorption of ionic impurities than the inorganic precipitates.

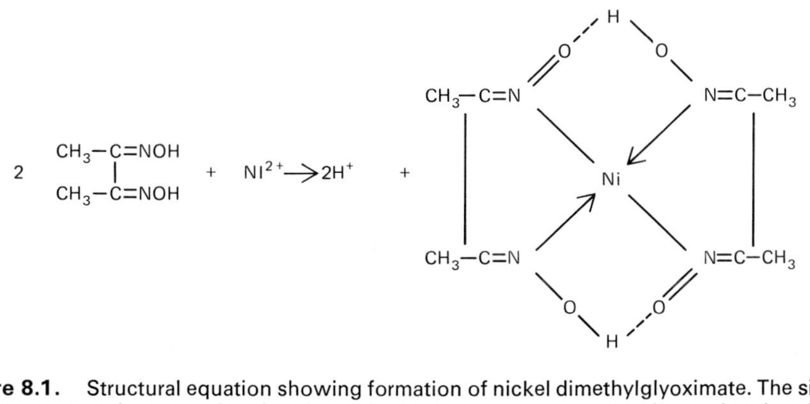

Figure 8.1. Structural equation showing formation of nickel dimethylglyoximate. The single line represents a single covalent bond, the double line represents a double covalent bond, the arrow represents a coordinate bond in which the arrow points away from the atom which furnishes the electron pair, and the dashed line represents a hydrogen bond.

In the determination by precipitation from homogeneous solution, the *p*H of the solution must be adjusted to a value just below incipient precipitation. A *p*H of approximately 2.6 to 2.8 is satisfactory. A *p*H greater than 2.8 will allow the precipitate to form immediately upon addition of dimethylglyoxime, while a *p*H much below 2.6 will require too long a time before precipitation begins. The dimethylglyoxime is dissolved in 1-propanol rather than ethanol to decrease loss by evaporation during the heating.

The *p*H is raised slowly and uniformly throughout the solution by the hydrolysis of urea (see Chapter 7). The hydrolysis is slow at room temperature but is moderate at temperatures near the boiling point of water so that the necessary *p*H rise occurs over a reasonable length of time.

Dimethylglyoxime is insoluble in water so that solutions in organic solvents must be used. Since the precipitate is also more soluble in organic reagents, the amount of reagent and thus the amount of dimethylglyoxime added are controlled so that neither will the dimethylglyoxime precipitate nor will the precipitation be incomplete due to solubility in the organic solvent. Use of a volatile solvent such as methanol or ethanol allows too much loss by evaporation with consequent precipitation of the excess precipitating agent.

The gravimetric factor of 0.2032 is very favorable since it requires almost a five milligram error in the precipitate to cause a one milligram error in the amount of nickel.

B. Precipitation Titrimetry

Precipitation titrations are titrations in which at least one product is an insoluble solid. While there are many reactions in which a substance is quantiatively precipitated, only a few are known for which satisfactory indicators are available. Precipitation methods are used primarily in the titration of chloride with silver, though others are known and used. This section will deal only with the chloride titration methods.

There are three distinct methods for the titration of chloride with silver based upon the different types of indicators used in the methods. The three types of indicators are (1) the formation of a colored solid as used in the Mohr method, (2) the formation of a colored soluble complex as in the Volhard method, and (3) the change of color associated with the adsorption of the indicator on the surface of a solid particle, which is the basis of the Fajans method.

In any titration there must be a very great change in the concentration of at least one ion at the equivalence point, and to show such changes graphically, the *p* system of expressing concentrations is used. In this system, the molar concentration of an ion is expressed in exponential form, and the *p* value is the exponent with the sign changed. Thus a molar concentration of chloride of 5.00×10^{-3} is also equal to $10^{-2.30}$ for which the *p*Cl would be 2.30. (See Appendix 10 concerning the use of logarithms and the expression of numbers in the exponential form.) A typical graph for the variation of *p*Cl with volume of titrant is shown in Fig. 8-2. It should be noted that there is a break of 4 *p*Cl units from 3 to 7 with two to four drops of titrant. The molar concentration change represented by this break at the end point is from 10^{-3} to 10^{-7}, which is a 10,000-fold change.

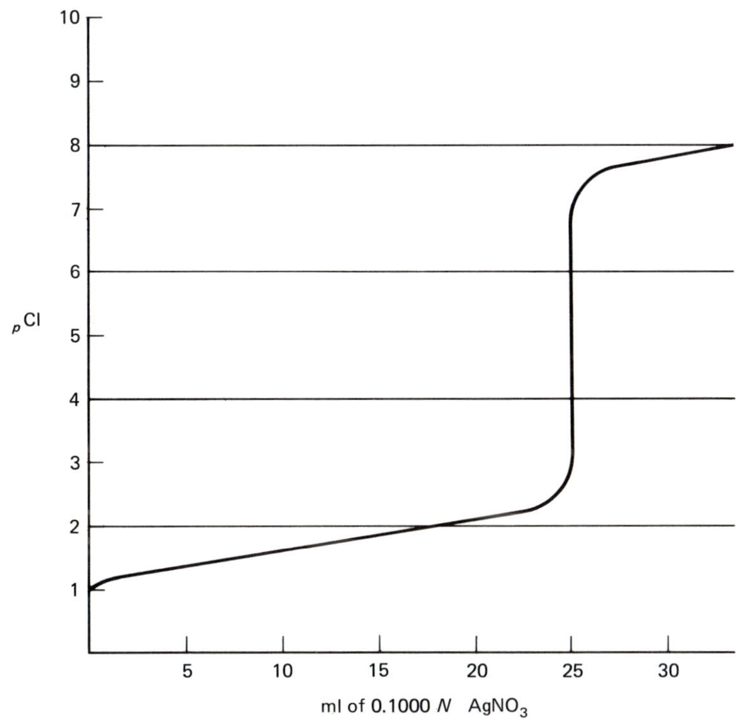

Figure 8.2. Titration curve for the titration of 25.00 ml of 0.1000 *N* NaCl with 0.1000 *N* $AgNO_3$. $M_{Cl^-} = 10^{-pCl^-}$.

1. The Mohr Method

In the Mohr method, potassium chromate is added to the chloride solution to act as the indicator, with the end point being the change in color due to the formation of reddish-brown silver chromate. As the titrant is added to the chloride solution, the silver ion is removed by reaction with the chloride ion until the chloride has been quantitatively precipitated as AgCl. The Ag_2CrO_4 then begins to precipitate as the silver ion concentration increases due to the complete precipitation of chloride. The end point depends upon the relative solubility of AgCl and Ag_2CrO_4 or upon the fractional precipitation of chloride in the presence of chromate.

For the method to be sufficiently accurate for use in analysis, the silver chromate must not begin precipitation until the chloride has been reduced to a negligible value, as shown symbolically in eqs. (5) and (6).

$$Ag^+ + \begin{matrix} Cl^- \\ CrO_4^{2-} \end{matrix} \longrightarrow AgCl \quad \text{(until } p\text{Cl is approximately 6)} \tag{5}$$

$$2Ag^+ + CrO_4^{2-} \longrightarrow Ag_2CrO_4 \quad \text{(only after } p\text{Cl is approximately 6)} \tag{6}$$

The calculations which show that the chloride is reduced to a neglible value are given in Chapter 6, section D4 and should be reviewed by the student.

A blank is necessary in this method since the Ag_2CrO_4 actually begins to precipitate slightly beyond the equivalence point, which occurs at $p\text{Cl} = 5$, and since some silver ion must be used to produce enough silver chromate to give a visible change. The latter error can be offset in part by using a potassium chromate indicator solution which is saturated with silver chromate.

The solution must be neutral or slightly basic (pH from 7 to 9) since chromate reacts with hydrogen ion in acid solutions to form hydrogen chromate ions, which reduces the concentration of chromate. At higher pH there are enough hydroxide ions present to cause the

$$CrO_4^{2-} + H^+ \longrightarrow HCrO_4^- \tag{7}$$

$$2Ag^+ + 2OH^- \longrightarrow 2AgOH \longrightarrow Ag_2O + H_2O \tag{8}$$

formation of silver hydroxide or oxide. Even though chromic acid is a strong acid, the second ionization is definitely weak since K_2 is 3.2×10^{-7}. Consequently, at pH values below 7, an appreciable fraction of the indicator is converted to the hydrogen chromate ion. Most directions offset errors due to pH by adjustment to the neutral point of phenolphthalein before titration by adding phenolphthalein and making the solution basic with 0.5 N NaOH followed by addition of one to two drops of 0.5 N H_2SO_4 to remove the color.

Most directions also call for the volume at the end point to be approximately 100 ml in order to keep the chromate ion concentration essentially equal in all titrations.

2. The Fajans Method (Adsorption Indicator Method)

The Fajans method is based upon the fact that some organic compounds which adsorb strongly on certain precipitates and undergo a change in color upon adsorption can be used as indicators for selected titrations.

Fajans and his coworkers found that fluorescein and several of its derivatives are strongly adsorbed on silver chloride precipitates and change in color from a yellow with a greenish fluorescence to a pinkish red upon being adsorbed. Adsorption indicators can be used only for those titrations in which conditions are proper for the adsorption to occur or disappear at or near the equivalence point.

In the titration of chloride by silver the precipitated silver chloride is formed in a solution which contains excess chloride and consequently adsorbs chloride ion in the primary layer to become negatively charged. These charged particles then attract cations in the secondary layer (see Fig. 7-4), which can be symbolized as $AgCl : Cl^- : Na^+$. The particles will remain negatively charged until the exact equivalence point at which they become neutral and tend to coagulate. With the first excess of silver, however, the particles adsorb silver ion strongly and become positively charged. The positive particles attract anions into the secondary layer, which can be symbolized as $AgCl : Ag^+ : NO_3^-$. As a consequence there is a sharp change in charge on the particles just beyond the equivalence point. If the chloride is added to a silver solution, the particles are positively charged before the equivalence point and become negatively charged just beyond the equivalence point.

Fluorescein is a weak acid ($HC_{20}H_{11}O_5$ or HFl as a short abbreviated form) which is dissolved as the sodium salt and exists in solution as the fluoresceinate ion Fl^-. The fluoresceinate ion is attracted to the positively charged silver chloride particles and repelled by the negatively charged particles. In the titration of chloride with silver, the indicator anion is repelled by the particles until just beyond the equivalence point, when the charge changes to positive. The attraction of the fluoresceinate ion into the secondary layer, which can be symbolized as $AgCl : Ag^+ : Fl^-$, causes a change in structure which produces a reddish pink color on the surface of the particles. The color change can be explained as being due to the precipitation of silver fluoresceinate upon the surface due to the increased concentration of fluoresceinate ions in the counterion region. Alternatively, the pink color has been explained as a structural change in the fluoresceinate ion due to the energy relationships in the adsorption process on the particle surface. With either explanation the color change must be preceded by the adsorption of the indicator into the secondary layer. Since the molecular or nonionized form of fluorescein is not attracted to the particles and will not be adsorbed, the indicator must be in the ionic form to be effective.

The ratio of ionized to nonionized form for weak acids is a function of hydrogen ion concentration, and as a consequence the *p*H of the solution is important to the proper functioning of the indicator. For fluorescein, the solution must be essentially neutral or slightly basic (*p*H in the range of 6–10) in order for sufficient ions to be present. Dichlorofluorescein, which is a stronger acid, can be used down to a *p*H of approximately 4 satisfactorily.

Since the indicator works by adsorption, it is important that the surface area be kept as large as possible, which means that the particle size should be as small as possible. To offset coagulation of the particles before the equivalence point, a protective colloid such as dextrin is usually added to the solution being titrated. The dextrin forms a permeable coating around the particle which extends the depth of the counter ion region and retards coagulation.

Even though the end point occurs with an excess of titrant, there is no satisfactory way to determine the end-point error since there is no convenient way to run a blank determination. The end-point error is usually small and can be neglected in most cases but can be estimated when necessary by titrating primary standard solutions of silver nitrate and sodium chloride of identical concentrations.

3. The Volhard Method

The Volhard method is an indirect or back titration method in which an excess of a standard solution of silver nitrate is added to a chloride solution. The excess silver nitrate is then determined by titration with a standard solution of potassium or ammonium thiocyanate with ferric ion as indicator. The actual titration involves the reaction of silver ion with thiocyanate ion, or

$$Ag^+ + SCN^- \longrightarrow AgSCN \tag{9}$$

The first excess of thiocyanate reacts with the ferric ion to form the intensely colored $FeSCN^{2+}$ complex, or

$$SCN^- + Fe^{3+} \longrightarrow FeSCN^{2+} \tag{10}$$

The indicator is usually a concentrated [40%(w/w) or almost saturated] solution of ferric alum [ferric ammonium sulfate hydrate, $FeNH_4(SO_4)_2 \cdot 12H_2O$] in 20% (v/v) nitric acid. The nitric acid helps to bleach the color of the ferric ion and also prevents the hydrolysis of the ferric ion to form the hydrous oxide.

The standard silver solution is added to the acidified chloride solution to form the precipitate of AgCl. The acid (usually 5 ml of HNO_3 for each 25 ml of chloride solution) speeds the coagulation of the particles and tends to displace adsorbed silver ions. Since AgSCN (K_{sp}, 1×10^{-12}) is less soluble (more insoluble) than AgCl (K_{sp}, 1×10^{-10}), the SCN^- can react with and dissolve the precipitated AgCl, or

$$AgCl_{(solid)} + SCN^- \longrightarrow AgSCN_{(solid)} + Cl^- \tag{11}$$

For this reason, the precipitated chloride must be removed from contact with the solution before titration with thiocyanate. The removal can be accomplished by filtration, but the resultant dilution causes indistinct end points due to the large volume of the solution. Also, some of the excess silver ion may not be washed completely out of the paper and would cause a negative error. Nitrobenzene is an insoluble organic liquid which will form a coating around the silver chloride particles and thus remove them from contact with the solution since the coating is impermeable to water and the solution. Practically all procedures now call for the addition of the nitrobenzene rather than filtration. It should be noted that the compound used must be insoluble in water and must coat the particles with an impermeable layer to keep the particles from contact with the solution even though they are still suspended in the titration vessel which contains the solution.

The Volhard method can also be used for determination of bromide and iodide without removal or protection of the precipitate since silver bromide and silver iodide are less soluble than silver thiocyanate. In the titration of iodide, the indicator should not be added until after the precipitation of the silver iodide to prevent oxidation of iodide ion by ferric ion. The Volhard method can also be used as a direct method for the determination of silver by titration with thiocyanate or for the determination of thiocyanate by titration with silver.

CHAPTER 9

Neutralization

A. Acids and Bases

For many years acids and bases were defined in terms of the typical properties which are associated with them. Thus acids are substances which turn litmus red, while bases turn this dye blue; acids react with bicarbonates (or carbonates) to form carbon dioxide, while bases react with carbon dioxide to form bicarbonates (or carbonates); acids have a sour taste, while bases feel slippery.

With the advent of the Arrhenius theory of ionization, the fact that acids react with metals to produce hydrogen and a salt led to the definition that acids are substances which dissociate into hydrogen ions and anions when dissolved in water or are substances which contain "replaceable" hydrogen. Since acids react with bases to form a salt and water, bases must be substances which dissociate into hydroxyl ions and cations when dissolved in water. These definitions are not completely satisfactory since they neglect the role played by the solvent and do not explain why solutions of salts often are acidic or basic rather than neutral.

Brønsted and Lowry independently proposed a more general definition in 1923 in which an acid is defined as any substance which can give up hydrogen ions (protons) to the solvent or to any base present and a base is any substance which can combine with hydrogen ions. Thus an acid is a proton donor, while a base is a proton acceptor. It should be realized that hydrogen ions or bare protons do not exist in water solution since the hydrogen ion is hydrated to form $H(H_2O)_x^+$. The value of x is usually assumed to be one and the hydrogen ion in solution is called the hydronium ion and is written as $H_3\dot{O}^+$. For simplicity, since it makes no major difference in most calculations or conceptions, the hydronium ion will be used only when necessary for clarity.

According to the Brønsted definition, acids and bases are related by the equation

$$\text{Acid} \longrightarrow H^+ + \text{base} \tag{1}$$

examples of which are

$$H_2O \longrightarrow H^+ + OH^- \tag{2}$$

$$HCl \longrightarrow H^+ + Cl^- \tag{3}$$

$$HC_2H_3O_2 \longrightarrow H^+ + C_2H_3O_2^- \tag{4}$$

$$H_3PO_4 \longrightarrow H^+ + H_2PO_4^- \tag{5}$$

$$H_2PO_4^- \longrightarrow H^+ + HPO_4^{2-} \tag{6}$$

$$NH_4^+ \longrightarrow H^+ + NH_3 \tag{7}$$

$$Al(H_2O)_6^{3+} \longrightarrow H^+ + AlOH(H_2O)_5^{2+} \tag{8}$$

These acids and bases are known as conjugate pairs, chloride being the conjugate base of hydrochloric acid and the ammonium ion being the conjugate acid of the base ammonia.

Acids may be classified according to charge as (1) neutral molecule acids such as acetic, (2) anion acids such as primary phosphate ($H_2PO_4^-$), and (3) cation acids such as the ammonium ion. Acids may also be classified according to the number of replaceable hydrogens as monoprotic (acetic, hydrochloric), diprotic (sulfuric), triprotic (phosphoric), and polyprotic (any acid with more than one replaceable hydrogen). Those which can act as either acid or base (primary phosphate, bicarbonate) are said to be amphiprotic.

To emphasize the role of the solvent, the general acid-base reaction can be written as

$$\text{Acid}_1 + \text{base}_2 \longrightarrow \text{acid}_2 + \text{base}_1 \tag{9}$$

and the reaction will take place quantitatively as written only if it is much easier to transfer a proton from acid$_1$ to base$_2$ than it is to transfer a proton from acid$_2$ to base$_1$. For example, the ionization of water takes place only to a limited extent, while the reverse of this reaction (the neutralization of a strong acid with a strong base) attains essential completion. These reactions may be shown as

$$H_2O + H_2O \longrightarrow H_3O^+ + OH^- \quad \text{(slight extent)} \tag{10}$$

$$H_3O^+ + OH^- \longrightarrow H_2O + H_2O \quad \text{(essential completion)} \tag{11}$$

Similarly, the ionization of hydrochloric acid or the dissolving of hydrogen chloride gas in water,

$$HCl + H_2O \longrightarrow H_3O^+ + Cl^- \tag{12}$$

takes place to a great extent because it is much easier to remove a proton from a chloride ion (or hydrogen chloride moelcule) in water than it is to remove a proton from a hydronium ion. Another way of stating this is that the chloride ion has a much smaller affinity for a hydrogen ion than does a water molecule.

Reactions such as these are not confined to water solutions only but may occur in any solvent which has acid or basic properties. The dissolving of hydrogen chloride gas in ethyl alcohol and in glacial acetic acid is shown by

$$HCl + C_2H_5OH \longrightarrow C_2H_5OH_2^+ + Cl^- \tag{13}$$

$$HCl + HC_2H_3O_2 \longrightarrow H_2C_2H_3O_2^+ + Cl^- \tag{14}$$

The extent to which these reactions take place depends upon the basic strength of the solvent. In water, hydrogen chloride is a strong acid since it is essentially completely dissociated, while in glacial acetic acid (which is a much weaker base than water) it is classified as a weak acid since it is dissociated only to a limited extent.

Since we are primarily concerned with aqueous solutions in analytical work, we shall confine our discussion to the strength of acids and bases in water. As previously stated, the value of the ionization constant is a measure of the strength of the acid or base, with small values indicating slight dissociation or weak acids and bases. For conjugate acid and base pairs, the strength of the base varies inversely as the strength of the acid. Thus Brønsted classifies the chloride ion as an infinitely weak base and the acetate ion as a strong base. Similarly, the sodium ion is classified as an infinitely weak acid and the aluminum ion as a strong acid.

B. Acidity and Basicity

Pure water is dissociated into hydrogen and hydroxyl ions so slightly that 1 liter of water contains only 0.0000001 g of hydrogen ion and 0.0000017 g of hydroxyl ion. As a result the concentration of nonionized molecules of water is a constant in pure water and in the dilute solutions ordinarily used in analysis. For this reason the ionization constant of water is usually stated as the ion product constant K_w. At 25°C (room temperature),

$$K_w = (H^+) \times (OH^-) = 1.0 \times 10^{-14} \tag{15}$$

Since we are disregarding the effect of ionic strength and assuming that active and molar concentrations are equal, this becomes

$$[H^+] \times [OH^-] = 1.0 \times 10^{-14} \tag{16}$$

In pure water, the hydrogen ion concentration equals the hydroxyl ion concentration since they are produced in equal numbers by ionization and

$$[H^+] = [OH^-] = \sqrt{K_w} = 1.0 \times 10^{-7}\ M \tag{17}$$

Pure water is said to be neutral since there is no excess of either hydrogen or hydroxyl ions. The dissolving of an acid in water produces hydrogen ions, and as a result the hydrogen ion concentration in an acid solution must be greater than it is in pure water or greater than 10^{-7} *M*. Similarly a solution of a base contains more hydroxyl ions than pure water, and the hydroxyl ion concentration is greater than 10^{-7} *M*. Due to the equilibrium existing between the water molecules and the ions, an increase in the concentration of either ion causes a corresponding decrease in the other. Consequently the hydroxyl ion concentration in an acid solution and the hydrogen ion concentration in a basic solution are both less than 10^{-7} *M*.

The molar hydrogen ion concentration in neutral or basic solutions is a very small number and consequently difficult and tedious to use. To simplify the expression of hydrogen ion concentrations, Sørensen suggested the use of the term *p*H, which he defined as the negative logarithm of the molar hydrogen ion concentration, or

$$pH = -\log[H^+] \tag{18}$$

Similarly, *p*OH is the negative logarithm of the molar hydroxyl ion concentration and *pK* the negative logarithm of the ionization constant. The measurement of hydrogen ion concentrations by potentiometric methods gives the hydrogen activity rather than the molar concentration, and as a result different symbolizations have been suggested to distinquish the type of concentrations referred to. Some of the symbolizations found in the literature are p_aH and pH for active concentrations, p_cH and p_MH for molar concentrations, and p_sH for potentiometric measurements. The practical pH scale in use today is defined in terms of measurements against pH standards set by the National Bureau of Standards in Washington, D.C. For our purposes, we shall use the symbols p_cH and p_cOH to emphasize the fact that we are neglecting the effect of ionic strength on the effective concentration of an ion.

If the term *acidity* is used to denote the hydrogen ion concentration in a solution, it may be misleading since the hydrogen ion concentration and the concentration of the acid in a solution are seldom identical except in the case of strong monoprotic acids. For this reason, the acidity of a solution is usually thought to be of three types: momentary, titratable (or total), and reserve (or residual) acidity. The momentary acidity is the actual hydrogen ion concentration or the p_cH of the solution. The titratable acidity is that which can be neutralized by addition of base. For strong acids the momentary and titratable acidity are essentially the same since the acid is completely ionized, but for weak acids which exist mainly as molecules in the solution, the momentary and total acidities differ greatly. The weak acids ionize to furnish more hydrogen ions as the base neutralizes those present and consequently constitute a reservoir of hydrogen ions which are available only after ionization. This acidity which is available only after ionization is known as the reserve or residual acidity. The sum of the momentary and reserve acidities equals the total or titratable acidity.

C. Relationships Among p_cH, $[H^+]$, p_cOH, $[OH^-]$, K_w and pK_w

The molar concentration and the p_cH are related by the expression

$$[H^+] = 10^{-p_cH} \tag{19}$$

or the p_cH is the exponent of the hydrogen ion concentration with its sign changed. Thus the p_cH of pure water is 7 since the hydrogen ion concentration is 10^{-7} M. For acid solutions in which the hydrogen ion concentration is greater than 10^{-7} M, the p_cH is a number less than 7, while for basic solutions in which the hydrogen ion concentration is less than 10^{-7} M, the p_cH is a number greater than 7.

To convert molar hydrogen ion concentration to p_cH, express the concentration as 10 raised to the correct power. The p_cH is this exponent with the sign changed. The calculation of the p_cH of a solution in which the hydrogen ion concentration is 3.45×10^{-9} *M* would be

$$[H^+] = 3.45 \times 10^{-9} = 10^{0.53} \times 10^{-9} = 10^{-8.47}$$
$$p_cH = 8.47$$

The reverse, or conversion of p_cH to hydrogen ion concentration, is performed as follows:

$$p_cH = 5.18$$
$$[H^+] = 10^{-p_cH} = 10^{-5.18} = 10^{0.82} \times 10^{-6} = 6.61 \times 10^{-6}\ M$$

The relationship between the p_cH, the p_cOH, and pK_w may be shown by converting the K_w expression to negative logarithms, or,

$$[H^+] \times [OH^-] = K_w = 10^{-14} \tag{20}$$

$$-\log[H^+] + -\log[OH^-] = -\log K_w = -\log 10^{-14} \tag{21}$$

so that by definition

$$p_cH + p_cOH = pK_w = 14 \tag{22}$$

and the sum of p_cH and p_cOH equals 14 in water and aqueous solutions.

D. Acid-Base Titrations

1. Conditions Existing in the Titrated Solution at the Equivalence Point

a. GENERAL. At the equivalence or stoichiometric point, equivalent quantities of the reactants have been brought into reaction so that the equivalents of base must be equal to the equivalents of acid. Another way of stating this is to say that the milliequivalents of titrant must be equal to the milliequivalents of the titrated substance at the equivalence point. Since the general reaction may be shown as

$$\text{acid} + \text{base} \longrightarrow \text{salt} + \text{water} \tag{23}$$

the solution at the equivalence point is a solution of the salt formed and is identical to a solution prepared by dissolving the salt in water. As a consequence the calculation of the p_cH at the equivalence point in a titration becomes the calculation of the p_cH of a solution of the salt formed.

b. CONCENTRATION OF SALT AT THE EQUIVALENCE POINT. The p_cH at the equivalence point in many cases is affected by the concentration of the salt. The concentration of the salt can be calculated from the concentration and volume (milliequivalents) of the acid or base and the final volume of the titrated solution at the equivalence point. The normal concentration must then be converted to molar concentration before the calculation for p_cH is made. As an example, calculate the molar concentration of the salt at the equivalence point in the titration of 40.00 ml of 0.1025 *N* hydrochloric acid with 0.2075 *N* barium hydroxide (1) assuming no dilution with water, (2) assuming dilution with 20.25 ml of water, and (3) assuming dilution to 100 ml with water. To solve such a problem, first determine the number of milliequivalents of the acid or of the base used since this will equal the number of milliequivalents of salt formed, or

$$\text{meqs of salt} = \text{ml} \times N \text{ (of either acid or base)}$$
$$= 40.00 \times 0.1025 = 4.100 \text{ meqs}$$

Next, calculate the total volume of the solution if it is not given. For (1) the volume will be 40.00 ml for the acid plus 19.75 ml for the base. The milliliters of base added can be found by dividing the milliequivalents of the acid by the normality of the base (in this case, $4.100/0.2075 = 19.75$). For (2) the volume will be 59.75 for the acid and base plus the 20.25 ml of water added, or 80.00 ml. The third step is to calculate the normal concentration of the salt as shown below:

$$N_{\text{salt}} = N_s = \text{meqs/ml} \tag{24}$$

for

$$(1)\quad N_s = 4.100/59.75 = 0.0686\ N$$
$$(2)\quad N_s = 4.100/80.00 = 0.0513\ N$$
$$(3)\quad N_s = 4.100/100.0 = 0.0410\ N$$

Then convert the normal concentration to molar concentration by

$$M_{\text{salt}} = M_s = N_s/\text{equiv./mole} \tag{25}$$

for

$$(1)\quad M_s = 0.0686/2 = 0.0343\ M$$
$$(2)\quad M_s = 0.0513/2 = 0.0257\ M$$
$$(3)\quad M_s = 0.0410/2 = 0.0205\ M$$

2. Solutions of Salts

a. HYDROLYSIS. When some salts are dissolved in water, the cation or the anion or both will react with water to some extent. This reaction is known as hydrolysis and is the opposite of neutralization. Since neutralization reactions go to essential completion, the reverse or hydrolysis reaction usually takes place only to a slight extent. For hydrolysis to occur, a weak acid or a weak base or both must be formed so that only the salts of weak acids and/or weak bases undergo hydrolysis. If acids and bases are classified as strong and weak, the four possible classifications of salts are:

1. Salt of strong acid and strong base such as NaCl
2. Salt of strong acid and weak base such as NH_4Cl
3. Salt of weak acid and strong base such as $NaC_2H_3O_2$
4. Salt of weak acid and weak base such as $NH_4C_2H_3O_2$

Of these, the first does not undergo hydrolysis since no weak acid or weak base can be formed, and as a consequence the solution is neutral with p_cH equal to 7. In the second case, the cation of the weak base reacts with water,

$$NH_4^+ + HOH \rightleftarrows NH_4OH + H^+ \tag{26}$$

to form the weak base and hydrogen ions, and so the solution is acidic with p_cH less than 7. For the third class, the anion of the weak acid reacts with water,

$$C_2H_3O_2^- + HOH \rightleftarrows HC_2H_3O_2 + OH^- \tag{27}$$

to form the weak acid and hydroxyl ions, and so the solution is basic with p_cH greater than 7. The salts of weak acids and weak bases react with water,

$$NH_4^+ + C_2H_3O_2^- + HOH \rightleftarrows NH_4OH + HC_2H_3O_2 \tag{28}$$

to form both the weak base and the weak acid, and the p_cH will depend upon the relative strengths of the acid and base. In most cases, the p_cH will be close to 7 but not necessarily exactly 7.

The same conclusions as to acidity or basicity of salt solutions can be deduced by use of the Brønsted theory since conjugate acids and bases have inverse strengths and since anions are bases and cations are acids. In the case of NaCl, the Na^+ ion is an infinitely weak acid since OH^- is the strongest possible base in aqueous solution, and the Cl^- ion is an infinitely weak base since hydrochloric acid is completely dissociated in solution and H^+ (or H_3O^+) is the strongest possible acid in aqueous solutions. As a consequence, the solution is neutral since both the acid and base are weaker than water.

In the case of NH_4Cl, the ammonium ion is stronger as an acid than chloride is as a base and consequently the solution is acidic, while for sodium acetate the solution is basic since acetate ion is a strong base. For salts of weak acids and weak bases such as $NH_4C_2H_3O_2$, the solution will contain a strong Brønsted acid (NH_4^+) and a strong Brønsted base ($C_2H_3O_2^-$), so the solution will be close to neutral, and the hydrogen ion concentration will depend upon the relative strengths of the strong acid and base present.

b. p_cH OF SOLUTIONS OF SALTS. (1) *Salts of strong acids and strong bases.* As previously explained, the p_cH of such solutions is 7.00 since it is the same as water. If the effect of ionic strength on K_w is taken into account the p_cH will deviate slightly from 7 in solutions of high ionic strength.

(2) *Salts of strong bases and weak acids.* Considering a solution of NaA, the sodium salt of the weak acid HA, the anion will react with water as follows:

$$A^- + HOH \rightleftharpoons OH^- + HA \tag{29}$$

for which

$$K_h = \frac{[OH^-][HA]}{[A^-]} \tag{30}$$

If this equation is multiplied by $[H^+]/[H^+]$, it becomes

$$K_h = \frac{[H^+][OH^-][HA]}{[H^+][A^-]} = \frac{K_w}{K_a} \tag{31}$$

Neglecting the small hydroxyl ion concentration produced from ionization of water, the hydroxyl ion concentration and the concentration of nonionized HA are equal since they are produced in the same amounts. Also, since the amount of anion which reacts is relatively small compared to the total amount present, the concentration of A^- can be considered to be equal to the molar concentration of the salt (M_s). Making these substitutions, eq. (30) may be written as

$$K_h = \frac{K_w}{K_a} = \frac{[OH^-]^2}{M_s} \tag{32}$$

or

$$[OH^-]^2 = \frac{K_w M_s}{K_a} \tag{33}$$

$$[OH^-] = \sqrt{\frac{K_w M_s}{K_a}} \tag{34}$$

Taking negative logs, this becomes

$$-\log[OH^-] = \tfrac{1}{2}(-\log K_w - \log M_s + \log K_a)$$

or

$$p_cOH = \tfrac{1}{2}(pK_w - pK_a - \log M_s) \qquad (35)$$

If $K_w/[H^+]$ is substituted for $[OH^-]$ in eq. (33), it becomes

$$\frac{K_w^2}{[H^+]^2} = \frac{K_w M_s}{K_a} \qquad (36)$$

or

$$[H^+]^2 = \frac{K_w^2 K_a}{K_w M_s} = \frac{K_w K_a}{M_s} \qquad (37)$$

and

$$[H^+] = \sqrt{\frac{K_w K_a}{M_s}} \qquad (38)$$

which becomes, upon taking negative logs,

$$p_cH = \tfrac{1}{2}(pK_w + pK_a + \log M_s) \qquad (39)$$

Any of these equations (34, 35, 38, or 39) may be used to calculate the p_cH of a solution of a salt of a monoprotic weak acid, the secondary salt of a diprotic weak acid, or the tertiary salt of a triprotic weak acid.

Example 1. Calculate the p_cH of a 0.020 M solution of sodium benzoate. Benzoic acid is monoprotic with pK_a equal to 4.20. (a) Using eq. (34),

$$[OH^-] = \sqrt{\frac{10^{-14} \times 2 \times 10^{-2}}{10^{-4.20}}} = \sqrt{\frac{10^{0.30} \times 10^{-16}}{10^{-4.20}}}$$

$$= \sqrt{\frac{10^{-15.70}}{10^{-4.20}}} = \sqrt{10^{-11.50}} = 10^{-5.75}\ M$$

$$p_cOH = 5.75$$

$$p_cH = 14.00 - 5.75 = 8.25$$

(b) Using eq. (35),

$$p_cOH = \tfrac{1}{2}[14.00 - 4.20 - (-1.70)] = 11.50/2 = 5.75$$

$$p_cH = 14.00 - 5.75 = 8.25$$

(c) Using eq. (38),

$$[\mathrm{H}^+] = \sqrt{\frac{10^{-14} \times 10^{-4.20}}{10^{-1.70}}} = \sqrt{10^{-16.50}} = 10^{-8.25}$$

$$p_c\mathrm{H} = 8.25$$

(d) Using eq. (39),

$$p_c\mathrm{H} = \tfrac{1}{2}(14.00 + 4.20 - 1.70) = 8.25$$

It can be seen from the above example that the use of eq. (39) involves the fewest calculations and consequently will be used in the remaining examples.

Example 2. Calculate the p_cH at the equivalence point in the titration of 25.00 ml of 0.1000 *N* acetic acid with 0.1000 *N* sodium hydroxide. The pK_a of acetic acid is 4.76. Assume no dilution with water. First, calculate M_s.

$$N_s = \frac{25.000 \times 0.1000}{50.00} = 0.0500\ N$$

$$M_s = 0.0500\ N/1 = 0.0500\ M = 10^{-1.30}$$

Then, using eq. (39),

$$p_c\mathrm{H} = \tfrac{1}{2}(14.00 + 4.76 - 1.30) = 17.46/2 = 8.73$$

(3) *Salts of weak bases and strong acids.* The cation of the salt MA of a weak base and strong acid will react with water,

$$\mathrm{M}^+ + \mathrm{HOH} \rightleftarrows \mathrm{H}^+ + \mathrm{MOH} \tag{40}$$

for which K_h equals

$$K_h = \frac{[\mathrm{H}^+][\mathrm{MOH}]}{[\mathrm{M}^+]} \tag{41}$$

If this equation is multiplied by $[\mathrm{OH}^-]/[\mathrm{OH}^-]$, it becomes

$$K_h = \frac{[\mathrm{H}^+][\mathrm{OH}^-][\mathrm{MOH}]}{[\mathrm{M}^+][\mathrm{OH}^-]} = \frac{K_w}{K_b} \tag{42}$$

Neglecting the small hydrogen ion concentration from the ionization of water, the hydrogen ion concentration and the concentration of nonionized molecules of the weak base MOH must be equal since they are produced in the same amounts by the reaction of the cation with water. Also, the concentration of the cation may be assumed to be equal to the molar concentration of the salt since the salt is completely ionic and the hydrogen ion concentration is small compared to the salt concentration. Making these substitutions, eq. (42) may be written as

$$K_h = \frac{K_w}{K_b} = \frac{[H^+]^2}{M_s} \tag{43}$$

or

$$[H^+] = \sqrt{\frac{K_w M_s}{K_b}} \tag{44}$$

and, taking negative logs,

$$\begin{aligned} -\log[H^+] &= \tfrac{1}{2}[-\log K_w - (-\log K_b) + (-\log M_s)] \\ p_cH &= \tfrac{1}{2}(pK_w - pK_b - \log M_s) \end{aligned} \tag{45}$$

Either eq. (44) or (45) may be used to calculate the p_cH of solutions of salts of weak bases and strong acids.

Example 3. Calculate the p_cH of a 0.100 M solution of silver nitrate. The pK_b of silver hydroxide is 3.96. (a) Using eq. (44),

$$[H^+] = \sqrt{\frac{10^{-14} \times 10^{-1}}{10^{-3.96}}} = \sqrt{10^{-11.04}} = 10^{-5.52}$$

$$p_cH = 5.52$$

(b) Using eq. (45),

$$p_cH = \tfrac{1}{2}(14.00 - 3.96 + 1.00) = 11.04/2 = 5.52$$

Example 4. Calculate the p_cH at the equivalence point in the titration of 40.00 ml of 0.0750 N NH_4OH with 0.1063 N hydrochloric acid. Assume that the solution has been diluted to 100 ml. The pK_b of NH_4OH is 4.74.

$$N_s = \frac{40.00 \times 0.0750}{100} = 0.0300\ N$$

$$M_s = N_s/1 = 0.0300\ M = 10^{-1.52}\ M$$

$$p_cH = \tfrac{1}{2}(14.00 - 4.74 + 1.52) = 10.78/2 = 5.39$$

(4) *Salts of weak bases and weak acids.* The titration of weak bases with weak acids is avoided in actual analysis since there is no sharp change in p_cH at the equivalence point. The p_cH of solutions of salts of weak acids and weak bases is close to 7 and may be calculated by

$$p_cH = \tfrac{1}{2}(pK_w + pK_a - pK_b) \tag{46}$$

This equation will not be derived, but it should be noted that it indicates that the p_cH is independent of the concentration of the salt.
Example 5. Calculate the p_cH of a 0.250 M solution of ammonium benzoate if pK_a is 4.20 and pK_b is 4.74.

$$p_cH = \tfrac{1}{2}(14.00 + 4.20 - 4.74) = 13.46/2 = 6.73$$

(5) *Amphiprotic salts.* Amphiprotic salts are formed at the equivalence in the titration of polyprotic acids with strong bases when dibasic acids are one-half neutralized and tribasic acids are one-third or two-thirds neutralized. The p_cH of the equivalence point solution is controlled by the amphiprotic ion since the salt is completely ionized into the amphiprotic ion and the cation which is an infinitely weak Brønsted acid. Considering a solution of NaHX, the amphiprotic salt of the diprotic acid H_2X, the HX^- anion can undergo two separate reactions:

$$HX^- \rightleftharpoons H^+ + X^{2-} \tag{47}$$

$$HX^- + H^+ \rightleftharpoons H_2X \tag{48}$$

which may be summed to give

$$2HX^- \rightleftharpoons H_2X + X^{2-} \tag{49}$$

The equilibrium constants for these equations are

$$K_{(47)} = \frac{[H^+][X^{2-}]}{[HX^-]} = K_2$$

$$K_{(48)} = \frac{[H_2X]}{[H^+][HX^-]} = \frac{1}{K_1}$$

$$K_{(49)} = \frac{[H_2X][X^{2-}]}{[HX^-]^2} = \frac{K_2}{K_1}$$

To get $[H^+]$ into the expression for $K_{(49)}$, we can substitute the value of $[HX^-]$ from the expression for $K_{(48)}$, or

$$[HX^-] = \frac{K_1[H_2X]}{[H^+]} \quad \text{or} \quad \frac{1}{[HX^-]} = \frac{[H^+]}{[H_2X]K_1}$$

and

$$\frac{1}{[HX^-]^2} = \frac{[H^+]^2}{[H_2X]^2K_1^2}$$

to obtain

$$[H_2X][X^-]\left(\frac{[H^+]^2}{[H_2X]^2K_1^2}\right) = \frac{[H^+]^2[X^{2-}][H_2X]}{[H_2X]^2K_1^2} = \frac{K_2}{K_1}$$

or

$$\frac{[H^+]^2[X^{2-}]}{[H_2X]} = \frac{K_2K_1^2}{K_1}$$

which may be simplified to

$$[H^+] = \sqrt{\frac{[H_2X]}{[X^{2-}]}}\sqrt{K_1K_2} \qquad (50)$$

Since reactions (47) and (48) usually take place to approximately the same extent, $[H_2X] \simeq [X^{2-}]$ and eq. (50) becomes

$$[H^+] = \sqrt{K_1K_2} \qquad (51)$$

and, upon taking negative logs,

$$p_cH = \tfrac{1}{2}(pK_1 + pK_2) \qquad (52)$$

Both eqs. (51) and (52) are applicable to solutions of primary salts of weak diprotic acids and the primary and secondary salts of weak triprotic acids. The hydrogen ion concentration is independent of the molar concentration of the salt provided the molar concentration is large enough to be greater than $3K_1$ and also greater than $2K_w/K_2$. These conditions are necessary for the concentrations of the nonionized acid and the secondary anion to be approximately equal.

Example 6. Calculate the p_cH of 0.068 M potassium hydrogen malate. Malic acid is diprotic with pK_1 equal to 3.40 and pK_2 equal to 5.05.

To solve such a problem, first check to see if the simplified formula (51) can be used, that is, that M_s is greater than $3K_1$ and greater than $2K_w/K_2$.

$$\begin{aligned} M_s &= 0.068 = 10^{-1.17} \\ 3K_1 &= 3 \times 10^{-3.40} = 10^{-2.92} \\ 2K_w/K_2 &= 2 \times 10^{-14}/10^{-5.05} = 10^{-13.70}/10^{-5.05} \\ &= 10^{-8.65} \end{aligned}$$

M_s satisfies both necessary conditions so that eq. (51) or (52) may be used.

$$\begin{aligned} p_c\text{H} &= \tfrac{1}{2}(pK_1 + pK_2) = \tfrac{1}{2}(3.40 + 5.05) \\ &= 8.45/2 = 4.23 \end{aligned}$$

3. p_cH of Solutions of Salts Containing Excess Acid or Base

a. GENERAL. The presence of a salt of a strong acid or base has no effect on the p_cH of the solution of a strong acid or base other than the effect of ionic strength on the practical pK_w. The presence of a salt of a weak acid or a weak base in a solution of the acid or base will affect the p_cH. Since slight to moderate additions of strong acids or strong bases do not appreciably affect the p_cH of such solutions, they are known as buffered solutions and will be discussed in more detail later.

b. SOLUTIONS OF A WEAK ACID AND ONE OF ITS SALTS. The hydrogen ion concentration in a solution containing a weak acid is governed by the ionization constant of the weak acid and the ratio of concentrations of the nonionized molecules and of the anion of the acid, or

$$[\text{H}^+] = K_a \frac{[\text{HA}]}{[\text{A}^-]} \tag{53}$$

Since the salt is completely ionic and since the anions of the salt react with hydrogen ions to form nonionized molecules, the hydrogen ion concentration is less in such a solution than in a solution of the acid itself. In such a solution, when pK_a is greater than 4, the molar concentrations of the nonionized acid and of the anion are both very large numbers compared to the hydrogen ion concentration and also very large compared to the amount of the salt which reacts with hydrogen ions to form nonionized molecules. As a result, the concentrations of nonionized molecules are essentially equal to the molar concentration of the acid (M_a), and the anion concentration is essentially equal to the molar concentration of the salt (M_s).

Using these terms, eq. (53) becomes

$$[H^+] = K_a \frac{M_a}{M_s} \tag{54}$$

Substituting millimoles per milliliter for molar concentration and cancelling the millilitres since the volume is the same for both the acid and the salt,

$$[H^+] = K_a \frac{\text{mmoles of acid/ml}}{\text{mmoles of salt/ml}} = K_a \frac{\text{mmoles of acid}}{\text{mmoles of salt}} \tag{55}$$

Converting to negative logs, eq. (54) and (55) become

$$\begin{aligned} -\log[H^+] &= -\log K_a + (-\log M_a) - (-\log M_s) \\ &= -\log K_a + (-\log \text{mmoles of acid}) - (-\log \text{mmoles of salt}) \end{aligned}$$

or

$$\begin{aligned} p_c\text{H} &= pK_a - \log M_a + \log M_s \\ &= pK_a - \log \text{mmoles of acid} + \log \text{mmoles of salt} \end{aligned} \tag{56}$$

Either molar concentrations or millimoles of acid and salt can be used in the calculation. It is often preferable to use millimoles rather than molar concentrations since the logarithms of millimoles will be positive numbers, while the logarithms of the molar concentrations are usually negative numbers.

Example 7. Calculate the p_cH of a solution which is 0.50 M in propionic acid and 0.25 M in sodium propionate. The pK_a for propionic acid is 4.87.

$$0.50 = 5.0 \times 10^{-1} = 10^{0.70} \times 10^{-1} = 10^{-0.30}$$

and

$$\log 0.50 = -0.30$$

$$0.25 = 2.5 \times 10^{-1} = 10^{0.40} \times 10^{-1} = 10^{-0.60}$$

and

$$\log 0.25 = -0.60$$

$$\begin{aligned} p_c\text{H} &= 4.87 - \log 0.50 + \log 0.25 = 4.87 - (-0.30) + (-0.60) \\ &= 4.87 + 0.30 - 0.60 = 4.57 \end{aligned}$$

Example 8. Calculate the p_cH of a solution prepared by addition of 20.00 ml of 0.1000 *M* sodium hydroxide to 30.00 ml of 0.1500 *M* acetic acid. The pK_a for acetic acid is 4.76. (a) Calculate the millimoles of salt formed. The millimoles of salt formed equals the millimoles of base added since acetic acid and sodium hydroxide are both univalent compounds, so that

$$\text{mmoles of salt} = 20.00 \times 0.1000 = 2.000$$

(b) Calculate the millimoles of acid left unreacted.

$$\text{mmoles of acid at start } (30.00 \times 0.1500) = 4.500$$

$$\text{mmoles of acid which reacted (mmoles base added)} = 2.000$$

$$\text{mmoles of acid left unreacted} = 2.500$$

(c) Calculate p_cH.

$$\begin{aligned} p_c\text{H} &= pK_a - \log \text{mmoles of acid} + \log \text{mmoles of salt} \\ &= 4.76 - \log 2.50 + \log 2.00 \\ &= 4.76 - 0.40 + 0.30 = 4.66 \end{aligned}$$

Example 9. Calculate the p_cH of a solution which contains 2.72 g of KH_2PO_4 and 3.48 g of K_2HPO_4 in 100 ml of solution. The pK_2 of phosphoric acid is 7.21. MWs: KH_2PO_4, 136, K_2HPO_4, 174. (a) Calculate mmoles of acid and salt.

$$\text{mmoles of acid } (KH_2PO_4) = 2.72/0.136 = 20.0$$

$$\text{mmoles of salt } (K_2HPO_4) = 3.48/0.174 = 20.0$$

(a) Calculate p_cH.

$$\begin{aligned} p_c\text{H} &= pK_a - \log \text{mmoles of acid} + \log \text{mmoles of salt} \\ &= 7.21 - \log 20.0 + \log 20.0 = 7.21 \end{aligned}$$

c. SOLUTION OF A WEAK BASE AND ONE OF ITS SALTS. The hydroxyl ion concentration (and thus the hydrogen ion concentration) in a solution of a weak base and one of its salts will be governed by the ionization constant (K_b) of the base and the ratio of concentrations of the nonionized molecules and of the cation of the base, or

$$[OH^-] = \frac{K_b[MOH]}{[M^+]} \qquad (57)$$

As in solutions of weak acids and salts, the hydroxyl ion concentration in solutions of weak bases and salts is less than in the solution of the base alone. If the pK_b of the base is 4 or more, the molar concentration of the nonionized molecules may be assumed to be equal to the molar concentration of the base, and the molar concentration of the cation may be assumed to be equal to the molar concentration of the salt. Making these substitutions, eq. (57) becomes

$$[OH^-] = \frac{K_b M_b}{M_s} \tag{58}$$

which, upon taking negative logarithms, becomes

$$p_cOH = pK_b - \log M_b + \log M_s \tag{59}$$

and millimoles can be substituted for molar concentration since the two compounds are contained in the same solution, so that eq. (59) may be written

$$p_cOH = pK_b - \log \text{mmoles of base} + \log \text{mmoles of salt} \tag{60}$$

Example 10. Calculate the p_cH of a solution which is 0.50 M in NH_4OH and 0.30 M in NH_4Cl. The pK_b of NH_4OH is 4.74.

$$\begin{aligned} p_cOH &= 4.74 - \log 0.50 + \log 0.30 \\ &= 4.74 - (-0.30) + (-0.52) \\ &= 4.74 + 0.30 - 0.52 = 4.52 \\ p_cH &= 14.00 - 4.52 = 9.48 \end{aligned}$$

Example 11. Calculate the p_cH in a solution that contains 2.00 mmoles of aconitine and 4.00 mmoles of aconitine chloride in 50.0 ml. The pK_b of aconitine is 5.89.

$$\begin{aligned} p_cOH &= 5.89 - \log 2.00 + \log 4.00 \\ &= 5.89 - 0.30 + 0.60 = 6.19 \\ p_cH &= 14.00 - 6.19 = 7.81 \end{aligned}$$

Example 12. How many grams of solid ammonium chloride (MW, 53.5) should be added to 100 ml of 1.00 M ammonium hydroxide to produce a solution with p_cH equal to 9.30? Assume no change in volume due to addition of the solid NH_4Cl. The pK_b of NH_4OH is 4.74. Since it is easier to calculate p_cOH than p_cH in solutions of a weak base and salt, first convert to p_cOH by subtraction from 14.00, or

$$p_c\text{OH} = 14.00 - p_c\text{H} = 14.00 - 9.30 = 4.70$$

Using eq. (58), calculate the molar concentration of the salt M_s.

$$[\text{OH}^-] = K_b \frac{M_b}{M_s}$$

or

$$M_s = M_b \frac{K_b}{[\text{OH}^-]} = 1.00 \frac{10^{-4.74}}{10^{-4.70}} = 10^{-0.04}$$
$$= 10^{0.96} \times 10^{-1} = 9.12 \times 10^{-1} = 0.912$$

Next, calculate the weight of NH_4Cl in 100 ml of a solution of this concentration, or

$$\text{g} = \text{MW} \times M \times \text{liters} = 53.5 \times 0.912 \times 0.100 = 4.88 \text{ g}$$

4. Titration Curves

a. GENERAL. The p_cH of the titrated solution may be calculated after the addition of various amounts of titrant and the results graphed as p_cH versus milliliters of titrant added. Such graphs are called titration curves. The actual shape of a titration curve depends upon the concentration and pK values of the acid and base used and upon which is the titrant.

b. TITRATION CURVE FOR STRONG ACID BY STRONG BASE. As an example, we shall use the titration of 25.00 ml of 0.1000 N HCl with 0.1000 N NaOH and calculate the p_cH at various points on the titration curve. In each case we shall assume no dilution with water.

(1) *Before addition of any* NaOH. The p_cH will be that of 0.1 N HCl which is completely ionic and monoprotic so that N equals M and the p_cH at the beginning of the titration will be 1.00.

$$[\text{H}^+] = 10^{-1.00}\,M$$
$$p_c\text{H} = 1.00$$

(2) *After addition* of 5.00 *ml of* NaOH. The added base will neutralize an equal number of milliequivalents of acid to produce a solution containing unneutralized acid and NaCl. Since NaCl is the salt of a

strong acid and strong base, it will have no effect on the p_cH, so that the calculation involves only the determination of the concentration of the unneutralized acid and the conversion to p_cH. In our example, milliequivlents equal millimoles since both the acid and base are uni-univalent compounds and the millimoles of base added will be equal to the millimoles of acid reacted.

$$\text{mmoles of acid at start} = 25.00 \times 0.1000 = 2.500$$

$$\text{mmoles of acid reacted (mmoles of base added)} = 5.00 \times 0.1000 = 0.500$$

$$\text{mmoles of acid left unreacted} = 2.500 - 0.500 = 2.000 \text{ mmoles}$$

$$\text{Volume of solution} = 25.00 + 5.00 = 30.00 \text{ ml}$$

$$[H^+] = 2.00/30.0 = 0.067 = 10^{-1.17}\ M$$

$$p_cH = 1.17$$

(3) *After addition of* 20.00 *ml of* NaOH.

$$\text{mmoles acid at start} = 2.500$$

$$\text{mmoles acid reacted (base added)} = 2.000$$

$$\text{mmoles acid left unreacted} = 0.500$$

$$\text{volume of solution } (25.00 + 20.00) = 45.00 \text{ ml}$$

$$[H^+] = 0.500/45.00 = 0.0111 = 10^{-1.95}$$

$$p_cH = 1.95$$

Note that the change in p_cH in going from one-fifth to four-fifths neutralized is less than one unit.

(4) *After addition of* 24.00 *ml of* NaOH.

$$\text{mmoles acid at start} = 2.500$$

$$\text{mmoles acid reacted} = 2.400$$

$$\text{mmoles acid left unreacted} = 0.100$$

$$\text{Volume } (25.00 + 24.00) = 49.00 \text{ ml}$$

$$[H^+] = 0.100/49.00 = 0.00204 = 10^{-2.69}$$

$$p_cH = 2.69$$

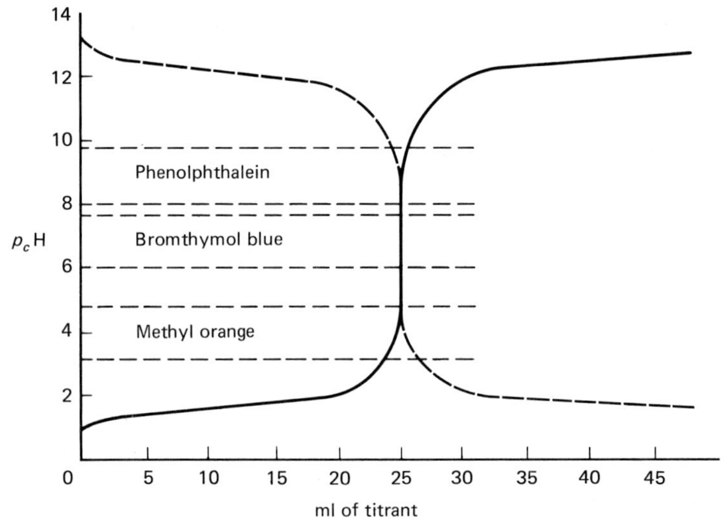

Figure 9.1. Titration of 25.00 ml of 0.1000 *N* HCl with 0.1000 *N* NaOH. The dashed line represents the reverse titration in which the acid is added to the base.

(5) *After addition of* 24.50 *ml of* NaOH.

mmoles of acid at start $= 2.500$

mmoles of acid reacted $= 2.450$

mmoles of acid left unreacted $= 0.050$

Volume $(25.00 + 24.50) = 49.50$ ml

$$[H^+] = 0.050/49.50 = 0.00101 = 10^{-3.00}$$

$$p_cH = 3.00$$

(6) *After addition of* 24.90 *ml* (*two drops before equivalence point*).

mmoles acid at start $= 2.500$

mmoles acid reacted $= 2.490$

mmoles acid left unreacted $= 0.010$

volume $(25.00 + 24.90) = 49.90$ ml

$$[H^+] = 0.010/49.9 = 0.000200 = 10^{-3.70}$$

$$p_cH = 3.70$$

(7) *After addition of* 24.95 *ml of* NaOH (*one drop from equivalence point*).

mmoles of acid at start = 2.500

mmoles of acid reacted = 2.495

mmoles of acid left unreacted = 0.005

volume (25.00 + 24.95) = 49.95 ml

$[H^+] = 0.005/49.95 = 0.000100 = 10^{-4.00}$

$p_cH = 4.00$

(8) *After addition of* 25.00 *ml of* NaOH (*the equivalence point*). Since there is no acid or base left unreacted, this solution is the same as a 0.05 *M* solution of NaCl in which the p_cH is 7.00. Note that there is a jump of 3 p_cH units due to addition of this one drop of base.

(9) *After addition of* 25.05 *ml of* NaOH (*one drop beyond equivalence point*). Beyond the equivalence point the solution will contain sodium chloride and the excess base. As a consequence, calculate the molar hydroxyl ion concentration and p_cOH instead of the molar hydrogen ion concentration.

mmoles base added (25.05 × 0.1) = 2.505

mmoles base reacted (mmoles acid at start) = 2.500

mmoles base unreacted = 0.005

volume (25.00 + 25.05) = 50.05 ml

$[OH^-] = 0.005/50.05 = 0.0000999 = 10^{-4.00}\ M$

$p_cOH = 4.00; \qquad p_cH = 14.00 - 4.00 = 10.00$

This represents a change of 3 p_cH units due to addition of the one drop of base in excess beyond the equivalence point, or a change of 6 p_cH units within one drop on either side of the equivalence point.

(10) *Continuing with similar calculations.* The p_cH at 25.10 ml is 10.30, at 25.50 ml is 11.00, at 26.00 ml is 11.29, at 30.00 ml is 11.95, and at 45.00 ml is 12.46. A graph of these results is shown in Fig. 9-1.

c. TITRATION CURVES FOR WEAK ACIDS BY STRONG BASE. As an example, we shall use the titration of 25.00 ml of 0.1000 *M* acetic acid with 0.1000 *M* NaOH and shall assume that the volumes are additive and that there has been no dilution with water. The pK_a of acetic acid is 4.76.

(1) *Before addition of any base.* This will be a solution of a weak acid, since pK_a is greater than 4,

$$[H^+] = \sqrt{K_a M_a} \qquad \text{or} \qquad p_cH = \tfrac{1}{2}(pK_a - \log M_a)$$

$$\log M_a = \log 10^{-1.00} = -1.00$$

$$p_cH = \tfrac{1}{2}(4.76 + 1.00) = 5.76/2 = 2.88$$

(2) *After addition of* 5.00 *ml of base.* The solution now contains unreacted acid and the salt formed by the reaction of the base so that it is necessary to calculate the millimoles of salt and acid present.

mmoles of acid at start $(25.00 \times 0.1000) = 2.500$

mmoles of salt formed (or base added) $(5.00 \times 0.1000) = 0.500$

mmoles of acid left unreacted $= 2.000$

$$p_cH = pK_a - \log \text{mmoles of acid} + \log \text{mmoles of salt}$$
$$= 4.76 - \log 2.0 + \log 0.5 = 4.76 - 0.30 + (-0.30) = 4.16$$

(3) *After addition of* 12.50 *ml of* NaOH. Since the acid is half neutralized, the concentration of the acid and salt are equal so that the $p_cH = pK_a$, or

$$p_cH = 4.76.$$

This is an important fact to remember, that is, that

$$p_cH = pK_a$$

at the half-neutralization point.

(4) *After addition of* 20.00 *ml of* NaOH. This would be calculated similar to (2), or

mmoles of acid at start $= 2.500$

mmoles of salt formed $= 2.000$

mmoles of acid left unreacted $= 0.500$

$$p_cH = pK_a - \log 0.5 + \log 2.0$$
$$= 4.76 + 0.30 + 0.30 = 5.36$$

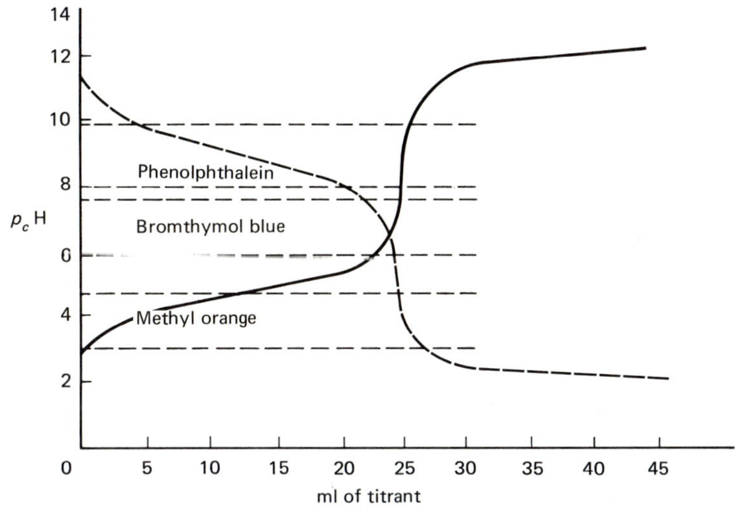

Figure 9.2. Titration of 25.00 ml of 0.1000 *N* $HC_2H_3O_2$ with 0.1000 *N* NaOH. The dashed line represents the titration of 25.00 ml of 0.1000 *N* NH_4OH with 0.1000 *N* HCl.

(5) *After addition of* 24.00 *ml of* NaOH.

mmoles salt formed = 2.400

mmoles acid left unreacted = 0.100

$$p_c\text{H} = 4.76 - (-1.00) + 0.38 = 6.14$$

(6) *After addition of* 24.90 *ml of* NaOH (*two drops before equivalence point*).

mmoles salt formed = 2.490

mmoles acid unreacted = 0.010

$$p_c\text{H} = 4.76 - (-2.00) + 0.40 = 7.16$$

(7) *After addition of* 25.00 *ml of* NaOH (*the equivalence point*). The solution now contains only sodium acetate so that the p_cH can be calculated from eq. (39).

$$M_s = \frac{25.00 \times 0.1000}{25.00 + 25.00} = \frac{2.500}{50.00} = 0.05\ M = 10^{-1.30}$$

$$p_c\text{H} = \tfrac{1}{2}(pK_w + pK_a + \log M_s) = \tfrac{1}{2}(14.00 + 4.76 - 1.30)$$
$$= 17.46/2 = 8.73$$

(8) *After addition of* 25.10 *ml of* NaOH (*two drops beyond equivalence point*). The solution now contains sodium acetate and excess

strong base. Since the salt has no effect on the p_cH in a solution of excess base, the calculation is the same as that for the titration of a strong acid by a strong base, or

$$\text{mmoles base in excess} = 2.510 - 2.500 = 0.010$$

$$\text{Volume} = 25.00 + 25.10 = 50.10 \text{ ml}$$

$$[OH^-] = 0.010/50.10 = 0.0001996 = 10^{-3.70}$$

$$p_cOH = 3.70$$

$$p_cH = 10.30$$

The p_cH values in the titrated solution for further additions of base are identical with those calculated for the titration of a strong acid with a strong base. The titration curve for this type of titration is shown in Fig. 9-2.

d. TITRATION CURVE FOR WEAK BASE BY STRONG ACID. As our example we shall use the titration of 25.00 ml of 0.1000 *M* NH_4OH with 0.1000 *M* HCl and assume that the volumes are additive and that there is no dilution with water. The pK_b for NH_4OH is 4.74.

(1) *Before addition of any* HCl. This will bc a solution of a weak base so that

$$[OH^-] = \sqrt{K_b M_b} \qquad \text{or} \qquad p_cOH = \tfrac{1}{2}(pK_b - \log M_b)$$

$$\log M_b = 10^{\ 1.00}$$

$$p_cOH = \tfrac{1}{2}(4.74 + 1.00) = 5.74/2 = 2.87$$

$$p_cH = 11.13$$

(2) *After addition of* 5.00 *ml of* HCl. This solution will now contain unreacted NH_4OH together with the NH_4Cl formed by the reaction of the acid and base. The p_cH may be calculated from eq. (60) as follows:

$$\text{mmoles base at start } (25.00 \times 0.1000) = 2.500$$

$$\text{mmoles salt formed (acid added)} = 5.00 \times 0.1000 = 0.500$$

$$\text{mmoles of base left unreacted} = 2.000$$

$$p_cOH = pK_b - \log \text{mmoles of base} + \log \text{mmoles of salt}$$

$$= 4.74 - \log 2.00 + \log 0.5$$

$$= 4.74 - 0.30 + (-0.30) = 4.74 - 0.60$$

$$= 4.17$$

$$p_cH = 14.00 - p_cOH = 14.00 - 4.17 = 9.83$$

All other p_cH values for additions of acid before the equivalence point is reached can be calculated in the same manner.

(3) *After addition of* 25.00 *ml of acid* (*the equivalence point*). At the equivalence point, the solution will contain the salt with no excess of either acid or base, and the p_cH can be calculated by use of eq. (45) as follows:

$$M_s = \frac{25.00 \times 0.1000}{25.00 + 25.00} = 0.05\ M = 10^{-1.30}$$

$$p_cH = \tfrac{1}{2}(pK_w - pK_b - \log M_s)$$

$$= \tfrac{1}{2}[14.00 - 4.74 - (-1.30)] = 10.66/2 = 5.33$$

(4) *After addition of* 30.00 *ml of* HCl. Beyond the equivalence point, the solution contains the salt and excess strong acid. Since the salt of a strong acid does not affect the p_cH of solutions of that acid, the p_cH is calculated from the millimoles of strong acid and the volume of the solution as follows:

mmoles strong acid added $(30.00 \times 0.1000) = 3.000$

mmoles acid reacted (mmoles of base at start) $= 2.500$

mmoles of strong acid unreacted (in excess) $= 0.500$

Volume $(25.00 + 30.00) = 55.00$ ml

$[H^+] = 0.500/55.00 = 0.00909 = 10^{-2.04}$

$p_cH = 2.04$

5. Feasibility of Titrations

For a titration to be feasible in the laboratory, there must be a sufficiently sharp break in the pH at the equivalence point and a suitable indicator must be available. The pH break depends upon the ionization constant of the weak acid or base and upon the concentration of the titrant. For most acid-base titrations, there must be a sharp break of between 1 and 2 pH units with 0.2–0.4 per cent (usually two to four drops) of the volume of titrant needed to attain the equivalence point. For titrations using 0.1 N strong acid or base as the titrant, such a break will occur if the ionization constant is greater than 10^{-8}. For other concentrations of titrant, the product of the concentration of titrant and ionization constant (MK_a) must exceed 10^{-9}, as shown in Fig. 9-3. This is only a general rule and

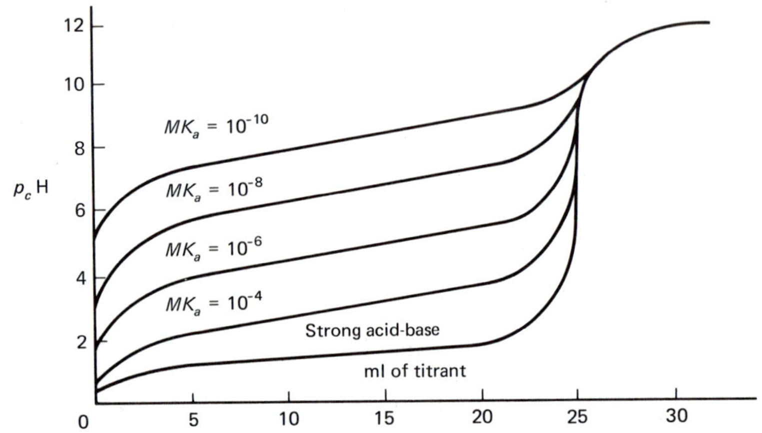

Figure 9.3. Titration curves for various values of MK_a, where M is the molar concentration of the NaOH titrant and K_a is the ionization constant of the weak acid.

should be applied with caution. Feasibility also depends upon the solubilities and upon the availability of a satisfactory indicator in the *p*H region of the equivalence point. Also, titrant concentrations less than 0.001 *N* are seldom used even though more dilute solutions are theoretically satisfactory.

6. Titration of Polyprotic Acids and Mixtures of Acids

Since polyprotic acids are neutralized by stepwise reaction with the hydrogen ions, two or more equivalence points should be possible in the titration of polyprotic acids. Thus, in the titration of maleic acid ($H_2C_4H_2O_4$), two equivalence points are theoretically possible according to the equations

$$H_2C_4H_2O_4 + NaOH \longrightarrow NaHC_4H_2O_4 + HOH \qquad (61)$$

$$NaHC_4H_2O_4 + NaOH \longrightarrow Na_2C_4H_2O_4 + HOH \qquad (62)$$

To be able to utilize such equivalence points in an acutal titration there must be a sufficiently sharp break in p_cH at both equivalence points. For titrations using 0.1 *N* strong base as titrant, satisfactory breaks will occur if the difference in the *pK* values is approximately 4 or more and if both *pK* values are 9 or less. For the case of maleic acid for which pK_1 is 2.00 and pK_2 is 6.26, both equivalence points may be used, but for carbonic acid with pK_1 equal to 6.37 and pK_2 equal to 10.25, only one of the equivalence points would give a large enough break in p_cH and this one is very seldom used due to the low solubility of carbonic acid in water. As shown in Fig. 9-4, curve A, only two equivalence points can be utilized with phosphoric acid (pK_1, 2.12; pK_2, 7.21; pK_3, 12.32). Similarly, only one end point is available with tartaric acid (pK_1, 3.02; pK_2, 4.54).

The same principles hold for the titration of mixtures of acids, and in general mixtures of strong and weak acids show two equivalence points, while mixtures of two strong acids or mixtures of two weak acids will show only one. Thus both acetic acid and hydrochloric acid can be determined in the presence of each other but acetic and propionic cannot.

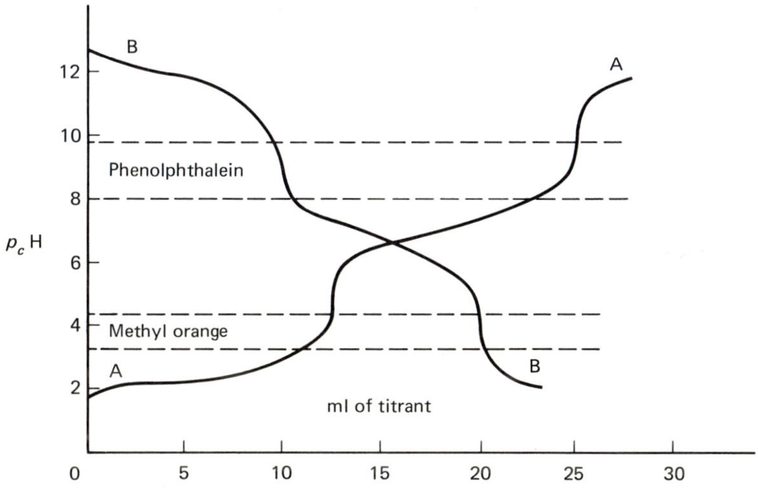

Figure 9.4. **A**, Titration of 12.50 ml of 0.1000 *M* H_3PO_4 with 0.1000 *M* NaOH. **B**, Titration of 10.00 ml of 0.1000 *M* Na_2CO_3 with 0.1000 *M* HCl.

In many cases, the salts of diprotic weak acids which show only one equivalence point can be titrated with strong acid to give two equivalence points which are readily utilizable in actual titrations. Thus Na_2CO_3 can be titrated with HCl according to the equations

$$Na_2CO_3 + HCl \longrightarrow NaHCO_3 + NaCl \tag{63}$$

$$NaHCO_3 + HCl \longrightarrow H_2CO_3 + NaCl \tag{64}$$

As shown in Fig. 9-4, curve B, the p_cH at the equivalence point for reaction (63) is approximately 8, while that for reaction (64) is in the range of 4.

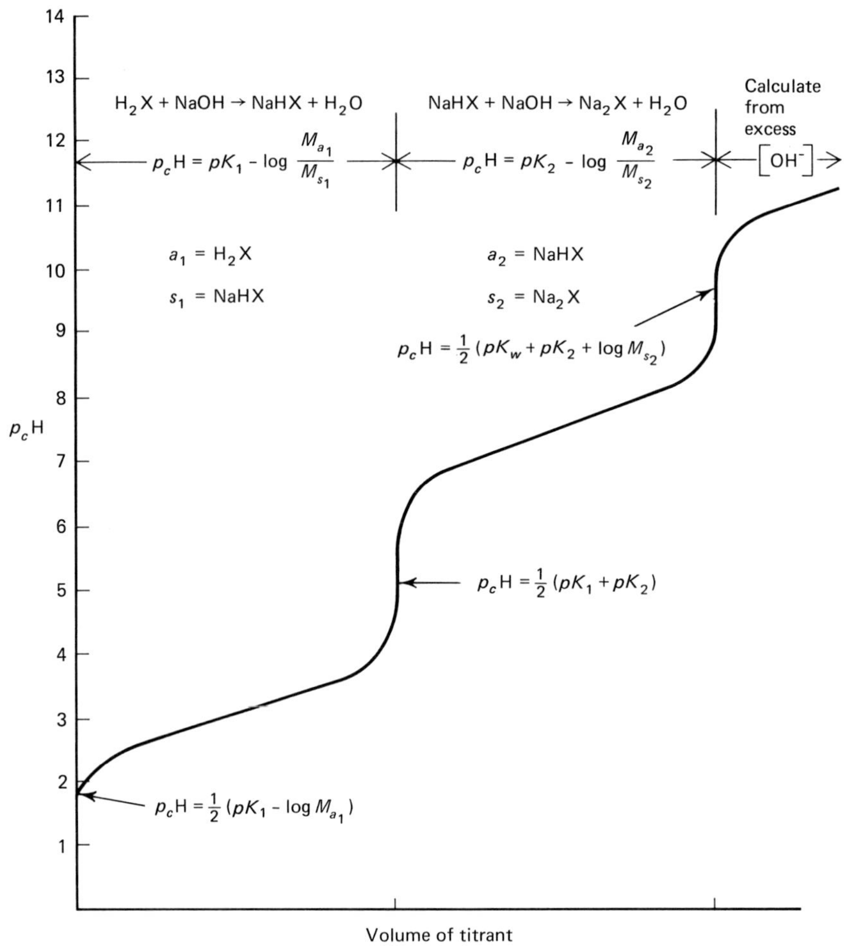

Figure 9.5. Summary of calculation of p_cH at various points on the titration curve for the titration of a dibasic acid H_2X with NaOH. M_a = analytical molar concentration of acid and M_s = analytical molar concentration of salt.

The p_cH break at both equivalence points is sufficiently large that either or both may be utilized in actual titrations.

In those cases where multiple equivalence points exist and can be used, potentiometric titration methods may be utilized if satisfactory indicators are not available.

Typical titration curves for weak dibasic acids and for weak diacid bases are shown in Figures 9-5 and 9-6, respectively. These figures also show the formulas for calculation of p_cH and p_cOH in various regions of the titration.

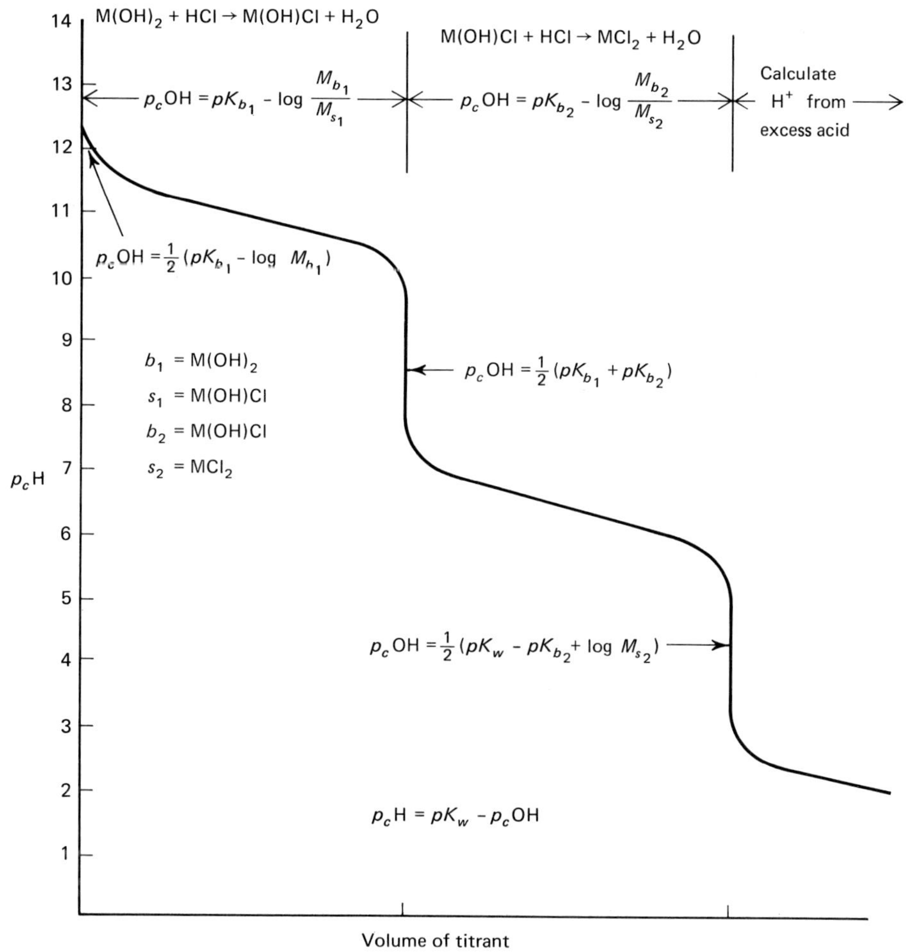

Figure 9.6. Summary of calculation of p_cOH and p_cH at various points on the titration curve for the titration of a diacid base, $M(OH)_2$, with HCl. M_b = analytical molar concentration of base, and M_s = analytical molar concentration of salt.

E. Indicators

1. General

Indicators for acid-base titrations are substances which change color over a definite p_cH range. Most neutralization indicators are weak organic acids or weak organic bases for which the molecules and ions have different colors. One form may be colorless and the other colored, or both may be colored. The color which exists at the lower p_cH is called the acid color and that at the upper p_cH the basic color, irrespective of the actual hydrogen ion concentrations at which they occur. The color which is seen when both colored forms are present is called the intermediate or transition color or virage. Thus for methyl orange the acid color is red and the basic color is yellow, so that when both forms are present the mixture of yellow and red appears to be orange.

2. Explanation of Indicator Action

The reason indicators show different colors at different hydrogen ion concentrations is usually explained on the basis of the effect that variations in the hydrogen ion concentration have on the ratio of concentrations of nonionized molecules and indicator ion.

For weak acids such as phenolphthalein,

$$\mathrm{HIn} \rightleftharpoons \mathrm{H^+ + In^-} \tag{65}$$

for which the ionization constant is

$$K_{In} = \frac{[\mathrm{H^+}][\mathrm{In^-}]}{[\mathrm{HIn}]} \tag{66}$$

the relationship may be expressed as

$$[\mathrm{H^+}] = K_{In}\frac{[\mathrm{HIn}]}{[\mathrm{In^-}]} = K_{In}\frac{[\text{acid form}]}{[\text{basic form}]} \tag{67}$$

For weak bases such as methyl orange,

$$\mathrm{InOH} \rightleftharpoons \mathrm{In^+ + OH^+}$$

the ionization expression may be written as

$$[\mathrm{OH^-}] = K_{In}\frac{[\text{basic form}]}{[\text{acid form}]} \tag{68}$$

Using the weak acids as an example, K_{In} is constant, and any variation in the hydrogen ion concentration will cause a corresponding change in the ratio $[\mathrm{HIn}]/[\mathrm{In^-}]$. At relatively large values of the hydrogen ion concentration, the indicator will exist mainly as molecules, while at relatively low values, the ions will predominate. Thus for weak acids, the color of the nonionized molecules will predominate in acid solutions and the color of the ions will appear in basic solutions. It should be noted, however, that both forms must be present in all solutions which contain the indicator. For weak bases the basic form is the nonionized molecule which predominates in basic solution.

The actual color changes of indicators are due to complex structural rearrangements which occur when the indicator reacts with an acid or base. Since most acid-base indicators are organic compounds which contain two or more benzene rings joined by one or more atoms of carbon or nitrogen, the most common change involves the interconversion of benzenoid and quinoid types of structure:

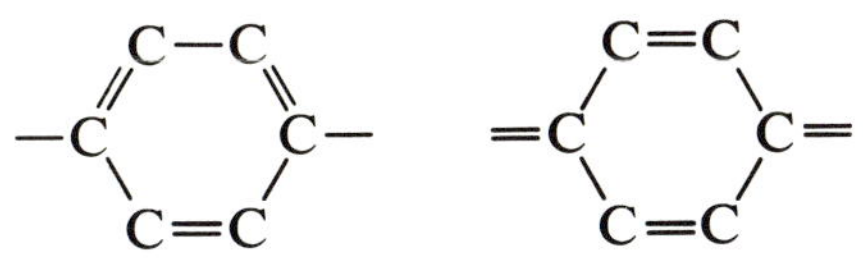

There are also other chromophoric groups which are associated with color and which are involved in the color changes of some indicators.

3. Range of Indicators

If the hydrogen ion concentration equals the ionization constant of the indicator, the ratio of concentrations must be unity and the two concentrations will be equal, as shown by eq. (67). Under these conditions, the eye sees the mixture of colors or the intermediate color of the indicator. At any other hydrogen ion concentration the eye will see some variation of the intermediate color or else will see only the acid color or only the basic color. If the two colored forms are equally easy to see, the eye can ordinarily distinguish the presence of one form with the other only if the concentration is at least one-tenth the total concentration of the indicator. Thus the eye will see some type of intermediate virage when the concentration ratio is between 1 : 10 and 10 : 1 of the acid to basic form. Outside this range the eye sees only the predominant form. Writing eq. (67) in negative logarithmic form,

$$-\log[\mathrm{H}^+] = -\log K_{In} + \left(-\log \frac{\text{acid form concentration}}{\text{basic form concentration}}\right) \tag{69}$$

$$p_c\mathrm{H} = pK_{In} - \log \frac{\text{acid form concentration}}{\text{basic form concentration}} \tag{70}$$

Substituting the ratios 1 : 10 and 10 : 1 in this equation, we obtain

$$p_c\mathrm{H} = pK_{In} - \log 10 = pK_{In} - 1$$

and

$$p_c\mathrm{H} = pK_{In} - \log 0.1 = pK_{In} - (-1) = pK_{In} + 1$$

TABLE 9.1. Acid-Base Indicators

Common name	Chemical name	*p*H range	pK_{In}	Wave length of maximum absorbance	Color change (acid-alkaline)
Cresol red (acid range)	*o*-cresolsulfonphthalein	0.2–1.8			Red to yellow
Methyl violet		0.5–1.5			Yellow to blue
Thymol blue (acid range)	Thymolsulfonphthalein	1.2–2.8	1.7	544	Red to yellow
Methyl yellow	Dimethylaminoazobenzene	2.8–4.0	3.3	508	Red to yellow
Bromphenol blue	Tetrabromophenolsulfonphthalein	3.0–4.6	3.8	592	Yellow to blue
Methyl orange	Dimethylaminoazobenzene-sulfonic acid	3.1–4.4	3.5	506	Red to orange-yellow
Bromocresol green	Tetrabromophenol-*m*-cresol-sulfonphthalein	4.0–5.6	4.7	614	Yellow to blue
Methyl red	Dimethylaminoazobenzene-*o*-carboxylic acid	4.2–6.2	5.0	533	Red to yellow
Chlorphenol red	Dichlorosulfonphthalein	4.8–6.4	6.0	573	Yellow to red
Bromthymol blue	Dibromothymolsulfonphthalein	6.0–7.6	7.1	617	Yellow to blue
Phenol red	Phenolsulfonphthalein	6.4–8.0	7.8	558	Yellow to red
Cresol red (base range)	See above	7.2–8.8	8.1	572	Yellow to red
Thymol blue (base range)	See above	8.0–9.6	8.9	596	Yellow to blue
Phenolphthalein		8.0–9.8	9.3	553	Colorless to red
Thymolphthalein		9.3–10.5	9.9	598	Colorless to blue
Alizarin yellow	*p*-Nitroaniline azo sodium salicylate	10.0–12.0	11.1		Yellow to violet
2,4,6-Trinitrotoluene		12.0–14.0			Colorless to orange

Thus the range of the color change of an indicator is approximately pK_{In} plus or minus 1 pH unit or a total of 2 pH units. In actual practice, the range is anywhere from 1 to 3 pH units and is seldom exactly 2, as shown in Table 9-1. This table lists some selected commonly used indicators giving the range, the acid and base colors, the pK_{In} values, and the wavelength of maximum absorbance. Of these the sulfonphthaleins are the most popular, together with methyl orange, methyl red, and phenolphthalein. Phenolphthalein is used in more titrations than any other indicator since it works well for most titrations in which sodium hydroxide is the titrant. Some dibasic indicators, such as thymol blue, have two color changes.

F. Differential Titration of Alkalies

Soda ash, which is often used as an inexpensive base in industry, is usually considered to be anhydrous sodium carbonate. It is seldom pure since it is contaminated by water and often is a mixture of sodium carbonate with either sodium hydroxide or sodium bicarbonate rather than sodium carbonate alone. As a consequence the total alkalinity is usually determined by titration with acid to the methyl orange end point using an end-point comparison standard or to a modified methyl orange end point. Sometimes it is necessary to know the composition as well as the total alkalinity. The composition can be determined if the sample is titrated to the phenolphthalein end point before the addition of methyl orange for the total alkalinity determination. It should be noted that the phenolphthalein end point for the titration of bases is the change from colored to colorless rather than the colorless to colored usually used. Colorless end points should be avoided when possible since the color change is usually gradual and there is no way to estimate how much the end point is overshot.

The basic reaction involved is the stepwise neutralization of sodium carbonate with hydrochloric acid shown in eq. (63) and (64). The end point for the titration to bicarbonate occurs at approximately pH 8 at which phenolphthalein becomes colorless, and the end point for the carbonic acid titration is near pH 4, which is in the methyl orange range. It should be noted that for samples which contain sodium carbonate only, the volume of acid required to the phenolphthalein end point should exactly equal the volume required to go from the phenolphthalein end point to the methyl orange end point. If the sample contains either bicarbonate or hydroxide, the volumes required for the two end points will not be equal.

Figure 9-7 shows the titration curves for samples containing only sodium carbonate and for samples which contain sodium carbonate together with either hydroxide or bicarbonate. Upon addition of acid, any

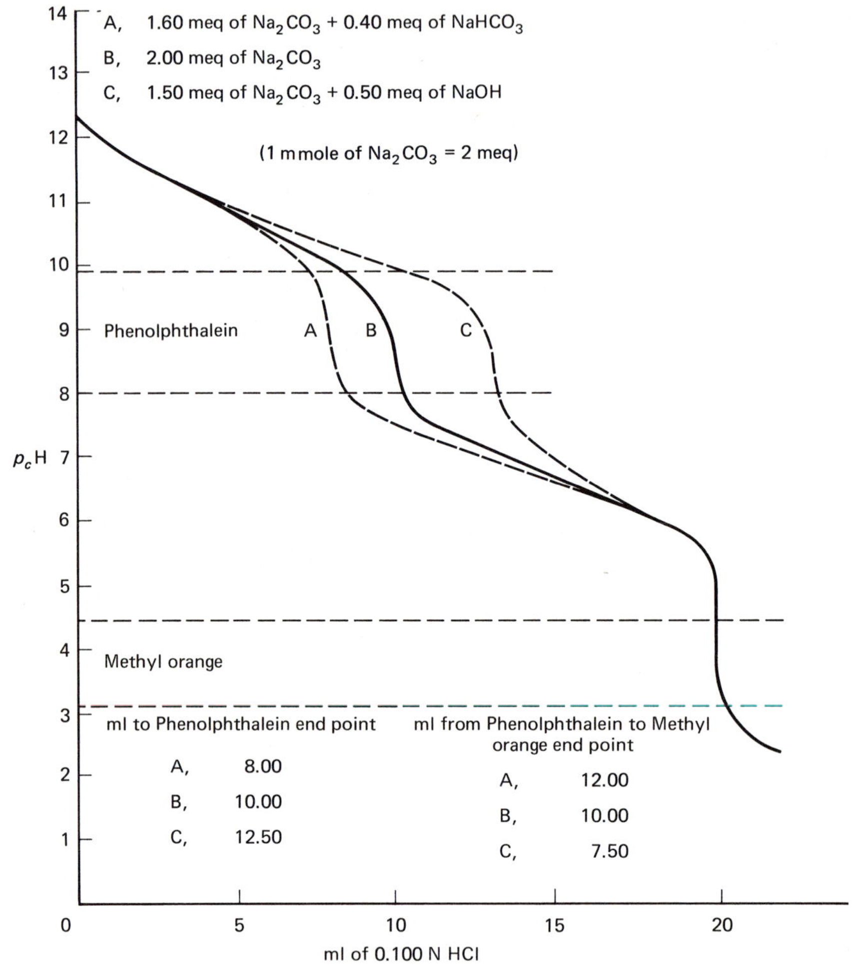

Figure 9.7. Differential titration of alkalies showing volume relations of titrant required to the phenolphthalein end point and from the phenolphthalein to the methyl orange end point for mixtures of various compositions.

strong base present is neutralized first. After this any sodium carbonate which is present is converted to sodium bicarbonate. Both these reactions are completed when the first end point is reached. Further addition of acid converts the sodium bicarbonate to carbonic acid. As a result, for mixtures of hydroxide and carbonate, the volume of acid required to the phenolphthalein end point will be greater than the volume required from the phenolphthalein to the methyl orange end point. For mixtures of sodium carbonate and sodium bicarbonate, the volume required to go from the phenolphthalein end point to the methyl orange end point will exceed that necessary to attain the phenolphthalein end point. For samples containing only sodium carbonate, the two volumes will be equal. Another way of stating these differences is that if more than half the total volume of acid required is utilized to reach the first end point, the sample contains sodium hydroxide and sodium carbonate; if exactly half the total volume is used in the first titration, the sample contains only sodium carbonate; if less than half the total volume is consumed in reaching the first end point, the sample contains sodium bicarbonate and sodium carbonate. Mixtures of sodium hydroxide and sodium bicarbonate are not possible since these two substances will react to form sodium carbonate and water. Examples of calculations for such titrations are given below.

Example 13. A 1.000-g sample of an industrial sodium carbonate alkali is dissolved and diluted to 100.0 ml. A 25.00-ml aliquot is titrated to the phenolphthalein end point and requires 24.76 ml of 0.1015 N hydrochloric acid. A second 25.00-ml aliquot is titrated to the methyl orange end point and requires 43.34 ml of the same acid. (a) State the components present and (b) calculate the percentage of each.

(a) Since the total volume required is 43.34 ml and the amount to the first end point is 24.76, or more than half the total, the sample contains sodium hydroxide and sodium carbonate.

(b) First calculate the volume of acid required to convert the sodium carbonate present to carbonic acid. This will be twice the volume of acid required to go from the phenolphthalein end point to the methyl orange end point, or

$$\begin{aligned} \text{volume required} &= 2(43.34-24.76) = 2 \times 18.58 \\ &= 37.16 \text{ ml} \end{aligned}$$

Then calculate the volume of acid required to neutralize the hydroxide. This will be the difference between the total volume required and the volume necessary for the sodium carbonate, or

$$\text{volume required} = (43.34 - 37.16) = 6.18$$

This last value may also be calculated by subtracting one-half the volume required for the sodium carbonate from the volume required in the titration to the phenolphthalein end point. Next calculate the percentage of sodium hydroxide (MW, 40.00) present.

$$\begin{aligned} \%\ \text{NaOH} &= \frac{\text{ml} \times N \times \text{meq wt} \times 100}{\text{wt of sample}} \\ &= \frac{6.18 \times 0.1015 \times 0.04000 \times 100}{1.000 \times 25.00/100.0} \\ &= 10.04\% \end{aligned}$$

Then calculate the percentage of Na_2CO_3 (MW, 106.0).

$$\% \, Na_2CO_3 = \frac{\text{ml} \times N \times \text{meq wt} \times 100}{\text{wt of sample}}$$

$$= \frac{37.16 \times 0.1015 \times 0.05300 \times 100}{1.000 \times 25.00/100.0}$$

$$= 79.96\%$$

The percentage of sodium carbonate can also be calculated from the volume required to go from the first to the second end point as follows:

$$\% \, Na_2CO_3 = \frac{18.58 \times 0.1015 \times 0.1060 \times 100}{1.000 \times 25.00/100.0}$$

$$= 79.96\%$$

It should be noted that in the second method of calculation of sodium carbonate that the volume of acid is one-half the volume for the first calculation and that the equivalent weight is twice the value used in the first calculation.

Example 14. An alkaline unknown contains inert material plus any one or any combination of sodium carbonate, sodium bicarbonate, and sodium hydroxide. A sample weighing 1.023 g was dissolved, phenolphthalein was added, and the sample titrated to loss of color with 0.1008 N HCl. Methyl orange was then added, and the sample was titrated with the same acid to the methyl orange point. The first titration required 30.69 ml and the second required an additional 40.92 ml. (a) State the components present and (b) calculate the percentage of each in the sample.

(a) Since a larger volume of acid is required to go from the phenolphthalein end point to the methyl orange end point than was required to attain the phenolphthalein end point, the sample contains sodium carbonate and sodium bicarbonate.

(b) First calculate the percentage of sodium carbonate from the data for the phenolphthalein end point. Since this represents the conversion of sodium carbonate (MW, 106.0) to sodium bicarbonate (MW, 84.01), the equivalent weight of the sodium carbonate will be the same as the molecular weight.

$$\% \, Na_2CO_3 = \frac{\text{ml} \times N \times \text{meq wt} \times 100}{\text{wt of sample}}$$

$$= \frac{30.69 \times 0.1008 \times 0.1060 \times 100}{1.023}$$

$$= 32.05\%$$

Next, determine the volume of acid required to react with the sodium bicarbonate present by subtracting the volume to the phenolphthalein end point from that required for the second titration, or

$$\text{ml for } NaHCO_3 = 40.92 - 30.69 = 10.23$$

This value can also be calculated by subtracting twice the volume to the phenolphthalein end point from the total volume required for both titrations, or

$$\text{ml for } NaHCO_3 = 71.61 - 61.38 = 10.23$$

Next, calculate the percentage of $NaHCO_3$ present.

$$\% \, NaHCO_3 = \frac{10.23 \times 0.1008 \times 0.08401 \times 100}{1.023}$$

$$= 8.47\,\%$$

Example 15. A 2.000-g sample is known to contain inert material plus any one or any combination of sodium carbonate, sodium bicarbonate, and sodium hydroxide. The sample was dissolved and titrated in the cold with 0.5000 *N* HCl. When phenolphthalein was the indicator, the end point was reached with 24.00 ml of the acid. Methyl orange was then added and a total of 48.00 ml of the acid was required to attain the end point.(a) State the components present and (b) calculate the percentage of each.

(a) The volume of acid required to the phenolphthalein end point (24.00 ml) is exactly half the total volume (48.00 ml) required for the complete titration. As a consequence sodium carbonate is the only basic component present.

(b) The percentage of sodium carbonate (MW, 106.0) may be calculated from either titration. Using the phenolphthalein end point,

$$\% \, Na_2CO_3 = \frac{24.00 \times 0.5000 \times 0.1060 \times 100}{2.000}$$

$$= 63.60\,\%$$

Using the methyl orange end point or the total titration,

$$\% \, Na_2CO_3 = \frac{48.00 \times 0.5000 \times 0.05300 \times 100}{2.000}$$

$$= 63.60\,\%$$

G. Buffers

1. Definitions

A solution which resists a change in the p_cH of the solution upon addition of slight to moderate amounts of acid or base is known as a buffered solution. Or, stated in another way, a solution is considered to be buffered if the addition of slight to moderate amounts of strong acid or strong base will not cause an appreciable change in the p_cH.

There are several terms which are often used interchangeably but which actually have slightly different meanings. Some of these terms are:

Buffer: A buffer is a compound or a mixture of compounds which has the ability to maintain the p_cH of a solution between narrow limits.

Buffer system: A buffer composed of two or more compounds is usually referred to as a buffer system.

Buffer action: The actual reaction by which a buffer system maintains the p_cH upon addition of acid or base is known as the buffer action.

Buffering solution: A buffering solution is a concentrated solution of a buffer system which may be added in small amounts to another solution to cause it to be buffered.

2. Types of Buffers

There are two types of buffers which can cause a solution to be buffered. One of these is a strong acid or strong base at a concentration of 0.1 *N* or more. In this case, the solution is buffered simply because it contains relatively large amounts of strong acid or strong base. The second type of buffer is a mixture of a weak acid or a weak base together with a salt of the acid or base.

3. Explanation of Buffer Action

When one mmole of a strong acid is added to 100 ml of water, the hydrogen ion concentration is 10^{-2} *M* and the p_cH is 2.00. For water, the addition of the one mmole to 100 ml causes a change of 5.00 p_cH units, and the same type of change (from 7.00 to 12.00) would be expected if one mmole of strong base were added to water.

A 0.1 *N* solution of a strong acid contains 10.0 mmoles of hydrogen ion per 100 ml of solution and has a p_cH of 1.00. If 1 mmole of hydrogen ion is added without appreciable change of volume, the concentration of the hydrogen ion is increased to 11.0 mmoles/100 ml and the p_cH is 0.96. Similarly, if 1 mmole of strong base is added without appreciable change in volume, the solution would contain 9.00 mmoles of hydrogen ion in 100 ml and the p_cH would be 1.05. Thus the addition of slight to moderate amounts of strong acid or strong base does not cause an appreciable change in the p_cH, and the solution is considered to be buffered. The same situation exists in a 0.1 *N* solution of a strong base.

If 1 mmole of hydrogen ion is added without appreciable change of volume to 100 ml of 0.00001 *N* strong acid, the p_cH changes from 5.00 to 2.00, and if 1 mmole of a strong base had been added instead, the p_cH would change from 5.00 to 12.00. Thus dilute solutions of strong acids or bases do not furnish enough hydrogen or hydroxyl ions to cause a solution to be buffered. If a buffered solution with p_cH between 1.00 and 13.00 is needed, simple solutions of strong acids or strong bases may not be used. In this region, buffer mixtures of a weak acid and one of its salts or a weak base and one of its salts must be used to obtain a buffered solution.

If 1 mmole of strong acid is added to 100 ml of a solution which is 0.100 *M* in acetic acid and 0.100 *M* in sodium acetate, the p_cH changes from 4.74 to 4.66, while the addition of 1 mmole of strong base causes the p_cH to change from 4.74 to 4.83. Evidently in solutions which contain a weak acid and its salt some mechanism exists which tends to neutralize the effect of the addition of strong acid or strong base. This action can be explained if we consider that the weak acid is present primarily as molecules, while the salt is completely ionic. Thus the solution contains a relatively large number of nonionized molecules and a relatively large number of a common ion.

The ionization of a weak acid may be shown as

$$HA \rightleftharpoons H^+ + A^- \tag{71}$$

for which the ionization constant is

$$K_a = \frac{[H^+][A^-]}{[HA]} \tag{72}$$

which can be written as

$$[H^+] = \frac{K_a[HA]}{[A^-]} = K_a \frac{[HA]}{[A^-]} \tag{73}$$

For weak acids it can be assumed that the acid is primarily present in the molecular form and that the salt is completely ionic so that the [HA] may be considered to be equal approximately to the molar concentration (analytical concentration, M_a) of the acid and $[A^-]$ to the molar concentration (analytical concentration, M_s) of the salt. Substituting, eq. (73) becomes

$$[H^+] = K_a \frac{M_a}{M_s} = K_a \frac{\text{mmole of acid}}{\text{mmole of salt}} \tag{74}$$

Since the ionization of a weak acid is an equilibrium, addition of strong acid or strong base simply shifts the equilibrium. A strong acid will react with the salt to form more weak acid molecules, while the strong base will react with the molecular acid to form more anions. Thus

$$\text{strong acid} + \text{anions} \longrightarrow \text{molecules of weak acid} \qquad (75)$$

and

$$\text{strong base} + \text{molecules of weak acid} \longrightarrow \text{anions} \qquad (76)$$

Applying this reasoning to the example of acetic acid and sodium acetate given, addition of strong acid converts acetate ion to acetic acid molecules, while addition of strong base changes acetic acid molecules to acetate ions. The calculations necessary to illustrate the p_cH changes may be shown by use of eq. (74) using 1.8×10^{-5} as the K_a of acetic acid.

$$[H^+] = 1.8 \times 10^{-5} \times \frac{0.10}{0.10} = 1.8 \times 10^{-5}$$

$$= 10^{0.26} \times 10^{-5} = 10^{-4.74}$$

$$p_cH = 4.74$$

The addition of 1.0 mmole of strong acid such as hydrochloric acid would convert 1.0 mmole of acetate ion to 1.0 mmole of acetic acid and form 1 mmole of sodium chloride. Since there are 10.0 mmoles of each originally present (100×0.10), after the addition there will be 11.0 mmoles of acetic acid and 9.0 mmoles of acetate ion, or the solution will be 0.11 *M* in acetic acid, 0.09 *M* in sodium acetate, and 0.01 *M* in sodium chloride. Applying these values in eq. (74),

$$[H^+] = 1.8 \times 10^{-5} \times \frac{0.11}{0.09} = 1.8 \times 10^{-5} \times 1.22$$

$$= 2.20 \times 10^{-5} = 10^{0.34} \times 10^{-5} = 10^{-4.66}$$

$$p_cH = 4.66$$

The addition of 1.0 mmole of a strong base such as sodium hydroxide would convert 1.0 mmole of acetic acid to 1.0 mmole of sodium acetate and 1.0 mmole of water, so that the concentration of acetic acid would be 0.09 *M* and that of sodium acetate would be 0.11 *M*. Substituting these values,

$$[H^+] = 1.8 \times 10^{-5} \times \frac{0.09}{0.11} = 1.8 \times 10^{-5} \times 0.82$$

$$= 1.48 \times 10^{-5} = 10^{0.17} \times 10^{-5} = 10^{-4.83}$$

$$p_cH = 4.83$$

The necessary conditions for a buffer, then, are either the presence of a large concentration of hydrogen or hydroxyl ions or else the presence of large concentrations of a slightly ionized material and a common ion.

4. Level and Range of Buffers

The level of a buffer is the actual $p_c\text{H}$ of the buffered solution containing the buffer, and the range is the $p_c\text{H}$ region in which it is effective as a buffer.

The level of a buffer depends upon the ionization constant (K_a) of a weak acid or base of the buffer system and upon the ratio of concentrations of the slightly ionized material and salt since it may be calculated using eq. (74) or eq. (54) through (59).

Inspection of Fig. 9-2 for the titration of acetic acid with sodium hydroxide shows that the $p_c\text{H}$ rises fairly rapidly at the beginning of the titration but then increases rather slowly until just preceding the equivalence point. This region of slow rise of $p_c\text{H}$ is known as the buffer region. The rise covers approximately two $p\text{H}$ units from the $p_c\text{H}$ at 10 per cent neutralization to the $p_c\text{H}$ at 90 per cent neutralization. This can be shown using eq. (56) and calculating the $p_c\text{H}$ at these two partial neutralization values.

$$p_c\text{H} = pK_a - \log \frac{M_a}{M_s}$$

For 10 per cent neutralization, $M_a = 0.90$ and $M_s = 0.10$, so that

$$p_c\text{H} = pK_a - \log \frac{0.90}{0.10} = pK_a - 0.95$$

For 90 per cent neutralization, $M_a = 0.10$ and $M_s = 0.90$, so that

$$p_c\text{H} = pK_a - \log \frac{0.10}{0.90} = pK_a - (-0.95)$$
$$= pK_a + 0.95$$

Consequently, the range of a buffer system is approximately $pK_a \pm 1.0$ $p\text{H}$ unit, or a total of two $p\text{H}$ units, centered at the value of pK_a.

5. Capacity of Buffers

The capacity of a buffer is a measure of the ability of the buffer system to neutralize added acid or base without a major change in $p_c\text{H}$. To make the definition more quantitative, the capacity is often defined as the number of millimoles of strong acid required per milliliter to cause a unit change in

p_cH. The capacity is also defined in terms of capacity to added base and called the beta value (β). The beta value is the number of millimoles of strong base required per milliliter to cause a unit rise in the p_cH.

The capacity of a buffer depends upon the ratio of concentrations and upon the total concentration. For any given ratio, the larger the total concentration, the greater the capacity.

Example 16. Calculate (a) the p_cH and (b) the capacity to added acid for a solution which contains 12.0 mmoles of acetic acid (pK_a, 4.74) and 8.0 mmoles of sodium acetate in 100 ml.

(a) Using eq. (56),

$$\begin{aligned} p_cH &= pK_a - \log \text{mmoles of acid} + \log \text{mmoles of salt} \\ &= 4.74 - \log 12.0 + \log 8.0 \\ &= 4.74 - 1.08 + 0.90 = 4.56 \end{aligned}$$

(b) To determine the capacity, first determine the ratio of millimoles of acid to salt at a p_cH one unit less than the p_cH of the solution.

$$\text{New } p_cH = 4.56 - 1.00 = 3.56$$

Using eq. (55),

$$[H^+] = K_a \frac{\text{mmoles acid}}{\text{mmoles salt}}$$

rearrange to calculate the ratio of mmoles

$$\frac{\text{mmoles acid}}{\text{mmoles salt}} = \frac{[H^+]}{K_a} = \frac{10^{-3.56}}{10^{-4.74}} = 10^{1.18} = 15.1$$

or

$$\text{mmoles acid} = 15.1 \text{ mmoles salt}$$

This value can then be substituted in the expression for the total number of millimoles present to calculate the millimoles of salt or acid after the addition of sufficient acid to lower the p_cH one unit, or

$$\begin{aligned} \text{mmoles acid} + \text{mmoles salt} = \text{total mmoles} &= 20.0 \\ 15.1 \text{ mmoles salt} + \text{mmoles salt} &= 20.0 \\ 16.1 \text{ mmoles salt} &= 20.0 \\ \text{mmoles salt} &= 1.24 \end{aligned}$$

Since there were 8.0 mmoles of salt in the buffered solution before the addition of any acid, and the added acid reacted with the salt to form molecular acetic acid, the difference between the millimoles in the original solution and those left after the addition of the acid will be a measure of the millimoles of acid added.

$$\text{mmoles of acid added} = \text{mmoles of salt at start} - \text{mmoles salt left}$$
$$= 8.0 - 1.2 = 6.8$$

The capacity is defined as the number of millimoles of hydrogen ion per milliliter required to cause a unit change in p_cH, and the solution involved here contains 100 ml and required 6.8 mmoles of hydrogen ion so that the capacity is

$$\text{capacity} = \frac{6.8}{100} = 0.068 \text{ mmoles/ml}$$

Note that for a solution which contained 10 times the concentration shown in this problem the capacity would be greater. Since the total millimoles would equal 200, the difference in millimoles before and after would be 68 rather than 6.8, and the capacity thus would be 0.68 mmoles/ml, or 10 times higher.

Example 17. Calculate (a) the p_cH and (b) the beta value of a solution which contains 2.5 mmoles of NaH_2PO_4 and 3.5 mmoles of Na_2HPO_4 in 100 ml. The pK_2 of phosphoric acid is 7.21.

(a) Calculate the p_cH using eq. (56).

$$p_cH = pK_2 - \log \text{mmoles acid} + \log \text{mmoles salt}$$

In this case, the NaH_2PO_4 is the acid and the Na_2HPO_4 is the salt, so that

$$p_cH = 7.21 - \log 2.5 + \log 3.5$$
$$= 7.21 - 0.40 + 0.54 = 7.35$$

(b) To determine the beta value, first determine the ratio of millimoles of acid to millimoles of salt at a p_cH one unit higher than the p_cH of the solution.

$$\text{new } p_cH = 7.35 + 1.00 = 8.35$$

Using eq. (55) rearranged as in Example 16,

$$\frac{\text{mmoles acid}}{\text{mmoles salt}} = \frac{[H^+]}{K_a} = \frac{10^{-8.35}}{10^{-7.21}} = 10^{-1.14}$$
$$= 10^{0.86} \times 10^{-2} = 7.3 \times 10^{-2}$$
$$= 0.073$$

so that

$$\text{mmoles acid} = 0.073 \times \text{mmoles of salt}$$

The total number of millimoles of acid and salt is 6.0, or

$$\text{mmoles acid} + \text{mmoles of salt} = 6.0$$

Substituting 0.073 mmoles of salt for millimoles of acid,

$$0.073 \text{ mmoles salt} + \text{mmoles salt} = 6.0$$

$$1.073 \text{ mmoles salt} = 6.0$$

$$\text{moles salt} = \frac{6.0}{1.073} = 5.6$$

The change in millimoles of salt will equal the amount of base added since 1 mmole of NaOH reacts with 1 mmole of acid to form 1 mmole of salt. Accordingly,

$$\text{increase in mmoles salt} = 5.6 - 3.5 = 2.1$$

or 2.1 mmoles of NaOH would be required per 100 ml to cause a unit increase in p_cH. The beta value then is

$$\text{beta value} = \frac{2.1 \text{ mmoles}}{100 \text{ ml}} = 0.021 \text{ mmoles/ml}$$

Both of the solutions in Examples 16 and 17 will show different capacities to added acid and added base since the number of millimoles of acid is not equal to the millimoles of salt. The maximum capacity to added acid or base for any given total concentration occurs when the concentrations of the salt and acid are the same. Solutions which contain more acid than salt will have a higher capacity to added base than to added acid, while solutions containing more salt than acid will have a higher capacity to added acid than to added base.

6. Preparation of Buffer Solutions of Definite p_cH and Definite Total Molar Concentration

It is often necessary to prepare a buffer of known p_cH for use in a given determination. Since the capacity as well as the p_cH is usually important in any procedure, the total molar concentration must also be specified so that the buffer system will have sufficient capacity to keep the p_cH within narrow limits during the determination. For optimum capacity to either acid or base, the pK of the weak acid or base must be close to the p_cH desired so that the molar ratio of acid or base to salt will be close to unity. The acid or base and salt selected must also be sufficiently soluble to prepare a solution with the desired capacity. The calculations necessary for the preparation of a buffer of definite p_cH and definite total molar concentration include the calculation of the molar ratio of the acid or base and the salt followed by use of this ratio and the total molar concentration to determine the actual concentration of each.

Example 18. Calculate (a) the molar concentration of acid and salt, and (b) the volumes of 1.00 M solutions of each which must be mixed and diluted to prepare 500 ml of a solution of sodium acetate and acetic acid which has a p_cH of 4.65 and a total molar concentration of 0.500 M. The pK_a of acetic acid is 4.74. First, calculate the molar ratio of acid and salt using eq. (74) rearranged as follows:

$$\frac{[H^+]}{K_a} = \frac{\text{mmoles of acid}}{\text{mmoles of salt}} = \frac{10^{-4.65}}{10^{-4.74}} = 10^{+0.09}$$

$$\text{mmoles acid} = 1.23 \text{ mmoles of salt}$$

The total number of millimoles equals 500 × 0.500 or 250, so that

$$\text{mmoles of acid} + \text{mmoles of salt} = 250$$

Substituting the value for millimoles of acid in this expression gives

$$1.23 \text{ mmoles salt} + \text{mmoles salt} = 250$$

$$2.23 \text{ mmoles salt} = 250$$

$$\text{mmoles salt} = \frac{2.50}{2.23} = 112.1 = 112$$

$$\text{mmoles acid} = 250 - 112 = 138$$

Next, calculate the volumes of 1.00 M solutions of acid and base which contain these numbers of millimoles.

$$\text{for the acid,}\quad \text{ml} = \frac{\text{mmoles}}{M} = \frac{138}{1.00} = 138 \text{ ml}$$

$$\text{for the salt,}\quad \text{ml} = \frac{\text{mmoles}}{M} = \frac{112}{1.00} = 112 \text{ ml}$$

To prepare the solution, mix 138 ml of 1.00 M acetic acid with 112 ml of 1.00 M sodium acetate and dilute to 500 ml.

Example 19. How many grams of ammonium chloride (MW, 53.5) and how many milliliters of concentrated ammonium hydroxide (16.0 *M*) should be dissolved, mixed, and diluted to 100 ml to prepare a solution with a p_cH of 9.00 and a total molar concentration of 5.00? The pK_b of NH_4OH is 4.76. Since the slightly ionized material is a base, it is preferable to work with p_cOH rather than p_cH, so first convert p_cH to p_cOH.

$$p_cOH = 14.00 - p_cH = 14.00 - 9.00 = 5.00$$

Next, determine the ratio of millimoles of base and salt using eq. (58).

$$\frac{[OH^-]}{K_b} = \frac{\text{mmoles base}}{\text{mmoles salt}} = \frac{10^{-5.00}}{10^{-4.76}} = 10^{-0.24}$$

$$= 10^{0.76} \times 10^{-1} = 5.76 \times 10^{-1} = 0.576$$

and

$$\text{mmoles base} = 0.576 \text{ mmoles salt}$$

The total number of millimoles equals 100×5.00 or 500, so that

$$\text{mmoles base} + \text{mmoles salt} = 500$$

and, substituting in this expression,

$$0.576 \text{ mmoles salt} + \text{mmoles salt} = 500$$

$$1.576 \text{ mmoles salt} = 500$$

$$\text{mmoles salt} = \frac{500}{1.576} = 317 \text{ mmoles}$$

$$\text{mmoles base} = 500 - 317 = 183$$

The grams of NH_4Cl is given by

$$g = \text{mmoles} \times \text{mMW} = 317 \times 0.0535 = 17.0 \text{ g}$$

The milliliters of concentrated NH_4OH is given by

$$\text{ml} = \frac{\text{mmoles}}{M} = \frac{183}{16.0} = 11.3 \text{ ml}$$

To prepare the solution, dissolve 17.0 g of NH_4Cl in a small amount of water, add 11.3 ml of concentrated NH_4OH, and dilute to 100 ml.

H. Nonaqueous Titrations

Water is not a satisfactory solvent for the titration of many organic acids and bases since a large number of organic compounds are not soluble in water and those which are soluble are often very weak acids or bases with ionization constants less than 10^{-8}. Since compounds which have very low ionization constants do not give a satisfactory break when titrated in water solution, only a very few organic acids and bases can be titrated in water. It has long been known that the strength of basic compounds, such as amines, is larger in glacial acetic acid than in water since acetic acid is less basic than water. Similarly some strong aqueous acids such as hydrochloric are actually weak acids in acetic acid, while some acids such as perchloric are essentially completely ionized in both acetic acid and water.

In water, the strongest acid is the solvated proton known as the hydronium ion which is formed by the ionization of water.

$$2H_2O \rightleftharpoons H_3O^+ + OH^- \qquad (77)$$

Similarly, in acetic acid the solvated proton formed by the ionization of acetic acid is also the strongest possible acid.

$$2HC_2H_3O_2 \rightleftharpoons H_2C_2H_3O_2^+ + C_2H_3O_2^- \qquad (78)$$

Since perchloric acid is essentially completely ionized in acetic acid, a solution of perchloric acid in acetic acid is usually chosen as the titrant solution for bases. Water interferes, and since the usual concentrated perchloric acid contains approximately 30 per cent water, acetic anhydride is added to such solutions to remove the water by converting it to acetic acid.

$$(CH_3CO)_2O + H_2O \rightleftharpoons 2HC_2H_3O_2 \qquad (79)$$

The general equation for the titration of a base in acetic acid is

$$H_2C_2H_3O_2^+ + B \longrightarrow BH^+ + HC_2H_3O_2 \qquad (80)$$

Nonaqueous titration in acetic acid is used often for aliphatic and aromatic amines such as aniline, acridine, brucine, butylamine, tribenzylamine, and nicotinamide. Also, salts such as choline chloride, codeine phosphate, thiamine nitrate, potassium biphthalate, sodium acetate, sodium bromide, sodium nitrite, sodium sulfate, and sodium tungstate may be titrated in acetic acid using perchloric acid as the titrant. In fact, potassium biphthalate is often used as a primary standard base in nonaqueous titrations since it can be obtained readily in stated purity. In aqueous work, this compound is probably the most popular primary standard acid used.

Other solvents which are used for titration of bases include dioxane, acetonitrile, nitrobenzene, chloroform, carbon tetrachloride, petroleum ether, and nitromethane. The perchloric acid titrant is usually dissolved in acetic acid for use with the inert solvents but may be dissolved in the solvent being used in some cases.

Weak organic acids may also be titrated in nonqueous media, but these titrations are not as popular as the weak base titrations. Titrants are usually sodium methoxide or sodium ethoxide in the alcohol or in benzene-methanol (1 : 6), but alcoholic potassium and sodium hydroxides are also used. Anhydrous ethylenediamine as solvent and sodium aminoethoxide in ethylenediamine-ethanolamine as the titrant are used in the titration of phenols and other very weak acids.

Most nonaqueous titrations are performed potentiometrically using silver-silver chloride and glass electrodes, with the glass electrode being the indicator electrode. The saturated calomel electrode may be utilized as the reference electrode but requires the use of a salt bridge. An ordinary *p*H meter is the usual measuring device. The *p*H reading should not be considered as the hydrogen ion concentration but only as a measure of the voltage or potential change during the titration.

Visual indicators may also be used. Crystal and methyl violet are the most popular for titrations in acetic acid, but 1-naphtholbenzein and benzoylauramine are also used. The indicator is usually dissolved in an inert solvent such as chlorobenzene, chloroform, nitrobenzene, or acetonitrile. The color change with crystal and methyl violet is from violet to blue to green, with blue being the end point. Most directions state that the end point is the disappearance of the last tinge of violet in the solution. The end point for most titrations is actually very sharp and can be obtained with less than one drop of titrant when using 0.1 *N* solutions. However, water interferes and must be absent since, in the presence of very small amounts of water, the end point is very difficult to determine. The water prolongs the disappearance of the violet tinge and makes the end point indistinct.

I. Problems

1. List the conjugate acids of the following bases:
   (a) $H_2AsO_4^-$
   (b) CO_3^{2-}
   (c) $Fe(OH)(H_2O)_5^{2+}$
   (d) $C_6H_5NH_2$ (aniline)
   (e) $C_6H_5O^-$ (phenolate)
   (f) $HC_6H_5O_7^{2-}$ (hydrogen citrate)

2. List the conjugate bases for the following acids:
   (a) HNO_3 (d) NH_4^+
   (b) H_2MoO_4 (e) $Ni(H_2O)_6^{2+}$
   (c) $H_2C_6H_5O_7^-$ (dihydrogen citrate) (f) $HCrO_4^-$
3. Calculate the p_cH of the following solutions:
   (a) 0.0025 *M* HNO_3 (d) 0.00050 *M* H_2SO_4
   (b) 0.0093 *N* H_2SO_4 (e) $10^{-4.15}$ *N* $HClO_4$
   (c) 0.0153 *M* HCl

   Ans. (b) 2.03
4. Calculate the p_cH of the following solutions:
   (a) 0.036 *N* NaOH (d) 0.0048 *M* $Ba(OH)_2$
   (b) 0.250 *N* $Ba(OH)_2$ (e) 0.0000077 *M* $Ca(OH)_2$
   (c) 1.00 *N* KOH

   Ans. (b) 13.40
5. How many millimoles of HCl would have to be added to 250 ml of water to produce solutions with the following p_cH values: (a) 2.43, (b) 0.64, (c) 4.33, (d) 5.09, and (e) 1.12.

   Ans. (b) 57.3
6. How many millimoles of NaOH would have to be added to 333 ml of water to produce solutions with the following p_cH values: (a) 9.45, (b) 12.67, (c) 13.75, (d) 8.90, and (e) 10.40.

   Ans. (b) 15.6
7. How many millimoles of NaOH would have to be added to 500 ml of water to produce solutions with the following p_cOH values: (a) 1.49, (b) 4.73, (c) 6.10, (d) 0.50, and (e) 2.50.

   Ans. (b) 0.0093
8. Calculate the molar concentration of hydrochloric acid solutions which have the following p_cH values: (a) 1.23, (b) 2.34, (c) 3.45, (d) 4.56, and (e) 5.67.

   Ans. (b) 0.0046 *M*
9. Calculate the molar concentration of sodium hydroxide solutions which have the following p_cH values: (a) 7.89, (b) 13.00, (c) 8.97, (d) 12.23, and (e) 10.12.

   Ans. (b) 0.10 *M*
10. Calculate the molar concentration of the salt at the equivalence point in the following titrations assuming (1) no dilution with water, (2) dilution with water to 100.0 ml, and (3) dilution with 20.00 ml of water:
    (a) 25.00 ml of 0.1017 *N* $HC_2H_3O_2$ with 0.0983 *N* NaOH
    (b) 20.00 ml of 0.2408 *N* H_2SO_4 with 0.1013 *N* KOH
    (c) 10.00 ml of 0.1525 *M* H_3PO_4 to Na_2HPO_4 with 0.1123 *M* NaOH
    (d) 50.00 ml of 0.0505 *N* HCl with 0.0894 *N* NaOH
    (e) 25.00 ml of 0.2500 *N* NH_4OH with 0.2075 *N* HCl

    Ans. (b) (1) 0.0357 *M*, (2) 0.0241 *M*, (3) 0.0275 *M*
11. State whether the p_cH of the following solutions will be 7, less than 7, more than 7, or approximately 7:
    (a) NH_4NO_3 (c) Na_2SO_4 (e) NH_4ClO_2
    (b) $Ca(NO_2)_2$ (d) $AlCl_3$ (f) $Fe(C_2H_3O_2)_3$
12. Calculate the p_cH of the following solutions of salts:
    (a) 0.0250 *M* $Ca(C_2H_3O_2)_2$
    (b) 0.1050 *M* NaClO
    (c) 0.2134 *M* $KC_7H_5O_2$ (potassium benzoate)

(d) 0.1000 M $LiC_3H_5O_2$ (lithium propionate)
(e) 0.1250 M $Cs_2C_4H_4O_5$ (cesium malate)

Ans. (b) 10.28

13. Calculate the p_cH of the following solutions of salts:
(a) 0.1000 M NH_4NO_3
(b) 0.0100 M NH_4Cl
(c) 0.2150 M $AgNO_3$
(d) 0.1250 M $(NH_4)_2SO_4$
(e) 0.0500 M $CuSO_4$

Ans. (b) 5.63

14. Calculate the p_cH at the equivalence points in the titrations for problems 10(a), 10(c), and 10(e).

Ans. (a) (1) 8.73, (2) 8.59, (3) 8.66

15. Calculate the p_cH of the following solutions of salts:
(a) 0.0125 M $NH_4C_7H_5O_2$ (ammonium benzoate)
(b) 0.0019 M $AgC_3H_5O_2$ (silver propionate)
(c) 0.5000 M $NH_4C_2H_3O_2$
(d) 0.2000 M $(NH_4)_2CO_3$
(e) 0.1033 M $AgNO_2$

Ans. (b) 7.46

16. Calculate the p_cH of the following solutions of salts:
(a) 0.2500 M $NaHCO_3$
(b) 0.0525 M $NaHC_4H_4O_5$ (soidium hydrogen malate)
(c) 0.074 M NaH_2PO_4
(d) 0.1000 M Na_2HPO_4
(e) 0.1250 M $Na_2H_2C_{10}H_{12}O_8N_2$ (disodium dihydrogen EDTA)

Ans. (b) 4.23

17. Calculate the p_cH for the following solutions:
(a) 0.50 M propionic acid and 0.50 M sodium propionate
(b) 0.010 M sodium benzoate and 0.015 M benzoic acid
(c) 0.15 M disodium hydrogen phosphate and 0.25 M sodium dihydrogen phosphate
(d) 0.22 M sodium dihydrogen phosphate and 0.18 M trisodium phosphate
(e) 0.25 M NH_4OH and 0.050 M NH_4Cl

Ans. (b) 4.02

18. Calculate the p_cH in the titrated solution assuming no dilution with water for the titration of 25.00 ml of 0.1000 N acid with 0.1000 N base after the addition of (1) 5.00 ml, (2) 12.50 ml, (3) 17.50 ml, and (4) 30.00 ml of the base for the following titrations: (a) hydrochloric acid with NaOH, (b) benzoic acid with NaOH, (c) hydrochloric acid with ammonium hydroxide, (d) sodium dihydrogen phosphate with NaOH to form disodium hydrogen phosphate, and (e) disodium hydrogen phosphate with HCl to form sodium dihydrogen phosphate.

Ans. (b) (1) 3.60, (2) 4.20, (3) 4.56, (4) 11.96

19. Calculate the p_cH of the following solutions:
(a) A solution which contains 7.5 g of Na_2CO_3 and 6.0 g of $NaHCO_3$ in 400 ml
(b) A solution which contains 6.85 millimoles of sodium acetate and 3.43 millimoles of acetic acid in 250 ml

(c) A solution prepared by adding 50.0 ml of 0.250 M NaOH to 50.0 ml of 0.150 M H_3PO_4

(d) A solution prepared by adding 10.0 ml of 3.00 N HCl to 90.0 ml of 1.00 N NH_4OH

(e) A solution which contains 48 millimoles of tartaric acid and 56 millimoles of potassium hydrogen tartrate in 330 ml

Ans. (b) 5.06

20. Calculate (1) the capacity to added acid and (2) the beta value for the solutions listed in problem 17. Assume 100 ml of each solution.

Ans. (b) (1) 0.0084, (2) 0.01172

21. Calculate (1) the capacity to added acid and (2) the beta value for the solutions listed in problem 19.

Ans. (b) (1) 0.0206, (2) 0.0118

22. A 2.000-g sample of an industrial sodium carbonate alkali is dissolved and diluted to 100.0 ml. One 25.00-ml aliquot is titrated to the phenolphthalein end point with HCl and a second 25.00-ml aliquot is titrated with the same acid to the methyl orange end point. Using the volumes and normalities shown below, (1) state the compounds present, and (2) calculate the percentage of each present:

Milliliters to phenolphthalein end point	Milliliters to methyl orange end point	N of acid
(a) 35.48	49.93	0.1936
(b) 24.63	49.26	0.1450
(c) 15.34	35.42	0.2250

Ans. (b) (1) Na_2CO_3 only, (2) 75.71 per cent Na_2CO_3

23. An alkaline unknown contains inert material plus any one or any combination of sodium hydroxide, sodium carbonate, and sodium bicarbonate. A given weight of sample is dissolved and titrated to the phenolphthalein end point with acid. Methyl orange is then added and the titration continued to the methyl orange end point with the same acid. For the data given below, (1) state the components present, and (2) calculate the percentage of each present:

Weight of sample (g)	Milliliters to phenolphthalein end point	Milliliters from phenolphthalein end point to methyl orange end point	N of acid
(a) 0.998	25.27	20.63	0.1500
(b) 1.002	27.89	35.69	0.2250
(c) 0.500	23.33	23.33	0.2000

Ans. (b) (1) Na_2CO_3, $NaHCO_3$; (2) Na_2CO_3, 66.39 %; $NaHCO_3$, 14.71 %

24. How many grams of each constituent must be dissolved and diluted to prepare the buffer solutions of given p_cH and total molar concentration listed below:

	Constituents	Volume of solution (ml)	p_cH	Total molar concentration (M)
(a)	$HC_2H_3O_2 + NaC_2H_3O_2$	400	4.35	0.250
(b)	$NH_4OH + NH_4Cl$	250	9.50	1.33
(c)	$NaHCO_3 + Na_2CO_3$	1000	10.50	0.500
(d)	$NaH_2PO_4 + K_2HPO_4$	500	7.00	1.25
(e)	$KHC_8H_4O_4 + K_2C_8H_4O_4$ (phthalate)	100	5.10	1.00

Ans. (b) NH_4OH, 6.49 g; NH_4Cl, 7.40 g.

CHAPTER 10

Oxidation-Reduction Theory

There are a very large number of oxidation-reduction reactions which are used as the basis for analytical methods. Oxidation was defined originally as the reaction of a substance with oxygen, and the substance which reacted was said to be oxidized. Thus the rusting of iron, which was considered to be the reaction of iron with the oxygen in the air to form iron(III) oxide, was referred to as the oxidation of iron. Similarly, the term *reduction* was used to designate the removal of oxygen from a material since many metal ores were oxides, and, in the metallurgical process, the ore was said to be reduced to the metal. As a consequence, the removal of oxygen became known as reduction. As the knowledge of the electronic structure of atoms increased, it was realized that in the reaction with iron, oxygen atoms actually removed electrons from iron atoms to form charged particles called ions, so that oxidation was really the removal of electrons from an atom. It was soon realized that other reactions, such as the reaction of iron and chlorine, also involved loss and gain of electrons even though oxygen was not a reactant. Since these reactions caused an increase in positive charge or oxidation state as did the reaction with oxygen, they were also termed oxidation. Similarly, the removal of oxygen from an oxide represented a gain of electrons by the metal ion and a decrease in oxidation state, and all such reactions were termed *reductions*, whether or not oxygen was involved.

Oxidation is now defined as the loss of electrons by a substance, and *reduction* is defined as the gain of electrons by a substance. Since electrons cannot be created or destroyed in ordinary chemical reactions, any oxidation must be accompanied by a corresponding reduction. The loss of electrons causes an increase in oxidation state so that oxidation is also defined as an increase in the oxidation state of an atom or ion. Similarly the gain of electrons causes a decrease in oxidation state, and reduction is defined as the decrease in the oxidation state on an atom or ion.

A substance which causes oxidation to occur in another is called an oxidizing agent and, since it must accept electrons from the substance it oxidizes, is reduced during the reaction. Also, a substance which causes reduction to occur is called a reducing agent and is itself oxidized during the reaction. All oxidation-reduction reactions then can be considered to be the reaction of an oxidizing agent with a reducing agent. In all such reactions the oxidizing agent is reduced and the reducing agent is oxidized.

A. Balancing Oxidation-Reduction Reactions

There are several satisfactory methods for balancing oxidation-reduction reactions, each of which has certain advantages over the other systems. Since analytical chemists are interested primarily in the actual species which react in a solution and in the net reaction which occurs, most analysts prefer the partial ion equation method which is often called the half-cell reaction, half reaction, or electrode reaction method. In this method only the actual ions or molecules which are involved in the reaction between the oxidizing agent and the reducing agent are shown. If a molecular equation is necessary, it can be obtained from the net ionic reaction, which is the result of the partial ion equation method of balancing reactions.

The steps taken to balance an equation include (1) the writing and balancing of the ionic reaction of the oxidizing agent, (2) the writing and balancing of the ionic reaction of the reducing agent, (3) the multiplication of one or both of these equations, if necessary, to balance the number of electrons, and (4) the summation of the two equations to obtain the net ionic reaction. The following rules are utilized in balancing the half reactions:

1. Show each substance as the major portion occurs in the solution. Ionic substances (strong electrolytes) are shown as ions, and weak and nonelectrolytes are shown as molecules. The student should remember that the symbol of an ion includes its charge and that the symbol of an element represents an atom, not an ion, of the element. Similarly, $Cr_2O_7^{2-}$ is the symbol of the dichromate ion, while Cr_2O_7 (without the charge) is the formula of chromium heptoxide.

2. The equations must be balanced both as to atoms and charges.

3. The number of electrons involved in each half reaction must be equal, or the half reactions must be multiplied by numbers which make the number of electrons involved equal.

4. The molecular equation may be obtained from the ionic equation by adding the ions which occur with the reactant ions and distributing them among the product ions. Any ions left over must represent whole numbers of molecules for the substances represented.

5. Water or its ions may be required to obtain material balance. It must be remembered, however, that hydrogen ions cannot be shown as reactant or product in a basic solution, and hydroxide ions may not be shown as reactant or product in an acidic solution.

Several examples will be shown as illustrations of this method of balancing equations.

Example 1. Balance the reaction between $K_2Cr_2O_7$ and $FeCl_2$ in HCl solution to form $FeCl_3$, $CrCl_3$, and other products. Since salts and strong acids are completely ionic, the unbalanced ionic equation may be represented as

$$Cr_2O_7^{2-} + Fe^{2+} + H^+ \longrightarrow Cr^{3+} + Fe^{3+} + H_2O \quad (1)$$

The unbalanced ionic half reaction of the oxidizing agent is

$$Cr_2O_7^{2-} + H^+ \longrightarrow Cr^{3+} + H_2O \quad (2)$$

To obtain material (atom) balance, there must be 2 Cr^{3+} ions, and to balance the oxygen from the $Cr_2O_7^{2-}$, there must be 7 molecules of water, which, in turn, would require 14 H^+ ions. The half reaction then becomes

$$Cr_2O_7^{2-} + 14H^+ \longrightarrow 2Cr^{3+} + 7H_2O \quad (3)$$

The number of electrons involved may now be obtained by balancing the charge on both sides of the equation. That is, the total charge on the products must equal the total charge on reactants. The total charge on the reactants is 12 positive and on the products is 6 positive. To reduce 12 positive to 6 positive requires the addition of six electrons as reactant, or

$$Cr_2O_7^{2-} + 14H^+ + 6e^- \longrightarrow 2Cr^{3+} + 7H_2O \quad (4)$$

and the equation is balanced both as to atoms and as to charge.

The unbalanced ionic equation for iron is

$$Fe^{2+} \longrightarrow Fe^{3+} \quad (5)$$

This equation is balanced as to atoms and requires only balancing as to charge. Since an increase in oxidation state represents a loss or removal of electrons, it will require the loss of one electron by the Fe^{2+} to become Fe^{3+}, or

$$Fe^{2+} - 1e^- \longrightarrow Fe^{3+} \qquad (6)$$

The electrons can actually be shown on either side of the equation and eq. (6) could be written as

$$Fe^{2+} \longrightarrow Fe^{3+} + 1e^- \qquad (7)$$

Since the number of electrons lost by iron must equal the number of electrons gained by the dichromate, eq. (6) must be multiplied by 6 before the two equations can be added to obtain the net ionic reaction, or

$$\begin{array}{llll} Cr_2O_7^{2-} + 14H^+ + 6e^- & \longrightarrow & 2Cr^{3+} + 7H_2O & (4) \\ 6Fe^{2+} \qquad\qquad - 6e^- & \longrightarrow & 6Fe^{3+} & (8) \\ \hline Cr_2O_7^{2-} + 6Fe^{2+} + 14H^+ & \longrightarrow & 2Cr^{3+} + 6Fe^{3+} + 7H_2O & (9) \end{array}$$

If the molecular equation is desired, the ions occurring with the reactants can be added and distributed among the products, or

$$\begin{array}{llll} Cr_2O_7^{2-} + 6Fe^{2+} + 14H^+ & \longrightarrow & 2Cr^{3+} + 6Fe^{3+} + 7H_2O & (9) \\ 2K^+ \quad + 12Cl^- + 14Cl^- & \longrightarrow & 6Cl^- \quad + 18Cl^- + 2K^+ + 2Cl^- & \\ \hline K_2Cr_2O_7 + 6FeCl_2 + 14HCl & \longrightarrow & 2CrCl_3 + 6FeCl_3 + 7H_2O + 2KCl & \\ & & & (10) \end{array}$$

As shown in this example, the ions left over after distribution among the products must represent a whole number of molecules of substances.

Example 2. Balance the reaction for the titration of calcium oxalate by permanganate in the analysis of limestone by the oxalate-permanganate method. In this method, the precipitated calcium oxalate is dissolved in sulfuric acid and heated before being titrated with the permanganate. The oxalate will be present mainly as bioxalate due to the acidity of the solution. The unbalanced ionic equation then becomes

$$MnO_4^- + HC_2O_4^- + H^+ \longrightarrow Mn^{2+} + CO_2 + H_2O \qquad (11)$$

The unbalanced ionic half reaction for MnO_4^- is

$$MnO_4^- + H^+ \longrightarrow Mn^{2+} + H_2O \qquad (12)$$

To obtain material balance, four molecules of water will be formed, and this will require eight hydrogen ions, or

$$MnO_4^- + 8H^+ \longrightarrow Mn^{2+} + 4H_2O \quad (13)$$

To obtain charge balance, the total charge on the reactants is 7 positive and upon the products is 2 positive. To convert a charge of 7 positive to 2 positive requires the addition of 5 electrons, or

$$MnO_4^- + 8H^+ + 5e^- \longrightarrow Mn^{2+} + 4H_2O \quad (14)$$

The unbalanced ionic half reaction for bioxalate is

$$HC_2O_4^- \longrightarrow CO_2 + H^+ \quad (15)$$

Material balance requires two molecules of CO_2, or

$$HC_2O_4^- \longrightarrow 2CO_2 + H^+ \quad (16)$$

Charge balance requires the conversion of a one negative charge to a 1 positive charge and this requires the removal of two electrons, or

$$HC_2O_4^- - 2e^- \longrightarrow 2CO_2 + H^+ \quad (17)$$

Before the addition of the two half-cell reactions, the minimum number of electrons which will satisfy both equations is 2 × 5, or 10, so that eq. (14) must be multiplied by 2 and eq. (17) by 5, or

$$2MnO_4^- + 16H^+ + 10e^- \longrightarrow 2Mn^{2+} + 8H_2O \quad (18)$$

$$5HC_2O_4^- - 10e^- \longrightarrow 10CO_2 + 5H^+ \quad (19)$$

$$2MnO_4^- + 5HC_2O_4^- + 16H^+ \longrightarrow 2Mn^{2+} + 10CO_2 + 8H_2O + 5H^+ \quad (20)$$

By convention, the same ion or molecule is not shown as both reactant and product, so the five H^+ are canceled into the $16H^+$, or $5H^+$ are subtracted from both sides of the equation to obtain

$$2MnO_4^- + 5HC_2O_4^- + 11H^+ \longrightarrow 2Mn^{2+} + 10CO_2 + 8H_2O \quad (21)$$

Example 3. Balance the reaction between cupric ion and iodide ion in the determination of copper by the iodimetric method. In this reaction, the products are triiodide ion and cuprous iodide. The unbalanced ionic reaction is

$$Cu^{2+} + I^- \longrightarrow Cu_2I_2 + I_3^- \quad (22)$$

The unbalanced ionic half reaction for cupric ion is

$$Cu^{2+} \longrightarrow Cu_2I_2 \quad (23)$$

Material balance requires two cupric ions and two iodide ions as reactants, or

$$2Cu^{2+} + 2I^- \longrightarrow Cu_2I_2 \quad (24)$$

Charge balance requires converting a two positive charge to zero by the addition of two electrons, or

$$2Cu^{2+} + 2I^- + 2e^- \longrightarrow Cu_2I_2 \quad (25)$$

The unbalanced ionic reaction for iodide is

$$I^- \longrightarrow I_3^- \quad (26)$$

Material balance requires three iodide ions, or

$$3I^- \longrightarrow I_3^- \quad (27)$$

and charge balance requires conversion of a three negative charge to a one negative charge by the removal of two electrons, or

$$3I^- - 2e^- \longrightarrow I_3^- \quad (28)$$

Addition of eqs. (25) and (28) gives the balanced ionic net reaction, or

$$\begin{array}{llll} 2Cu^{2+} + 2I^- + 2e^- & \longrightarrow & Cu_2I_2 & (25) \\ 3I^- \qquad\quad - 2e^- & \longrightarrow & I_3^- & (28) \\ \hline 2Cu^{2+} + 5I^- & \longrightarrow & Cu_2I_2 + I_3^- & (29) \end{array}$$

In Examples 1, 2, and 3 it should be noted that there is both charge balance and material balance in the net ionic reaction. As a consequence, the student should check the charge balance of the net ionic reaction each time an oxidation-reduction reaction is balanced to be sure that the balancing has been done correctly.

B. Equivalent Weights in Oxidation-Reduction Reactions

The method of calculating the equivalent weights in any oxidation-reduction reaction was discussed in Section C.2 of Chapter 3 and should be reviewed by the student. It should be emphasized that the equivalent weight is a function of the electron change and not the oxidation state of the ion, atom, radical, or molecule. As a consequence the student must know the actual products of the reaction in order to determine the electron change. Since most oxidizing agents and reducing agents can produce different products under difference conditions, information as to the products to be expected is usually given in the beginning course in quantitative analysis. For the more common oxidants, MnO_4^- is always reduced to Mn^{2+} in acid solution, but MnO_2 is the product in solutions which are near neutrality, and manganate (MnO_4^{2-}) is the product in strongly basic solutions. Dichromate in acid solution is reduced to Cr^{3+}. Usually the student is expected to know the product for dichromate and permanganate titrations under various conditions.

C. The Electrochemical Theory of Oxidation-Reduction

The electric current which flows through wires and furnishes the energy for lighting, heating, and cooling homes and for industrial uses is simply a flow of electrons along the wire. The flow is characterized by three units—the volt, the ampere, and the ohm. The volt (V) is the unit of driving force known as the potential or the electromotive force (emf) which causes the electrons to flow. The ampere (A) is the unit of current and is a measure of the number of electrons which pass a given point in a given unit of time. The ohm (Ω) is the unit of resistance to the flow of the electrons through the conductor. These three factors are related by Ohm's law, which can be stated as

$$E = IR \tag{30}$$

in which E is the potential in volts, I is the current in amperes, and R is the resistance in ohms. Each of these quantities has an absolute definition, but they are usually defined in terms of each other. A volt is the potential necessary to cause a current of one ampere to flow through a resistance of one ohm. An ampere is the amount of current which passes a given point when a potential of one volt operates through a resistance of one ohm. The ohm is the amount of resistance offered to a current of 1 A under a potential of one volt.

When an oxidation-reduction reaction takes place in a solution there is a transfer of electrons through space from one atom or ion to another. In the reaction between metallic zinc and copper sulfate solution, which produces metallic copper and zinc sulfate solution, a copper ion must come in very close contact with a zinc atom for the transfer to take place since electrons do not flow freely through a solution. The ionic reaction may be shown as

$$Cu^{2+} + Zn \longrightarrow Cu + Zn^{2+} \qquad (31)$$

and the individual half reactions are

$$Cu^{2+} + 2e^- \longrightarrow Cu \qquad (32)$$

$$Zn - 2e^- \longrightarrow Zn^{2+} \qquad (33)$$

If some device could be designed which would separate the reactants and cause the electron transfer to take place along a wire, an electric current would be produced as the electrons flowed through the wire. A galvanic or voltaic cell is such a device. A galvanic cell is composed of two half-cells, each of which must contain the two forms of an oxidation and reduction pair and a metallic conductor called the electrode. The electrode can be one of the two forms of the oxidation-reduction pair or it may be an inert conductor such as platinum which does not enter into the half-cell reaction. A galvanic cell which takes advantage of the copper-zinc reaction of eq. (31) could be designed as shown in Fig. 10-1. A container which has a bar of

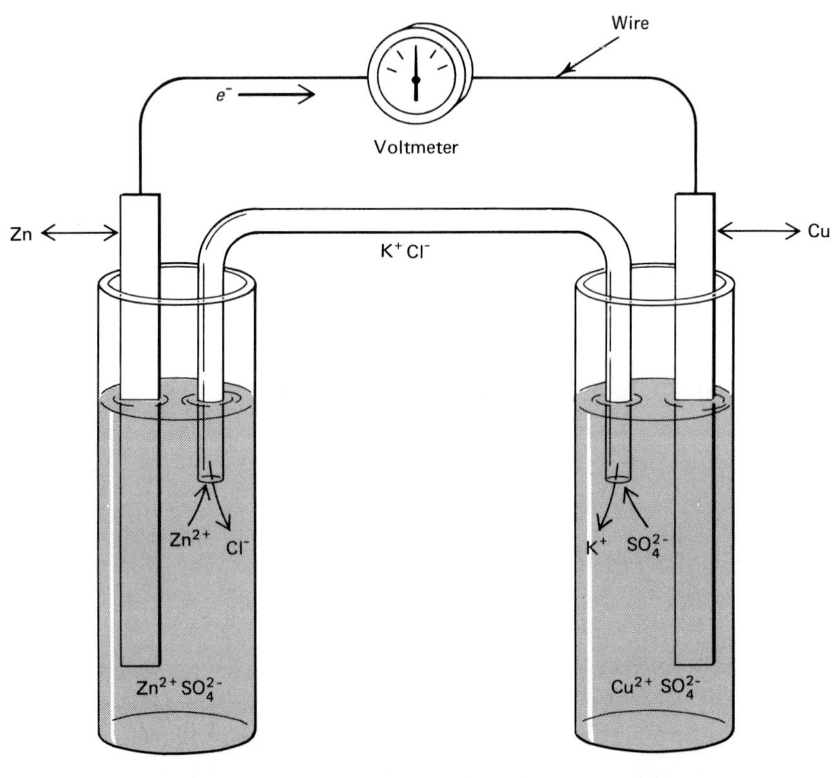

Figure 10.1. A galvanic cell showing a zinc-zinc sulfate half-cell and a copper-copper sulfate half-cell connected by a potassium chloride salt bridge.

zinc metal immersed in a solution of zinc sulfate is connected by a U-tube filled with a concentrated solution of an electrolyte to a second container which has a bar of copper immersed in a solution of copper sulfate. The two metal bars (electrodes) are connected by a wire. The U-tube filled with electrolyte is known as a salt bridge and allows ions to move from one container to the other to maintain electrical neutrality. It is usually filled with a concentrated solution of KCl embedded in a gelatin matrix to prevent free syphoning from one solution to the other. A voltmeter and/or an ammeter may be incorporated into the circuit to measure the potential created and the amount of current which flows.

Upon completion of the circuit, either by connecting the wire or by inserting the salt bridge, oxidation-reduction reactions occur at the electrodes and electrons flow through the wire. The direction of flow of electrons depends upon the composition of the two half-cells. Each half-cell exerts a potential or driving force which is related to the ease of reduction of the metal ions and which causes electrons to flow from it to another half-cell. However, each half-cell can also accept electrons. As a consequence the half-cell with the greater driving force or potential will force the other half-cell to accept electrons. In the zinc-copper cell, zinc has a greater driving force than copper since copper ion is more easily reduced than zinc ion and electrons flow through the wire from the zinc half-cell to the copper half-cell. In the zinc half-cell, zinc atoms go into solution to form zinc ions leaving two electrons on the metal bar. These two electrons then flow through the wire to the copper bar where they are given to a copper ion, which deposits as elementary copper. As a consequence the zinc bar will lose weight, while the copper bar gains weight. The reactions which occur at the electrodes may be shown as

$$\mathrm{Zn} - 2e^- \longrightarrow \mathrm{Zn}^{2+}$$

$$\mathrm{Cu}^{2+} + 2e^- \longrightarrow \mathrm{Cu}$$

which are the same as eqs. (32) and (33). Summation of these two reactions gives

$$\mathrm{Zn} + \mathrm{Cu}^{2+} \longrightarrow \mathrm{Cu} + \mathrm{Zn}^{2+} \qquad (34)$$

which is the same as eq. (31) for the reaction which takes place when zinc is placed in a copper sulfate solution.

The zinc electrode in the above example is called the negative pole since it has a negative charge due to the electrons deposited on it as zinc atoms are converted to zinc ions. The copper electrode is considered to be the positive electrode since it has a deficiency of electrons caused by the conversion of a copper ion to an atom. The zinc electrode, having the higher driving force, is considered to be a region of high electron density, while the copper electrode with a weaker driving force has a lower electron density, and since electrons flow freely along a metallic conductor from a region of high electron density to a region of lower electron density, the electrons flow through the wire from the negative half-cell to the positive half-cell, or from the zinc to the copper.

The salt bridge maintains electrical neutrality in the two solutions by transport of ions. The deposition of a copper ion leaves an excess of two negative charges in the copper solution, while the formation of a zinc ion causes an excess of two positive charges in the zinc solution. These charges are neutralized by the movement of ions into, out of, or through the salt bridge. It really makes no difference whether two Cl^- ions move from the bridge into the zinc solution and two K^+ ions move into the copper solution or whether the zinc and sulfate ions migrate into the salt bridge as long as electrical neutrality is preserved. In reality, probably both types of migration occur.

The driving force, or potential, of a half-cell cannot be measured alone but only by comparison to other half-cells. Another way of stating this is that differences in potential can be measured but individual half-cell potentials cannot be measured. As a consequence, it is necessary to adopt some particular electrode as the standard electrode and measure all other potentials against it. The hydrogen half-cell has been designated as the standard half-cell and assigned a potential of zero when the hydrogen ion is at unit activity.

Half-cells which cause the hydrogen half-cell to accept electrons (reduce hydrogen ions to hydrogen gas) are assigned a negative potential, while those half-cells which accept electrons from the hydrogen half-cell (convert hydrogen gas to hydrogen ions) are assigned a positive potential. In this system, the strong oxidizing agents such as permanganate are assigned positive potentials and the strong reducing agents such as zinc are assigned negative potentials. This system has been adopted by the International Union of Pure and Applied Chemistry (IUPAC) to offset confusion in the assignment of positive and negative potentials to half-cells. There is another convention which assigns electrode potentials exactly opposite to the IUPAC convention and consequently causes confusion. In the IUPAC convention, the half-cell reaction is written to show the gain of electrons or to show reduction, while tables which use the alternative convention show the half-cell reaction as oxidation or the loss of electrons by the reactant.

D. The Nernst Equation

The actual potential of any given half-cell depends not only upon the components of the half-cell reaction system but also upon the concentrations of all the ions which are involved in the half-cell reaction. For example, a cell composed of two zinc half-cells will produce a current if the concentrations of the zinc ions differ in the two half-cells.

The equation which shows the relationship between the actual potential of a half-cell and the active concentration of the reactants and products of the half-cell reaction is known as the Nernst equation. For the general half-cell reaction,

$$aA + bB + ne^- \rightleftharpoons cC + dD \tag{35}$$

the Nernst equation[1] is

$$E = E^\circ + \frac{2.3RT}{nF} \log \frac{(A)^a(B)^b}{(C)^c(D)^d} \tag{36}$$

in which E is the actual potential in volts produced by the half-cell, E° is the standard reduction potential of the half-cell, R is the universal gas constant (8.31 V-Coulombs/°K/mole), T is the absolute temperature, F is the faraday (96,500 Coulombs), and n is the number of electrons in the balanced half-cell reaction. The value of $2.3RT/F$ is 0.060 at 30°C[2] and 0.059 at 25°C. It should be noted that the activity quotient expression in the logarithm term is identical to the reciprocal of the equilibrium constant for the reaction as written so that eq. (36) may be shown as

$$E = E^\circ + \frac{0.060}{n} \log \frac{1}{K_{eq}} \tag{37}$$

For a half-cell composed of a metal and metal ion such as the zinc-zinc ion half-cell, the equation becomes

$$\begin{aligned} E_{Zn} &= E^\circ_{Zn} + \frac{0.060}{2} \log \frac{(Zn^{2+})}{(Zn_{solid})} \\ &= E^\circ_{Zn} + \frac{0.060}{2} \log (Zn^{2+}) \end{aligned} \tag{38}$$

---

[1] The Nernst equation for reduction reactions of this type is often written in the form

$$E = E^\circ - \frac{2.3RT}{nF} \log \frac{(C)^c(D)^d}{(A)^a(B)^b}$$

in which the symbols have the same meaning as given for eq. (36). The two expressions are equivalent since

$$\log \frac{(A)^a(B)^b}{(C)^c(D)^d} = -\log \frac{(C)^c(D)^d}{(A)^a(B)^b}$$

or

$$\log \frac{1}{K_{eq}} = -\log K_{eq}$$

[2] The value at 30°C will be utilized in all problems to simplify calculations.

since solids are considered to be at unit activity. It should be noted that the concentrations specified in the Nernst equation are the active (effective) concentrations or activities. In many calculations, molar concentrations can be utilized in place of the activities without major error.

For a half-cell reaction such as permanganate in acid solution for which the equation is

$$MnO_4^- + 8H^+ + 5e^- \rightleftharpoons Mn^{2+} + 4H_2O \qquad (14)$$

the Nernst equation, using molar concentrations, is

$$E_{MnO_4^-/Mn^{2+}} = E^\circ_{MnO_4^-/Mn^{2+}} + \frac{0.060}{5} \log \frac{[MnO_4^-][H^+]^8}{[Mn^{2+}]} \qquad (39)$$

Note that eq. (14) exemplifies reduction and that the logarithm term utilizes the reciprocal of the equilibrium constant for the reduction reaction as written, or the equilibrium constant for the reverse reaction.

Complete tables of standard reduction potentials, E°, are listed in various handbooks and textbooks. An abbreviated table of E° values for certain half-cells is given in Appendix 9. These E° values are those obtained when the concentrations of all ions which are involved in the half-cell reaction are at unit activity so that the logarithmic term becomes zero as the logarithm of one and has no effect on the half-cell potential. The conducting portion of the half-cell is assumed to be platinum or some other inert conductor for those half-cells which do not involve a metal as one member of the oxidation-reduction pair, such as the permanganate-manganous ion half-cell and the ferric-ferrous ion half-cell.

Example 4. Calculate the actual potential of the ferric-ferrous ion half-cell in the titration with dichromate before the addition of any dichromate. In most such titrations, the concentration of the ferrous ion is adjusted to approximately 0.1 M before the addition of dichromate and the ferric ion concentration is of course negligible. Analysts usually consider 10^{-6} M (or less) to be a negligible concentration, and 10^{-6} M will be used to represent a negligible concentration in this text whenever the actual value is not known. The E° value for the ferric-ferrous ion half-cell is 0.77 V.

$$E_{Fe^{3+}/Fe^{2+}} = E^\circ_{Fe^{3+}/Fe^{2+}} + \frac{0.060}{n} \log \frac{[Fe^{3+}]}{[Fe^{2+}]} \qquad (40)$$

$$= 0.77 + \frac{0.060}{1} \log \frac{10^{-6}}{10^{-1}}$$

$$= 0.77 + 0.060 \log 10^{-5}$$

$$= 0.77 + 0.060\,(-5)$$

$$= 0.77 - 0.30 = 0.47 \text{ V}$$

Example 5. Calculate the actual potential of the ferric-ferrous ion half-cell at the end of the titration with dichromate. In this example assume that the concentrations are the reverse of the concentrations in Example 4 since the ferrous ion has been quantitatively converted to ferric ion. The calculation is

$$E_{Fe^{3+}/Fe^{2+}} = 0.77 + 0.060 \log \frac{[Fe^{3+}]}{[Fe^{2+}]}$$

$$= 0.77 + 0.060 \log \frac{10^{-1}}{10^{-6}}$$

$$= 0.77 + 0.060 \log 10^{+5}$$

$$= 0.77 + 0.060\ (5)$$

$$= 0.77 + 0.30 = 1.07\ \text{V}$$

Example 6. Calculate the actual potential in a solution in which the permanganate concentration is 10^{-1} *M*, the manganous concentration is 10^{-4} *M*, and the p_cH is 1. Also, calculate the half-cell potential if the p_cH is 3. The $E°$ value for the permanganate-manganous ion half-cell is 1.51. Substitution in eq. (39) gives

$$E_{MnO_4^-/Mn^{2+}} = 1.51 + \frac{0.060}{5} \log \frac{10^{-1} \times (10^{-1})^8}{10^{-4}}$$

$$= 1.51 + 0.012 \log 10^{-5}$$

$$= 1.51 + 0.012(-5)$$

$$= 1.51 - 0.060 = 1.45\ \text{V}$$

For p_cH 3, the calculation becomes

$$E_{MnO_4^-/Mn^{2+}} = 1.51 + 0.012 \log \frac{10^{-1} \times (10^{-3})^8}{10^{-4}}$$

$$= 1.51 + 0.012 \log 10^{-21}$$

$$= 1.51 + 0.012(-21) = 1.51 - 0.25$$

$$= 1.26\ \text{V}$$

E. The Effects of p_cH upon the Potential of Half-Cells

Example 6 shows that the potential of certain half-cells will be affected by a variation in the hydrogen ion concentration or p_cH. The extent of this effect can be estimated for various electrodes to see which is the most greatly affected by variations in p_cH. Only those half-cells in which hydrogen or hydroxide is involved in the half-cell reaction are affected greatly by variations in p_cH. Other half-cells may be affected by the hydrogen ion concentration but the extent cannot be predicted by the Nernst equation. Using the general oxidation-reduction half-cell reaction,

$$XO_{m/2}^{b-} + mH^+ + ne^- \longrightarrow X^{c+} + \frac{m}{2} H_2O \qquad (41)$$

the Nernst equation is

$$E = E^\circ + \frac{0.060}{n} \log \frac{[XO_{m/2}^{b-}][H^+]^m}{[X^{c+}]} \qquad (42)$$

which may be written as

$$E = E^\circ + \frac{0.060}{n} \log \frac{[XO_{m/2}^{b-}]}{[X^{c+}]} + \frac{0.060m}{n} \log [H^+] \qquad (43)$$

Consequently, at constant concentrations of $XO_{m/2}^{b-}$ and X^{c+}, the value of E is governed by the variations in the value of the second logarithmic term which contains m/n as a multiplier. The ratio of m/n can be calculated and used to predict the extent of the variation caused by a given change in p_cH.

As an example, determine which of the following half-cells will be most greatly affected and which will be least affected by a given change in p_cH: $Cr_2O_7^{2-}/2Cr^{3+}$; MnO_4^-/Mn^{2+}; MnO_4^-/MnO_2; IO_3^-/I^-; AsO_4^{3-}/AsH_3, for which the half-cell reactions are

$$Cr_2O_7^{2-} + 14H^+ + 6e^- \longrightarrow 2Cr^{3+} + 7H_2O \qquad (4)$$

$$MnO_4^- + 8H^+ + 5e^- \longrightarrow Mn^{2+} + 4H_2O \qquad (14)$$

$$MnO_4^- + 4H^+ + 3e^- \longrightarrow MnO_2 + 2H_2O \qquad (44)$$

$$IO_3^- + 6H^+ + 6e^- \longrightarrow I^- + 3H_2O \qquad (45)$$

$$AsO_4^{3-} + 11H^+ + 8e^- \longrightarrow AsH_3 + 4H_2O \qquad (46)$$

The m/n values are

$Cr_2O_7^{2-}/2Cr^{3+}$	$14/6 = 2.33$
MnO_4^-/Mn^{2+}	$8/5 = 1.60$
MnO_4^-/MnO_2	$4/3 = 1.33$
IO_3^-/I^-	$6/6 = 1.00$
AsO_4^{3-}/AsH_3	$11/8 = 1.38$

Of these five half-cells, $Cr_2Or_7^{2-}/2Cr^{3+}$ will be the most greatly affected by a given change in p_cH and IO_3^-/I^- will be the least affected.

F. The Relationship Between Standard Potentials and the Equilibrium Constant

The value of the equilibrium constant for an oxidation-reduction reaction depends upon the values of the standard potentials of the oxidizing and reducing agents and upon the number of electrons involved in the balanced equation for the reaction. To develop the expression which relates these factors, consider the general oxidation-reduction reaction

$$\mathrm{Ox}_{(1)} + \mathrm{Red}_{(2)} \longrightarrow \mathrm{Ox}_{(2)} + \mathrm{Red}_{(1)} \tag{47}$$

for which the equilibrium constant expression is

$$K_{eq} = \frac{[\mathrm{Ox}_{(2)}][\mathrm{Red}_{(1)}]}{[\mathrm{Ox}_{(1)}[\mathrm{Red}_{(2)}]} \tag{48}$$

Equation (47) should be read as "the oxidized form of substance 1 reacts with the reduced form of substance 2 to form the reduced form of substance 1 and the oxidized form of substance 2."

In any oxidation-reduction reaction, there must be electrical equilibrium whenever chemical equilibrium exists. Stated another way, the potential of the oxidation half-cell must equal the potential of the reduction half-cell whenever chemical equilibrium exists in an oxidation-reduction system, or

$$E_{(1)} = E_{(2)} \qquad \text{at equilibrium} \tag{49}$$

Using the subscript (1) to represent the oxidizing agent and (2) to represent the reducing agent, the Nernst equations for $E_{(1)}$ and $E_{(2)}$ are

$$E_{(1)} = E^\circ_{(1)} + \frac{0.060}{n} \log \frac{[\mathrm{Ox}_{(1)}]}{[\mathrm{Red}_{(1)}]} \tag{50}$$

$$E_{(2)} = E^\circ_{(2)} + \frac{0.060}{n} \log \frac{[\mathrm{Ox}_{(2)}]}{[\mathrm{Red}_{(2)}]} \tag{51}$$

Equations (50) and (51) can be equated since n is the same in both and equals the number of electrons in the balanced oxidation-reduction reaction.

$$E^\circ_{(1)} + \frac{0.060}{n} \log \frac{[\mathrm{Ox}_{(1)}]}{[\mathrm{Red}_{(1)}]} = E^\circ_{(2)} + \frac{0.060}{n} \log \frac{[\mathrm{Ox}_{(2)}]}{[\mathrm{Red}_{(2)}]} \tag{52}$$

Equation (52) may be rearranged to give

$$E^\circ_{(1)} - E^\circ_{(2)} = \frac{0.060}{n}\left(\log \frac{[\mathrm{Ox}_{(2)}]}{[\mathrm{Red}_{(2)}]} - \log \frac{[\mathrm{Ox}_{(1)}]}{[\mathrm{Red}_{(1)}]}\right) \qquad (53)$$

or

$$\frac{n}{0.060}(E^\circ_{(1)} - E^\circ_{(2)}) = \log \frac{[\mathrm{Ox}_{(2)}][\mathrm{Red}_{(1)}]}{[\mathrm{Red}_{(2)}][\mathrm{Ox}_{(1)}]} \qquad (54)$$

The logarithmic term is the same as eq. (48) for the equilibrium constant for the reaction, and K_{eq} may be substituted to give

$$\log K_{eq} = \frac{n}{0.060}(E^\circ_{(1)} - E^\circ_{(2)}) \qquad (55)$$

In eq. (55), the subscript (1) refers to the oxidizing agent and the subscript (2) refers to the reducing agent. As in other equilibria, the value of K_{eq} is a measure of the completeness of the reaction or a measure of the concentrations of the reactants left at equilibrium. The larger the value of K_{eq}, the more complete the reaction.

The equilibrium constant for any oxidation-reduction reaction can be calculated using eq. (55). However, a high value of the constant does not indicate that the reaction is satisfactory for analytical determinations since the kinetics of the reaction are not indicated by the equilibrium constant. Even though very negligible amounts of reactants might be left at equilibrium, a reaction is not suitable for an analytical determination if the attainment of equilibrium requires an unduly long period of time. An example of the calculation of K_{eq} using eq. (55) is the calculation of the equilibrium constant for the reaction of potassium dichromate and ferrous iron represented by eq. (9). The E° values for the two half-cells are 1.33 for the dichromate-chromic ion half-cell and 0.77 for the ferric-ferrous ion half-cell. The number of electrons in the balanced reaction is 6 so that

$$\begin{aligned}\log K_{eq} &= \frac{6}{0.060}(1.33 - 0.77) \\ &= 100(0.56) = 56\end{aligned} \qquad (56)$$

and

$$K_{eq} = 10^{+56}$$

Since the value of the equilibrium constant is large, the reaction goes to completion and the reaction is quantitative.

As a second example, calculate the equilibrium constant for reaction (29) for copper(II) ion plus iodide ion. The E° values are 0.86 and 0.54, and n equals 2.

$$\log K_{eq} = \frac{2}{0.060}(0.86 - 0.54) \qquad (57)$$
$$= 33.3(0.32) = 10.66$$

and

$$K_{eq} = 10^{10.66} = 4.6 \times 10^{10}$$

Even though the value for K_{eq} for the copper-iodide reaction is much less than that for the dichromate-iron reaction, the copper-iodide reaction is quantitative and can be used satisfactorily in analytical determinations. Since a reaction is assumed to be quantitative if the K_{eq} is 10^6 or greater, the minimum difference in E° values which allows quantitatively complete reaction is $0.36/n$. This value is obtained from eq. (55) as follows:

$$0.060 \log 10^6 = n(E^\circ_{(1)} - E^\circ_{(2)})_{\min} \qquad (58)$$
$$(E^\circ_{(1)} - E^\circ_{(2)})_{\min} = \frac{0.060 \times 6}{n} = \frac{0.36}{n}$$

For any reaction to take place as written, the K_{eq} value must be 1 or more, which means that the E° value of the oxidizing agent must be a larger positive value than the E° value of the reducing agent. A K_{eq} value less than 1 indicates that the reverse reaction takes place and, if the value is 10^{-6} or less, that the reverse reaction is quantitative. Also, for any reaction to take place, both an oxidizing agent and a reducing agent must be present since one reducing agent does not react with another reducing agent, nor do two oxidizing agents react with each other.

The value of $n(E^\circ_{(1)} - E^\circ_{(2)})$ can be used for several similar reactions to predict which will have the largest and which will have the smallest equilibrium constants. The reaction which has the largest value of n times the difference in E° values will have the largest equilibrium constant.

G. Typical Curve for Oxidation-Reduction Titrations

The change in potential during the course of an oxidation-reduction titration can be followed by immersing two electrodes in the solution during the titration. One of the two electrodes must be a reference electrode and the other must be an indicating electrode. A reference electrode produces a constant potential irrespective of changes in the composition of the solution in which it is immersed, but the potential exerted by an indicator electrode must vary quantitatively as the concentration of the measured ion varies in the solution. Since the concentration of the titrated ion varies as it reacts with the titrant, the potential difference between the two electrodes will vary with the volume of titrant which is added. A typical graph for the change in potential during the course of a titration is shown in Fig. 10-2, which represents the titration of 25.00 ml of 0.1000 *N* ferrous sulfate with 0.1000 *N* ceric sulfate in sulfuric acid solution. The saturated calomel electrode (SCE) used in this titration is probably the most popular of all reference electrodes. The potential measurement against a standard hydrogen half-cell (SHE) is also shown for comparison.

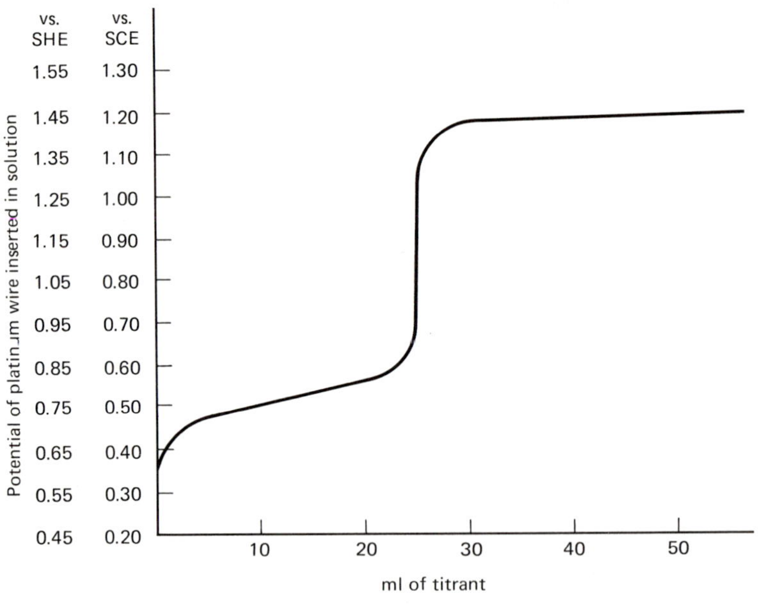

Figure 10.2. Variation of potential during the titration of 25.00 ml of 0.1000 *N* $FeSO_4$ with 0.1000 *N* $Ce(SO_4)_2$ in H_2SO_4 solution. SHE is the standard hydrogen electrode and SCE is the saturated calomel electrode.

The potential at the beginning of the titration up to the end point is controlled by the potential of the titrated ion, and after the end point it is controlled by the potential of the titrant ion. As a consequence the break at the equivalence point is a function of the difference in $E°$ values. In the iron-ceric ion reaction, the break is approximately 0.4 to 0.5 V. As with neutralization titrations, there must be a sharp break in potential at the equivalence point for the titration to be satisfactory.

H. Calculation of Potentials at Various Points on the Titration Curve

1. Before the Equivalence Point

The ceric sulfate-ferrous sulfate titration will be used as the example in the calculation of the potentials along the titration curve. Since it is difficult to estimate the actual ferric ion concentration in the solution at the beginning of the titration before the addition of any ceric sulfate, we shall calculate the potentials after the addition of 5.00, 12.50, 20.00, and 24.90 ml of the ceric sulfate.

a. AFTER ADDITION OF 5.00 ml OF TITRANT. The equation for the reaction of ceric ion with ferrous ion is

$$Ce^{4+} + Fe^{2+} \longrightarrow Ce^{3+} + Fe^{3+} \tag{59}$$

and there is an electron change of 1 on both ions. The E° value for ceric-cerous is 1.44 V, and for ferric-ferrous it is 0.77 V.

At the beginning of the titration there are 2.500 meqs of ferrous iron in the solution. Since there is an electron change of one, the equivalent weight is equal to the atomic weight for both cerium and iron and milliequivalents equal millimoles for both.

After addition of titrant, the solution will contain ferrous ions, ferric ions, cerous ions, and ceric ions. The ferrous and ferric concentrations can be calculated from the millimoles at start and the millimoles which have reacted, but there is no simple way to estimate the ceric ion concentration left unreacted from the data of the titration. As a consequence, the ferric-ferrous system should be used to calculate the potential in the solution before the equivalence point is reached.

$$\text{mmoles of } Fe^{2+} \text{ at start} = 25.00 \times 0.1000 = 2.500$$

$$\text{mmoles of } Fe^{3+} \text{ and } Ce^{3+} \text{ formed}$$

$$(= \text{mmoles of } Ce^{4+} \text{ added}) = 5.00 \times 0.1000 = \underline{0.500}$$

$$\text{mmoles of } Fe^{2+} \text{ left unreacted} = 2.000$$

Since the millimoles of ferrous and ferric are in the same solution, millimoles may be substituted for molar concentration in the Nernst equation so that the calculation of the potential becomes

$$E_{Fe^{3+}/Fe^{2+}} = 0.77 + \frac{0.060}{1} \log \frac{0.500}{2.000} \tag{60}$$

$$= 0.77 + 0.060 \log 0.25$$

$$= 0.77 + 0.060 \log 10^{-0.60}$$

$$= 0.77 + 0.060(-0.60) = 0.77 - 0.036$$

$$= 0.73 \text{ V}$$

b. AFTER THE ADDITION OF 12.50 ml OF TITRANT.

mmoles of Fe^{2+} at start = 2.500

mmoles of Fe^{3+} and of Ce^{3+} formed
(=mmoles Ce^{4+} added) = 1.250

mmoles of Fe^{2+} left unreacted = 1.250

$$E_{Fe^{3+}/Fe^{2+}} = 0.77 + 0.060 \log \frac{1.250}{1.250}$$

$$= 0.77 \text{ V}$$

c. AFTER ADDITION OF 20.00 ml OF TITRANT.

mmoles of Fe^{2+} at start = 2.500

mmoles of Fe^{3+} and Ce^{3+} formed
(=mmoles of Ce^{4+} added) = 2.000

mmoles of Fe^{2+} left unreacted = 0.500

$$E_{Fe^{3+}/Fe^{2+}} = 0.77 + 0.060 \log \frac{2.000}{0.500}$$

$$= 0.77 + 0.060 \log 4 = 0.77 + 0.060 \log 10^{0.60}$$

$$= 0.77 + 0.060(0.60) = 0.77 + 0.036$$

$$= 0.81 \text{ V}$$

d. AFTER ADDITION OF 24.90 ml OF TITRANT.

mmoles of Fe^{2+} at start = 2.500

mmoles of Fe^{3+} and Ce^{3+} formed
(=mmoles of Ce^{4+} added) = 2.490

mmoles of Fe^{2+} left unreacted = 0.010

$$E_{Fe^{3+}/Fe^{2+}} = 0.77 + 0.060 \log \frac{2.490}{0.010}$$
$$= 0.77 + 0.060 \log 249$$
$$= 0.77 + 0.060 \log 10^{2.40}$$
$$= 0.77 + 0.060(2.40) = 0.77 + 0.14$$
$$= 0.91 \text{ V}$$

2. At the Equivalence Point

The potential at the equivalence point for reactions of the general type

$$a\text{A} + b\text{B} \rightleftharpoons a\text{C} + b\text{D}$$

is given by

$$E_{\text{equiv}} = \frac{n_A E_A^\circ + n_B E_B^\circ}{n_A + n_B} \tag{61}$$

where the subscript A refers to the half-reaction of the oxidizing agent and the subscript B refers to the half-reaction of the reducing agent. Applied to the ferrous-ceric titration, this becomes

$$E_{\text{equiv}} = \frac{1.44 + 0.77}{1 + 1} = \frac{2.21}{2} = 1.105 \text{ V}$$

Equation (61) does not hold for more complicated titration reactions such as the permanganate-ferrous iron titration.

3. Beyond the Equivalence Point

Beyond the equivalence point, it is difficult to estimate the amount of ferrous ion left unreacted, but it is easy to calculate the concentrations of cerous and ceric ions if it is assumed that the millimoles of cerous formed equals the millimoles of ferrous at the start of the titration, that is, that the amount of ferrous ion left unreacted is negligible compared to the amount of cerous ion that is formed.

a. AFTER THE ADDITION OF 25.10 ml OF TITRANT.

mmoles of Fe^{2+} at start = mmoles of Ce^{3+} formed = 25.00 × 0.1000 = 2.500

mmoles of Ce^{4+} added = 25.10 × 0.1000 = 2.510

mmoles of Ce^{4+} left unreacted = 0.010

$$E_{Ce^{4+}/Ce^{3+}} = E^{\circ}_{Ce^{4+}/Ce^{3+}} + \frac{0.060}{n} \log \frac{\text{mmoles of } Ce^{4+}}{\text{mmoles of } Ce^{3+}} \quad (62)$$

$$E_{Ce^{4+}/Ce^{3+}} = 1.44 + \frac{0.060}{1} \log \frac{0.010}{2.500}$$

$$= 1.44 + 0.060 \log 0.004$$

$$= 1.44 + 0.060 \log 10^{-2.40}$$

$$= 1.44 + 0.060(-2.40) = 1.44 - 0.14$$

$$= 1.30 \text{ V}$$

b. AFTER ADDITION OF 30.00 ml OF TITRANT.

$$\text{mmoles of } Ce^{3+} \text{ formed} = 25.00 \times 0.1000 = 2.500$$

$$\text{mmoles of } Ce^{4+} \text{ added} = 30.00 \times 0.1000 = 3.000$$

$$\text{mmoles of } Ce^{4+} \text{ left unreacted} = 0.500$$

$$E_{Ce^{4+}/Ce^{3+}} = 1.44 + 0.060 \log \frac{0.500}{2.500}$$

$$= 1.44 + 0.060 \log 0.200$$

$$= 1.44 + 0.060 \log 10^{-0.70}$$

$$= 1.44 + 0.060(-0.70)$$

$$= 1.44 - 0.04 = 1.40 \text{ V}$$

c. AFTER FURTHER ADDITIONS OF TITRANT. By similar calculations the potential after 42.50 ml of titrant is 1.43, and after 50.00 ml it is 1.44.

I. Calculation of Concentrations at Various Points on the Titration Curve

1. Before the Equivalence Point

Using the ceric-ferrous reaction as the example, the concentrations of the ferrous, ferric, and cerous ion concentrations can be calculated as shown in Sections H.1.a, b, c, and d. The ceric ion concentration left unreacted cannot be calculated directly from the titration data, and, as a consequence, the ferric-ferrous half-cell is used to calculate the potentials before the equivalence point. Since the potential of the two half-cells must be equal, the calculated potential can be used with the Nernst equation for the ceric-cerous half-cell to calculate the concentration left unreacted. As an example, calculate the ceric ion concentration left unreacted after the addition of

12.50 ml of 0.1000 *N* $Ce(SO_4)_2$ to 25.00 ml of 0.1000 *N* $FeSO_4$ in sulfuric acid solution. Assume that the solution has been diluted to 100 ml. The potential in the solution, as calculated in Section H.1.b, is 0.77 V and the E° potential for the ceric-cerous half-cell is 1.44 V.

The milliequivalents of cerous ion produced, assuming that the concentration of ceric ion left unreacted is negligible compared to the cerous ion concentration, will be equal to the milliequivalents of ceric added, or

$$\begin{aligned}\text{meqs of } Ce^{3+} \text{ produced} &= \text{meqs of } Ce^{4+} \text{ added}\\ &= 12.50 \times 0.1000 = 1.250 \text{ meqs}\end{aligned}$$

and the normal concentration of Ce^{3+} will be 1.250 meqs/100.0 ml or 0.0125 *N*.

Since there is an electron change of one, the normal and molar concentrations will be the same so that the molar Ce^{3+} concentration is 0.0125 *M*.

Substitution of these values in the Nernst equation,

$$E_{Ce^{4+}/Ce^{3+}} = E^\circ_{Ce^{4+}/Ce^{3+}} + \frac{0.060}{n} \log \frac{[Ce^{4+}]}{[Ce^{3+}]} \tag{62}$$

gives

$$\begin{aligned}0.77 &= 1.44 + 0.060 \log \frac{[Ce^{4+}]}{0.0125}\\ &= 1.44 + 0.060 \log \frac{[Ce^{4+}]}{10^{-1.90}}\end{aligned}$$

which can be rearranged as

$$\log 10^{1.90}[Ce^{4+}] = \frac{0.77 - 1.44}{0.060} = -11.17$$

or

$$10^{1.90}[Ce^{4+}] = 10^{-11.17}$$

and

$$[Ce^{4+}] = \frac{10^{-11.17}}{10^{1.90}} = 10^{-13.07} = 8.5 \times 10^{-14}\ M$$

2. At the Equivalence Point

The concentrations of reactants left unreacted at the equivalence point is a measure of the completeness of the reaction and can be calculated from the titration data and the equilibrium constant or the equivalence point potential. The equilibrium constant method should be used for all reactions which involve different numbers of electrons in the half reactions.

Examples of both methods of calculation will be shown for the ceric-ferrous reaction, and an example of calculation by the equilibrium constant method will be shown for the arsenious acid-potassium permanganate reaction.

In the ceric-ferrous reaction, the cerous and ferric ion concentrations can be calculated for both methods from the titration data if it is assumed that the reaction is quantitatively complete, that is, that the concentrations of reactants left unreacted at the equivalence point are negligible compared to the concentrations of the products formed. The milliequivalents of cerous ion formed will equal the milliequivalents of ceric ion added, and the milliequivalents of ferric ion formed will equal the milliequivalents of ferrous ion at the start of the titration. For the titration of 25.00 ml of 0.1000 N $FeSO_4$ with 0.1000 N $Ce(SO_4)_2$, assuming an equivalence point volume of 100.0 ml,

$$\text{meqs of } Ce^{3+} \text{ formed} = \text{meqs of } Ce^{4+} \text{ added}$$
$$= 25.00 \times 0.1000 = 2.500$$

and

$$\text{meqs of } Fe^{3+} \text{ formed} = \text{meqs of } Fe^{2+} \text{ at start}$$
$$= 25.00 \times 0.1000 = 2.500$$

The normal concentration of Ce^{3+} and Fe^{3+} will be

$$N = \frac{2.500}{100.0} = 0.025\ N$$

and, for both ions,

$$M = \frac{N}{\text{equivs/mole}} = \frac{0.025}{1} = 0.025\ M$$

Since there is an electron change of one in both half reactions, milliequivalents equal millimoles and normal concentration equals molar concentration for all reactants and products.

a. BY THE EQUIVALENCE POTENTIAL. The equivalence potential for the ceric-ferrous reaction is 1.105 V as calculated by eq. (61). Since the potential of each half-cell must be equal and must equal the equivalence potential at the equivalence point,

$$E_{equiv} = 1.105 = E^\circ_{Fe^{3+}/Fe^{2+}} + \frac{0.060}{n} \log \frac{[Fe^{3+}]}{[Fe^{2+}]} \tag{63}$$

and

$$E_{equiv} = 1.105 = E^{\circ}_{Ce^{4+}/Ce^{3+}} + \frac{0.060}{n} \log \frac{[Ce^{4+}]}{[Ce^{3+}]} \tag{64}$$

Using the ferric-ferrous half-cell first,

$$1.105 = 0.77 + 0.060 \log \frac{0.025}{[Fc^{2+}]}$$

or

$$\log \frac{0.025}{[Fe^{2+}]} = \frac{1.105 - 0.77}{0.060} = \frac{0.0335}{0.060} = 5.58$$

and

$$\frac{0.025}{[Fe^{2+}]} = 10^{5.58}$$

so that

$$[Fe^{2+}] = \frac{0.025}{10^{5.58}} = \frac{10^{-1.60}}{10^{5.58}} = 10^{-7.18}$$

$$= 6.6 \times 10^{-8}\ M$$

Using the ceric-cerous half-cell, the concentration of ceric ion left unreacted is calculated by eq. (64) as

$$1.105 = 1.44 + 0.060 \log \frac{[Ce^{4+}]}{[Ce^{3+}]}$$

$$= 1.44 + 0.060 \log \frac{[Ce^{4+}]}{0.025}$$

or

$$0.060 \log \frac{[Ce^{4+}]}{0.025} = 1.105 - 1.44 = -0.335$$

and

$$\log 40[Ce^{4+}] = -\frac{0.335}{0.060} = -5.58$$

$$10^{1.60}[Ce^{4+}] = 10^{-5.58}$$

$$[Ce^{4+}] = \frac{10^{-5.58}}{10^{1.60}} = 10^{-7.18} = 6.6 \times 10^{-8}$$

b. BY USE OF THE EQUILIBRIUM CONSTANT. The equilibrium constant for the ceric-ferrous reaction may be calculated by eq. (55) as

$$\log K_{eq} = \frac{1}{0.060}(1.44 - 0.77) = \frac{0.67}{0.060} = 11.17$$

$$K_{eq} = 10^{11.17}$$

The equilibrium constant expression for the ceric-ferrous reaction is

$$\frac{[Ce^{3+}][Fe^{3+}]}{[Ce^{4+}][Fe^{2+}]} = 10^{11.17} \qquad (65)$$

$[Ce^{3+}] = [Fe^{3+}] = 0.025 = 10^{-1.60}$ M for the titration of 25.00 ml of 0.1000 N $FeSO_4$ with 0.1000 N $Ce(SO_4)_2$ assuming an equivalence point volume of 100 ml. Since the milliequivalents of Ce^{4+} left unreacted must equal the milliequivalents of Fe^{2+} left unreacted, $[Ce^{4+}] = [Fe^{2+}]$. Substitution of $[Ce^{4+}]$ for $[Fe^{2+}]$ in eq. (65) gives

$$\frac{10^{-1.60} \times 10^{-1.60}}{[Ce^{4+}]^2} = 10^{11.17}$$

or

$$[Ce^{4+}]^2 = \frac{10^{-3.20}}{10^{11.17}} = 10^{-14.37}$$

and, extracting the square root,

$$[Ce^{4+}] = 10^{-7.19} = 6.5 \times 10^{-8}\ M$$

and, since $[Ce^{4+}] = [Fe^{2+}]$,

$$[Fe^{2+}] = 6.5 \times 10^{-8}\ M$$

The small difference in the values calculated from the equivalence potential and from the equilibrium constant arises from the rounding off of the exponents to two decimal places.

c. CALCULATION OF THE EQUIVALENCE CONCENTRATIONS FOR THE PERMANGANATE-ARSENIOUS ACID REACTION. Since both half-cells for the ceric-ferrous reaction have an electron change of one, the calculation of the equivalence point concentrations is simpler than for reactions in which the electron changes of the half-cells differ, such as the permanganate-arsenious acid reaction. The balanced reaction for permanganate and arsenious acid in hydrochloric acid solution is

$$2MnO_4^- + 5H_3AsO_3 + 6H^+ \longrightarrow 2Mn^{2+} + 5H_3AsO_4 + 3H_2O \quad (66)$$

Since the number of electrons involved in the two half reactions differs, the equilibrium constant should be used for the calculation of concentrations. The equilibrium constant, calculated by eq. (55), is

$$\log K_{eq} = \frac{10}{0.060}(1.53 - 0.56) = \frac{9.7}{0.060} = 161.67$$

For the titration of 25.00 ml of 0.1000 N H_3AsO_3 with 0.1000 N $KMnO_4$, assuming a p_cH of one in an equivalence point volume of 100.0 ml, the concentrations of the Mn^{2+} ion and H_3AsO_4 can be calculated from the titration data since the milliequivalents of these substances (assuming quantitative reaction) will be equal to the milliequivalents of the reactants in the titration, or

$$\begin{aligned}\text{meqs of } Mn^{2+} \text{ formed} &= \text{meqs of } MnO_4^- \text{ added}\\ &= 25.00 \times 0.1000 = 2.500\end{aligned}$$

and

$$\begin{aligned}\text{meqs of } H_3AsO_4 \text{ formed} &= \text{meqs of } H_3AsO_3 \text{ at start}\\ &= 25.00 \times 0.1000 = 2.500\end{aligned}$$

The normal concentration of each of these in the equivalence point solution is

$$\frac{2.500 \text{ meqs}}{100.0 \text{ ml}} = 0.0250\ N$$

Since molar concentrations are required in the equilibrium constant, the normal concentrations can be converted to molar by dividing by the number of equivalents per mole, or

$$[H_3AsO_4] = \frac{N}{2} = \frac{0.0250}{2} = 0.0125 = 10^{-1.90}\ M$$

and

$$[Mn^{2+}] = \frac{N}{5} = \frac{0.0250}{5} = 0.0050 = 10^{-2.30}\ M$$

The equilibrium constant expression for the permanganate-arsenious acid reaction is

$$K_{eq} = \frac{[Mn^{2+}]^2[H_3AsO_4]^5}{[MnO_4^-]^2[H_3AsO_3]^5[H^+]^6} = 10^{161.67} \quad (67)$$

The molar concentrations of arsenic acid and manganous ion are $10^{-1.90}$ and $10^{-2.30}$ M, respectively, as shown previously. There is no simple way to calculate the concentrations of permanganate and arsenious acid left unreacted from the titration data, and, as a consequence, the relationship between the molar concentrations of the two reactants must be determined so that one can be expressed in terms of the other for substitution in the equilibrium constant expression.

Since the milliequivalents of MnO_4^- and H_3AsO_3 left unreacted are equal, the normal concentrations are equal, or

$$N_{MnO_4^-} = N_{H_3AsO_3}$$

For permanganate, the molar concentration is

$$[MnO_4^-] = \frac{N_{MnO_4^-}}{5}$$

or

$$N_{MnO_4^-} = 5[MnO_4^-]$$

For arsenious acid,

$$[H_3AsO_3] = \frac{N_{H_3AsO_3}}{2}$$

or

$$N_{H_3AsO_3} = 2[H_3AsO_3]$$

and

$$5[MnO_4^-] = 2[H_3AsO_3]$$

or

$$[H_3AsO_3] = \tfrac{5}{2}[MnO_4^-] = 2.5[MnO_4^-] = 10^{0.40}[MnO_4^-]$$

Substitution in eq. (67) gives

$$\frac{[10^{-2.30}]^2 \times [10^{-1.90}]^5}{[MnO_4^-]^2 \times (10^{0.40}[MnO_4^-])^5 \times (10^{-1})^6} = 10^{161.67}$$

or

$$[MnO_4^-]^2 \times 10^{2.00}[MnO_4^-]^5 = \frac{10^{-4.60} \times 10^{-9.50}}{10^{-6} \times 10^{161.67}}$$

and

$$10^{2.00}[MnO_4^-]^7 = 10^{-169.77}$$

$$[MnO_4^-]^7 = \frac{10^{-169.77}}{10^{2.00}} = 10^{-171.77}$$

$$[MnO_4^-] = 10^{-171.77/7} = 10^{-24.54} = 2.9 \times 10^{-25}\ M$$

The arsenious acid concentration is

$$[H_3AsO_3] = 2.5[MnO_4^-] = 10^{0.40} \times 10^{-24.54}$$
$$= 10^{-24.14} = 7.2 \times 10^{-25}\ M$$

J. Oxidation-Reduction Indicators

In some oxidation-reduction titrations, a colored reactant or product can act as the indicator. For example, permanganate ion has a deep purple color, while the color of manganous ion is a very pale pink, so pale, in fact, that it is essentially colorless in the concentrations found at the equivalence point in permanganate titrations. As a result, the first excess of permanganate colors the solution providing the other products are colorless. The appearance or disappearance of the color of the triiodide ion in iodimetric procedures can also serve as an end-point indicator.

For reactions in which the reactants or products do not indicate the end point, indicators must be added to the solution. Such indicators may form a colored complex with a reactant or product or else must be reversible oxidation-reduction systems in which the two members of the oxidation-reduction pair have different colors. An example of a colored complex indicator is the use of thiocyanate ion for the titration of iron(III) (ferric) with titanium(III). The deep red color of the $FeSCN^{2+}$ complex colors the solution until all the iron is reduced. Another example is the use of starch in iodimetric titrations. Starch forms a blue complex with iodine and provides very sharp end points.

There are several oxidation-reduction indicators which are oxidation-reduction systems. It can be shown that the color change of such indicators occurs over a range of approximately 0.12 V if both colored forms are equally easy to see. The range is centered at the $E°$ value and is usually expressed as $E° \pm 0.06$ V. As a consequence, the $E°$ value of the indicator must lie between the $E°$ values of the oxidant and reductant and should be close to the equivalence potential.

Diphenylamine and diphenylamine sulfonate are used as indicators for the titration of iron(II) (ferrous) ion with dichromate in the presence of phosphoric acid. The color change is from colorless to lavender or purple. The indicator does not change color at the true end point unless phosphoric acid is present.

Ferroin (ferrous 1,10-phenanthroline) and modified ferroins such as nitroferroin (ferrous 5-nitro-1,10-phenanthroline) are often used as indicators for ceric titrations. The dark red ferrous complex is oxidized to the pale blue ferric complex. $E°$ for ferroin is 1.06 V, and for the modified ferroins it ranges from 0.81 to 1.25 V.

K. Problems

1. Balance the following equations to obtain the ionic net reaction and the molecular equation.
   (a) $K_2Cr_2O_7 + FeSO_4 + H_2SO_4 \rightarrow Cr_2(SO_4)_3 + Fe_2(SO_4)_3 + H_2O + K_2SO_4$
   (b) $KMnO_4 + Fe(NH_4)_2(SO_4)_2 + H_2SO_4 \rightarrow MnSO_4 + Fe_2(SO_4)_3 + (NH_4)_2SO_4 + K_2SO_4 + H_2O$
   (c) $Na_2S_2O_3 + KI_3 \rightarrow Na_2S_4O_6 + KI + NaI$
   (d) $KIO_3 + KI + HC_2H_3O_2 \rightarrow KI_3 + KC_2H_3O_2 + H_2O$
   (e) $KBrO_3 + SbCl_3 + HCl \rightarrow KBr + SbCl_5 + H_2O$
   (*f*) $KMnO_4 + AsH_3 + H_2SO_4 \rightarrow H_3AsO_4 + MnSO_4 + H_2O + K_2SO_4$
   (g) $H_3AsO_3 + KMnO_4 + H_2SO_4 \rightarrow H_3AsO_4 + MnSO_4 + K_2SO_4 + H_2O$
   (h) $Ce(SO_4)_2 + SnSO_4 \rightarrow Ce_2(SO_4)_3 + Sn(SO_4)_2$
   (i) $Ce(SO_4)_2 + H_2C_2O_4 \rightarrow Ce_2(SO_4)_3 + CO_2 + H_2SO_4$
   (j) $KMnO_4 + K_4Fe(CN)_6 + H_2SO_4 \rightarrow MnSO_4 + K_3Fe(CN)_6 + K_2SO_4 + H_2O$
   Ans. (b) $2KMnO_4 + 10Fe(NH_4)_2(SO_4)_2 + 8H_2SO_4 \rightarrow 2MnSO_4 + + 5Fe_2(SO_4)_3 + 10(NH_4)_2SO_4 + K_2SO_4 + 8H_2O$
2. Calculate the numerical values of the equivalent weights of the reactants and products which are involved in the electron change for the reactions in problem 1.
   Ans. (b) $KMnO_4$, 31.61; $MnSO_4$, 30.20; $Fe(NH_4)_2(SO_4)_2$, 284.1; $Fe_2(SO_4)_3$, 199.9
3. Balance the reactions and express the equivalent weights by showing the number of equivalents per mole for each of the substances listed on the right for the reactions listed on the left:

	Reaction	Substance
(a)	$MnO_4^- + H^+ \rightarrow Mn^{2+} + H_2O$	$KMnO_4$, $MnSO_4$, Mn
(b)	$MnO_4^- + H^+ \rightarrow MnO_2 + H_2O$	$KMnO_4$, MnO_4^-, MnO_2, Mn
		(*Ans.* (b) $\frac{KMnO_4}{3}$ $\frac{MnO_4^-}{3}$ $\frac{MnO_2}{3}$ $\frac{Mn}{3}$)
(c)	$H_2O_2 + H^+ \rightarrow H_2O$	H_2O_2, O, H_2O
(d)	$PbO_2 + H^+ \rightarrow Pb^{2+} + H_2O$	Pb, PbO_2, $PbSO_4$, PbS, PbO
(e)	$H_3AsO_4 + H^+ \rightarrow H_3AsO_3 + H_2O$	Na_3AsO_4, As_2O_5, As_2O_3, H_3AsO_4, As
(f)	$SO_4^{2-} + H^+ \rightarrow H_2SO_3 + H_2O$	K_2SO_4, Na_2SO_3, S, SO_3, SO_2

4. Calculate the actual potential of a $Cr_2O_7^{2-}/2Cr^{3+}$ half-cell at following p_cH and concentration values:

	p_cH	$[Cr_2O_7^{2-}]$	$[Cr^{3+}]$	
(a)	−1	10^{-7}	0.05	
(b)	1	0.05	0.01	(*Ans.* 1.22)
(c)	3	10^{-5}	10^{-5}	
(d)	6	10^{-1}	10^{-1}	
(e)	0	10^{-8}	0.025	

5. Repeat problem 4 for the MnO_4^-/Mn^{2+} half-cell substituting $[MnO_4^-]$ for $[Cr_2O_7^{2-}]$ and $[Mn^{2+}]$ for $[Cr^{3+}]$.
Ans. (b) 1.42
6. Repeat problem 4 for the H_3AsO_4/H_3AsO_3 half-cell substituting $[H_3AsO_4]$ for $[Cr_2O_7^{2-}]$ and $[H_3AsO_3]$ for $[Cr^{3+}]$.
Ans. (b) 0.52
7. Repeat problem 4 for the IO_3^-/I^- half-cell substituting $[IO_3^-]$ for $[Cr_2O_7^{2-}]$ and $[I^-]$ for $[Cr^{3+}]$.
Ans. (b) 1.03
8. Repeat problem 4 for the $2BrO_3^-/Br_2$ half-cell substituting $[BrO_3^-]$ for $[Cr_2O_7^{2-}]$ and $[Br_2]$ for $[Cr^{3+}]$.
Ans. (b) 1.46
9. Which one of the four half-cells in problem 3(a), (b), (c), and (d) will be affected (1) most greatly and (2) the least by a given change in p_cH?
Ans. (1) PbO_2/Pb^{2+}, (2) H_2O_2/H_2O
10. Which one of the half-cells in problems 4, 5, 7, and 8 will be most greatly affected and which will be least affected by a given change in p_cH? Prove your deduction using the answers to parts (c) and (e) of problems 4–8.
11. Calculate the equilibrium constants for the reactions in problem 1 except (f).
Ans. (b) 1.10×10^{125}
12. Balance the equations and calculate the equilibrium constants for the following reactions:
    (a) Cerium(IV) (ceric) sulfate plus chromium(II) (chromous) sulfate to form cerium(III) (cerous) sulfate and chromium(III) (chromic) sulfate
    (b) Potassium permanganate plus oxalic acid in the presence of sulfuric acid to form manganese(II) (manganous) sulfate and carbon dioxide
    (c) Potassium iodate plus tin(II) (stannous) chloride in the presence of hydrochloric acid to form potassium iodide and tin(IV) (stannic) chloride

(d) Iron(III) (ferric) chloride plus potassium iodide to form potassium triiodide and iron(II) (ferrous) chloride

(e) Arsenic acid plus sulfurous acid to form arsenious acid and sulfuric acid.

Ans. (b) 2.6×10^{338}

13. Without calculating the actual equilibrium constant values, predict which one of the reactions in problem 1 will have the largest equilibrium constant and which one will have the smallest equilibrium constant.

14. Without calculating the actual equilibrium constant values, predict which one of the reactions in problem 12 will have (a) the largest equilibrium constant and (b) the smallest equilibrium constant.

Ans. (a) (b), (b) (d)

15. Predict which of the following reactions will take place spontaneously as written, and without calculating the actual equilibrium constant values, predict which one will have the largest equilibrium constant. The equations are not balanced.

(a) $MnO_4^- + Fe^{2+} + H^+$
(b) $Ce^{4+} + Sn^{2+} + H_2SO_4$
(c) $Ce^{3+} + Fe^{3+} + H_2SO_4$
(d) $Cr_2O_7^{2-} + Ce^{4+} + H_2SO_4$
(e) $MnO_4^- + Sn^{2+} + H^+$
(f) $Mn^{2+} + Fe^{2+} + H^+$

16. Calculate the potential in the titrated solution after the addition of the indicated volume of 0.1000 N $KMnO_4$ to 25.00 ml of 0.1000 N $Sn(SO_4)$ in H_2SO_4. Assume in each case that the volume is 100.0 ml and that the hydrogen ion concentration is 1 M.

(a) 5.00 ml
(b) 12.50 ml
(c) 20.00 ml
(d) 24.95 ml
(e) 25.00 ml
(f) 25.10 ml
(g) 30.00 ml
(h) 40.00 ml
(i) 50.00 ml

Ans. (b) 0.15, (e) 1.12, (g) 1.50

17. Repeat problem 16 for the titration of 0.2000 N $FeSO_4$ with 0.2000 N $K_2Cr_2O_7$ in sulfuric acid solution assuming that $[H^+]$ is 2 M.

18. Repeat problem 16 for the titration of 1.8219 g of Sb_2O_3 which has been dissolved in HCl with 0.2500 N KIO_3 assuming that the p_cH is 1.0.

19. Repeat problem 16 for the titration of 0.2793 g of iron wire which has been dissolved in hydrochloric acid and reduced to the iron(II) (ferrous) state before titrating with 0.2000 N cerium(IV) (ceric) chloride. Note that the E° potential of the ceric-cerous half-cell is 1.28 in chloride solution.

20. (a) Calculate the molar concentration of the titrant ion left unreacted in the titrated solution before the equivalence point in problem 16 (a), (b), (c), and (d);

(b) Calculate the molar concentrations of the titrant ion and of the titrated ion left unreacted for problem 16(e);

(c) Calculate the molar concentration of the titrated ion left unreacted for problem 16 (f), (g), (h), and (i).

Ans. (a) (b) 1.2×10^{-116}; (b) titrant ion, 2.08×10^{-35} M; (c) (g) 1.25×10^{-47} M.

21. Repeat problem 20 using the titration in problem 17.
22. Repeat problem 20 using the titration in problem 18.
23. Repeat problem 20 using the titration in problem 19.

CHAPTER 11

Oxidation-Reduction Titrations

Oxidation-reduction titrations constitute a major portion of all titrimetric methods and have the advantage that the end points can often be determined potentiometrically for those titrations for which there is no satisfactory visual indicator. A representative example of each of the more important reagents will be discussed in this chapter. One of the most effective oxidizing agents used is potassium permanganate, which is a strong oxidizing agent that does not require the use of a separate indicator. The main disadvantages of permanganate are that it is difficult to prepare in stable form and that it cannot be used in acid solutions containing chloride unless special precautions are taken. Potassium dichromate is also used extensively since it can be obtained in primary standard grade and can be used in acid solutions containing chloride. Ceric sulfate is a strong oxidizing agent which also can be used in acid solutions containing chloride, but it has the disadvantage (or in some cases the advantage) that the strength of cerium(IV) solutions as oxidizing agents depends upon the anions present since the standard half-cell potential ranges from 1.28 V in chloride solutions to 1.70 V in perchlorate solutions. Ceric ion is usually used in sulfate solution for which the standard potential is 1.44 V. Iodine (as triiodide) is used as a weak oxidizing agent with strong reducing agents in direct titrations but has the disadvantage of losing strength slowly by volatilization of iodine from the solution. Many widely used iodimetric methods are indirect in that iodide ion is first oxidized to iodine (or triiodide), which is then titrated with thiosulfate.

A. Permanganate Methods

Permanganate is used for the determination of calcium as the oxalate, for the titration of iron, for the determination of arsenic as the arsenite, and for other substances such as hydrogen peroxide, antimony, ferrocyanide, and sulfur dioxide. Permanganate is usually used in strong acid solutions (approximately 1–2 *N*) in which the product is manganese(II)(manganous) ion. In neutral and slightly acidic solutions the product is manganese dioxide, while in strongly basic solutions manganate ion (MnO_4^{2-}) is produced. The interference of chloride ion can be offset partially by adding a solution containing phosphoric acid, sulfuric acid, and manganous sulfate.

1. Standardization

Potassium permanganate is not a primary standard, and permanganate solutions must be standardized against primary standard substances. The most popular primary standards are sodium oxalate, arsenious oxide, and high-purity (99.99%) iron wire. Mohr's salt (ferrous ammonium sulfate hexahydrate), $Fe(NH_4)_2(SO_4)_2 \cdot 6H_2O$, has been used as a primary standard in the past. It has the advantage of a large equivalent weight, but the fact that it cannot be dried without decomposition is a distinct disadvantage. Selection of the primary standard depends upon the titration for which the solution will be used. In general, it is preferable to standardize a solution by the same reaction which will be utilized in the analysis. As an example, sodium oxalate is preferable for solutions which will be used in the calcium determination, while iron wire would be selected for solutions utilized in the determination of iron.

2. Determination of Calcium in Limestone by the Oxalate-Permanganate Method

Magnesium usually occurs with calcium in natural materials such as limestone and other rock samples so that the method of analysis for calcium normally involves the separation of the magnesium before the actual determination of the calcium. One advantage of the oxalate method for calcium is the fact that oxalate can be used to separate the calcium from any magnesium which may be present since magnesium forms a soluble complex with excess oxalate, while calcium does not. Other heavy metals interfere and must be removed prior to the oxalate separation.

The limestone sample is usually dissolved carefully and slowly in hydrochloric acid in a covered beaker to ensure that none of the sample is lost through effervescence. Most procedures also call for a digestion period in the acid solution to ensure complete solution of the carbonate fraction. The acid-insoluble residue left is composed mainly of silica and insoluble silicates. For samples in which the acid-insoluble residue is composed of insoluble silicates, especially clays, the residue is filtered, dried, and fused with sodium carbonate, and the fused material is dissolved in a small amount of acid. This solution is then added to the original solution of the sample before continuation of the analysis. The insoluble residue remaining from fusion is weighed and reported as silica.

Limestone samples often contain iron and aluminum in amounts large enough to cause interference in the oxalate separation. The iron and aluminum are usually removed by making the solution basic with ammonia and filtering the precipitated hydrous oxides. Care must be exercised in this step to offset negative errors caused by adsorption of calcium and magnesium ions on the hydrous oxides if the *p*H is too high. The hydrous oxide particles are colloidal in size and consequently are charged due to adsorption of ions in the primary region. The particles are positively charged at *p*H values below 8 for aluminum and below 8.5 for iron. At higher *p*H values the particles become negatively charged due to adsorption of hydroxide ions and attract positive ions such as calcium and magnesium into the secondary layer. As a result, most procedures call for the addition of a very slight excess of ammonia in the precipitation step. The oxides can be ignited and weighed or can be determined separately by photometric methods.

The filtrate from the hydrous oxides is made acidic with hydrochloric acid, and methyl red is added as an indicator. An excess of oxalate is then added (usually as ammonium oxalate), and the *p*H is raised to a value between 5 and 6 by the slow addition of ammonia until the intermediate orange color of the indicator is attained. This *p*H ensures the quantitative precipitation of calcium oxalate and offsets the coprecipitation of calcium hydroxide which will occur if the *p*H exceeds 7. The magnesium remains in solution as a soluble complex if there is sufficient excess oxalate. The particles of calcium oxalate are very small and must be digested for approximately one hour before attaining sufficient size to allow satisfactory filtration. The digestion period must not be too long, however, since magnesium oxalate will coprecipitate if the precipitate is left in contact with the solution for extended periods of time.

The transfer and washing of the precipitated calcium oxalate also requires careful attention to detail since the oxalate is to be titrated rather than the calcium. As a consequence, all the excess oxalate must be washed out of the precipitate. Since calcium oxalate is comparatively soluble, extended washing will cause a negative error due to dissolving the precipitate. As a consequence, most procedures call for washing by decantation a specified number of times with specified volumes of cold water, followed by transfer and washing in the filtration device (paper or suction crucible) with a specified number of washes of a specified volume.

In spite of the difficulties, results are precise if the washing directions are followed carefully. The accuracy is also satisfactory if directions are followed since the factors which can cause positive errors are carefully balanced against the factors which cause negative errors. Any deviation from the directions, however, can be the cause of major errors in the determination, even though the precision will still be excellent as long as all samples are treated exactly the same.

The precipitated calcium oxalate is dissolved in dilute sulfuric acid and titrated with potassium permanganate at elevated temperatures. The reaction involved in the titration may be represented by the ionic equation

$$2MnO_4^- + 5HC_2O_4^- + 11H^+ \longrightarrow 2Mn^{2+} + 10CO_2 + 8H_2O \quad (1)$$

and the half reaction for bioxalate, which is equivalent to the calcium, is

$$HC_2O_4^- - 2e^- \longrightarrow 2CO_2 + H^+ \quad (2)$$

Each bioxalate ion requires two electrons, and since each bioxalate represents one calcium ion, as shown by eq. (3), the equivalent weight for calcium in the titration equals one-half the atomic weight.

$$CaC_2O_4 + H^+ \longrightarrow Ca^{2+} + HC_2O_4^- \quad (3)$$

The bioxalate solution must be heated almost to boiling and must be maintained at a temperature above 60°C during the titration to ensure that the reaction takes place at a reasonable rate. At room temperature, the reaction is so slow that the titration is not feasible. Even at elevated temperatures the reaction proceeds slowly until some manganous ion is formed. The manganous ion acts as a catalyst for the reaction, which will proceed rapidly at elevated temperatures in the presence of manganous ion. For this reason, the solution should be allowed to decolorize after the addition of the first few drops of titrant before the titration is continued.

Equation (1) indicates that hydrogen ions are used up in the reaction. If the titrant is allowed to run into the solution without agitation, the *p*H may be raised in the region of entry to a value at which manganese dioxide will be precipitated. Even though the manganese dioxide redissolves upon agitation, the reaction is not completely reversible and an error can result. Most procedures specify adequate agitation during the titration to offset this error.

The end point of the titration occurs when the first excess of permanganate colors the solution pink to lavender. Since the end point occurs beyond the equivalence point, a positive error is involved. This error can be offset by determination of the amount of permanganate required to produce the same color in a sulfuric acid solution of equal volume and by subtracting this blank from the titration values.

B. Dichromate Methods

Potassium dichromate is used primarily in the determination of iron in iron ores but is also used in the determination of uranium, sodium (after precipitation as $NaZn(UO_2)_3(C_2H_3O_2)_9 \cdot 6H_2O$), and organic compounds. Organic compounds are boiled with an excess of standard dichromate solution. The excess dichromate is then determined by addition of an excess of standard ferrous iron solution followed by titration of the excess ferrous iron by standard dichromate.

Potassium dichromate is obtainable in primary standard grade and standard solutions can be prepared by the direct method. The standard solutions so prepared are so stable that they can be boiled without decomposition and do not oxidize chloride ion in dilute (1–2 *N*) acid solution. The product in 1–2 *N* acid solution is always the Cr^{3+} ion since there are no other stable oxidation states between $Cr_2O_7^{2-}$ and Cr^{3+}. The half-cell reaction for dichromate in acid solution is

$$Cr_2O_7^{2-} + 14H^+ + 6e^- \longrightarrow 2Cr^{3+} + 7H_2O \tag{4}$$

so that the equivalent weight of potassium dichromate is one-sixth the molecular weight. A tenth normal solution of potassium dichromate is transparent, and the meniscus is read easily. Diphenylamine sulfonate is a satisfactory indicator for most dichromate titrations.

1. Determination of Iron in an Iron Ore by Titration with Potassium Dichromate

Many iron ores are directly soluble in hydrochloric acid with little or no acid-insoluble residue. Some of the ores which contain materials insoluble in hydrochloric acid can be determined without removal of the acid-insoluble residue, but fusion must be used in those cases in which the acid-insoluble material contains iron.

After solution in hydrochloric acid, the iron must be reduced to the iron(II) (ferrous) state before titration with potassium dichromate. The prior reduction of iron is an important and critical step in the procedure and must be performed with care. Before reduction, the solution is evaporated to low volume (usually about 5 ml) on a steam or electric hot plate. The iron is present as a chloro complex, which has a dark brown color which is used as the indicator for the reduction. If the sample goes to dryness on the hot plate, it must be redissolved in hydrochloric acid before being reduced. In some samples, a precipitate will form in the concentrated solution, especially if the volume has been reduced too low. Such precipitates usually can be dissolved by addition of a few milliliters of warm water to the warm solution.

The iron is reduced to the ferrous state by reaction with tin(II)(stannous) chloride and, since ferrous iron is oxidized readily in acid solution by the oxygen of the air, rapid work is necessary after the reduction step to prevent a negative error. For this reason, most procedures require that all chemicals be measured and ready for addition and that the burette be filled with the titrant before the reduction step is started.

The reaction of stannous ion with ferric ion is usually written as shown in equation (5).

$$Sn^{2+} + 2Fe^{3+} \longrightarrow Sn^{4+} + 2Fe^{2+} \qquad (5)$$

Since the iron and tin are present as chloro complexes the reaction actually is

$$SnCl_4^{2-} + 2FeCl_4^- \longrightarrow SnCl_6^{2-} + 2Fe^{2+} + 6Cl^- \qquad (6)$$

The $FeCl_4^-$ ion has a yellow color (dark brown at this concentration) while the Fe^{2+} ion is a pale green color and essentially colorless in the concentrations found in the reduced solution. As a consequence the fading color of the $FeCl_4^-$ ion can be used to follow the course of the reduction. The reaction is comparitively slow near the end point, and the stannous chloride must be added dropwise with a 10- to 15-sec interval between drops. It is important that all the iron be reduced to the ferrous state, but an excess of stannous chloride of more than one or two drops will cause errors and difficulty in the later steps in the analysis. The solution must be diluted before titration, and hydrochloric acid should be added before dilution to prevent precipitation of $Sn(OH)_4$.

Since potassium dichromate reacts with stannous tin, the excess stannous ion added in the reduction step must be removed before titration. The removal is accomplished by the addition of mercury(II) (mercuric) chloride. The stannous ion reacts with the mercuric ion to form mercury(I) (mercurous) chloride, a very insoluble substance which will separate from the solution as a silky, finely divided white precipitate. Since a finely divided precipitate is desired, the mercuric chloride is added rapidly with constant agitation. The desired reaction is

$$SnCl_4^{2-} + 2HgCl_2 \longrightarrow SnCl_6^{2-} + Hg_2Cl_2 \qquad (7)$$

The mercurous chloride is sufficiently insoluble that it will not be oxidized by the potassium dichromate and thus does not cause an error. If too large an excess of stannous ion is added in the reduction step, the mercuric chloride will react to form elementary mercury by

$$SnCl_4^{2-} + HgCl_2 \longrightarrow SnCl_6^{2-} + Hg^0 \tag{8}$$

The elementary mercury formed by this reaction does react with potassium dichromate and will cause an error in the determination. As a consequence, most procedures call for a finely divided, shiny white precipitate to form within 3 min after the addition of the mercuric chloride. If no precipitate forms, insufficient stannous chloride was added and all the ferric ion was not reduced. If a heavy white to gray precipitate forms, too much stannous chloride was added, and the sample should be discarded since the mercury formed will react with the titrant to give a positive error.

Immediately after the reduction step and the formation of the mercurous chloride precipitate, sulfuric acid, phosphoric acid, and the diphenylamine indicator should be added and the solution titrated as rapidly as possible to prevent air oxidation of the ferrous iron.

The sulfuric acid is added to ensure the correct acidity of the solution and because the indicator works better in the presence of sulfate ions. The phosphoric acid forms a colorless complex [probably $Fe(H_2PO_4)_6^{3-}$] with the ferric ions as they are produced in the reaction. The formation of the colorless complex prevents the tan color of the $FeCl_4^-$ from obscuring the end point. The complex also serves to depress the potential of the ferric-ferrous half-cell and keeps the potential at a lower value before the end point is reached. This causes a sharper potential break at the end point and allows the use of diphenylamine sulfonate as the indicator. The effect of the phosphoric acid upon the potential change during the titration is shown in Fig. 11-1. In the absence of phosphoric acid, the color change appears before the equivalence point.

The diphenylamine sulfonate serves as the indicator and is oxidized from colorless to a lavender shade. In the actual titration the presence of the complexed ferric ions and the green color of the chromic ions combine with the color of the indicator to form a gray to black color just before the end point, and the end point is a dark purple color which will remain at least 30 sec. Some dichromate is necessary to oxidize the indicator, and it is difficult or impossible to run a blank determination since the color change is not sharp in the absence of iron. A blank of 0.07 ml of 0.1000 N $K_2Cr_2O_7$ is usually subtacted from the titrant volume.

C. Cerium(IV) (Ceric) Methods

Cerium(IV) (ceric) compounds act as strong oxidizing agents in acid solution and are reduced to the colorless cerium(III) (cerous) ions. The

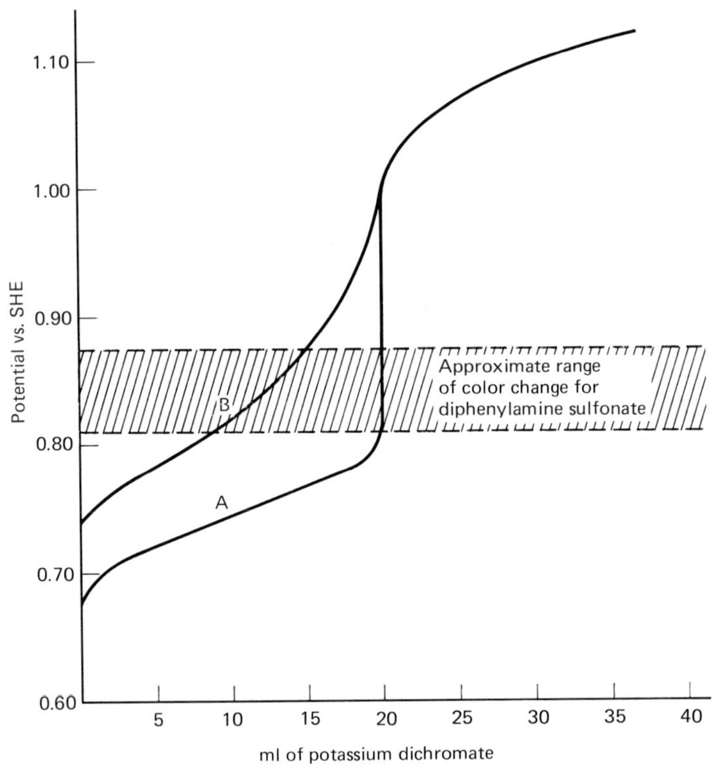

Figure 11.1. Effect of phosphoric acid on the potential during the titration of 20.00 ml of 0.1000 *N* $FeSO_4$ with 0.1000 *N* $K_2Cr_2O_7$. **A**, In 3 *M* H_3PO_4 and 1 *M* H_2SO_4. B, In 1 *M* H_2SO_4.

strength of the cerium(IV) solutions is dependent upon the anions present since the standard half-cell potential is 1.28 V in hydrochloric acid, 1.44 V in sulfuric acid, 1.61 V in nitric acid, and 1.70 V in perchloric acid. The variation in the standard half-cell potential is sometimes explained as being due to complex formation since cerium(IV) forms complexes with sulfate, chloride, and nitrate ions but does not form complexes with perchlorate ions. Also the stability constants of the complexes correlate with the standard potentials. Cerium(IV) is usually used in sulfuric or hydrochloric acid solutions for the titration of those substances which can be titrated with permanganate.

Cerium(IV) solutions must be standardized since the usual salts such as ceric ammonium sulfate [$(NH_4)_2Ce(SO_4)_3 \cdot 2H_2O$] are not primary standards. The solution is usually standardized against sodium oxalate, arsenious oxide, or high-purity iron wire. The disappearance of the yellow-orange color of the ceric ion can be used as the indicator, but ferroin [tris(1,10-phenanthroline) iron(II) sulfate] is an excellent indicator which changes from a dark red to a light blue (colorless in the actual titration) when it is oxidized to the iron(III) state.

Ceric solutions are stable indefinitely, can be used in the presence of chloride, and have the advantage that there is only one possible change in oxidation number, as shown by the half-cell equation,

$$Ce^{4+} + 1e^{-} \longrightarrow Ce^{3+} \tag{9}$$

Ceric solutions may be used for the determination of iron, oxalate, hydrogen peroxide, arsenite, antimony(III), uranium(IV), vanadium(IV), tin(II), and plutonium(III). Ferroin or similar phenanthroline indicators are usually satisfactory for most procedures.

D. Methods Involving Iodine

Elementary iodine is comparatively insoluble in water but reacts with iodide ion to form a soluble complex known as triiodide ion which has a dark brown color.

$$I_2^0 + I^- \longrightarrow I_3^- \tag{10}$$

As a consequence the triiodide ion is the reactive species in all methods which involve iodine in aqueous solution rather than elementary iodine. In this discussion, the two terms (iodine and triiodide) will be used interchangeably and will be taken to mean the triiodide ion when an actual reaction is being discussed.

Iodine is a weak oxidizing agent which is used for direct titration of strong reducing agents such as arsenite. Indirect titration methods in which an oxidizing agent is allowed to react with iodide ion to produce triiodide ion which is then titrated with sodium thiosulfate are probably more popular than the direct methods. Both direct and indirect methods must be used in neutral or acidic solutions since the reactions are not stoichiometric in basic solution due to side reactions. The appearance or disappearance of the color of the triiodide ion can be used to indicate the end point in both direct and indirect methods but starch is usually used as the indicator except in strongly acid solution. Iodine forms a blue complex with starch which constitutes a very sensitive method for the detection of the presence of iodine or triiodide ion. The actual color of the starch-iodine complex depends upon the relative concentration of triiodide ion. The color deepens from blue to dark blue to purple to black as the concentration of triiodide increases.

Iodine is limited in its use as a titrant in direct methods due to the low standard half-cell potential of 0.54 V and due to the fact that the concentration of iodine solutions changes slowly because of volatilization of the solute. Even though iodine can be obtained in a very pure state, actually primary standard grade, by successive sublimation, the concentrations of standard solutions must be checked periodically even though they were prepared originally by the direct method. The preparation of any standard iodine solution involves the solution of weighed amounts of iodine and potassium iodide in as small a volume of water as possible followed by dilution with water to the desired volume.

Arsenic(III), antimony(III), tin(II), hydrogen sulfide, sulfite, thiosulfate, and ferrocyanide are often determined by the direct method of titration with standard iodine using starch as the indicator. The substances ordinarily determined by the indirect method include the halogens and oxyanions of the halogens, strong oxidizing agents such as permanganate and dichromate, organic compounds which can be brominated quantitatively, and copper. The more active halogens displace iodide from potassium iodide to form triiodide ion, which is then titrated with thiosulfate as indicated in

$$Cl_2^0 + 3KI \longrightarrow KI_3 + 2KCl \qquad (11)$$

$$KI_3 + 2Na_2S_2O_3 \longrightarrow Na_2S_4O_6 + KI + 2NaI \qquad (12)$$

The general ionic reactions for all indirect methods are, assuming a two-electron change for the oxidizing agent,

$$Ox + 3I^- \longrightarrow Red + I_3^- \qquad (13)$$

$$I_3^- + 2S_2O_3^{2-} \longrightarrow 3I^- + S_4O_6^{2-} \qquad (14)$$

The determination of copper will be discussed in detail as an example of indirect iodine methods.

1. Indirect Iodine Methods Involving Titration with Thiosulfate

a. PREPARATION AND STANDARDIZATION OF THIOSULFATE SOLUTIONS. Thiosulfate solutions are usually prepared from sodium thiosulfate pentahydrate ($Na_2S_2O_3 \cdot 5H_2O$), which is not a primary standard, and, as a result, thiosulfate solutions must be standardized against suitable standard solids or solutions. The half-cell reaction of thiosulfate is represented by

$$2S_2O_3^{2-} - 2e^- \longrightarrow S_4O_6^{2-} \qquad (15)$$

The $S_4O_6^{2-}$ ion is known as tetrathionate. Equation (15) indicates that each thiosulfate radical requires one electron to be converted to a tetrathionate radical so that the equivalent weight of sodium thiosulfate pentahydrate equals the molecular weight.

Thiosulfate solutions are unstable since they are readily decomposed by bacteria and by acid. The bacterial decomposition may be offset by using recently boiled and cooled water to prepare the solution or by the addition of a germicidal agent such as chloroform or mercuric chloride. The acid decomposition is avoided by adding a small amount of sodium carbonate to the solution. The acid decomposition involves the disproportionation of the hydrogen thiosulfate ion ($HS_2O_3^-$) to the hydrogen sulfite ion (HSO_3^-) and sulfur. Freshly prepared solutions of thiosulfate change concentration slowly for a period of time so that the solution should be prepared and allowed to stand several days before standardization.

Thiosulfate solutions may be standardized against many oxidizing agents but potassium permanganate or potassium iodate solutions are preferable. Standardization against permanganate involves the addition of an aliquot of a standard solution of permanganate to an acidified solution of potassium iodide followed by titration with the thiosulfate solution using starch as the indicator. The reaction of permanganate with iodide is

$$2MnO_4^- + 15I^- + 16H^+ \longrightarrow 2Mn^{2+} + 5I_3^- + 8H_2O \qquad (16)$$

Permanganate is often used simply because previously standardized solutions are available. The reaction is rapid and the pale pink color of the manganous ion does not interfere with the color change at the end point.

Potassium iodate is also used for the standardization of thiosulfate. Potassium iodate may be obtained in primary standard grade and is readily soluble in water. In the standardization, an aliquot of a standard solution of iodate is added to a solution containing excess iodide. The solution is then acidified with HCl (1 : 10) and the iodate and iodide react to form triiodide ion by

$$IO_3^- + 8I^- + 6H^+ \longrightarrow 3I_3^- + 3H_2O \qquad (17)$$

This reaction is unusual in that it can be used to determine any one of the three reactants. In the presence of excess iodide and hydrogen ions, it is quantitative for iodate; in the presence of excess iodate and hydrogen ions, it is quantitative for iodide; and in the presence of excess iodate and iodide ions, it is quantitative for hydrogen ions and can be used to standardize acid solutions. The triiodide formed is titrated with thiosulfate. The balanced sequence of reactions involved in the standardization of thiosulfate is

$$IO_3^- + 8I^- + 6H^+ \longrightarrow 3I_3^- + 3H_2O \qquad (17)$$

$$3I_3^- + 6S_2O_3^{=} \longrightarrow 9I^- + 3S_4O_6^{2-} \qquad (18)$$

which may be summed to give

$$IO_3^- + 6S_2O_3^- + 6H^+ \longrightarrow I^- + 3S_4O_6^{2-} + 3H_2O \qquad (19)$$

Equation (19) shows that one IO_3^- reacts with six $S_2O_3^{2-}$ by the transfer of six electrons to form one I^-, three $S_4O_6^{2-}$, and three H_2O. As a consequence the equivalent weight for potassium iodate in the standardization of thiosulfate equals the molecular weight divided by 6, or one-sixth the molecular weight. Most students feel that the equivalent weight should be one-fifth the molecular weight since each iodate requires five electrons in the first reaction to form triiodide; however, in standardizations the complete balanced sequence of reactions must be utilized to determine the equivalent weight of a reactant or product.

Other substances used to standardize thiosulfate solutions include potassium bromate, potassium dichromate, oxalic acid, potassium hexacyanoferrate(III), iodine, and metallic copper.

b. DETERMINATION OF COPPER BY THE INDIRECT IODINE METHOD. Copper ores and other materials containing copper are usually determined by the indirect iodine method or by electrodeposition of the copper from a nitrate and sulfate solution. The indirect iodine method is usually faster, but the electrodeposition method is capable of greater accuracy.

Many copper ores contain metallic copper as well as a salt or oxide of copper, and as a consequence the sample is usually dissolved in nitric acid. The use of nitric acid creates several problems, however, since nitric acid and the oxides of nitrogen both react with iodide to form triiodide ion and thus cause positive errors. In the solution step, metallic copper reacts with the nitric acid to form oxides of nitrogen, as indicated by

$$Cu + 4HNO_3 \longrightarrow Cu(NO_3)_2 + 2NO_2 + 2H_2O \qquad (20)$$

and the copper salts or oxides react to form soluble salts, as indicated by

$$CuO + 2HNO_3 \longrightarrow Cu(NO_3)_2 + H_2O \qquad (21)$$

$$CuCO_3 + 2NHO_3 \longrightarrow Cu(NO_3)_2 + CO_2 + H_2O \qquad (22)$$

The excess nitric acid and the oxides of nitrogen must be removed prior to the determination of copper. Most procedures call for evaporation of the solution to low volume followed by dilution and further evaporation to remove the oxides of nitrogen and the excess nitric acid.

The oxides or salts of iron(III) often contaminate copper ores and will cause positive errors if allowed to remain in the solution since iron(III) (ferric) ions react with iodide in neutral or slightly acid solution to form triiodide ions. The interference due to iron is offset by the addition of ammonium bifluoride (NH_4HF_2), which reacts with ferric ions to form a soluble colorless complex, as shown by

$$Fe^{3+} + 6F^- \longrightarrow FeF_6^{3-} \tag{23}$$

The iron often precipitates as the hydrous oxide during the pH adjustment step but the formation constant of the hexafluoroferrate(III) ion is so large that any hydrous oxide precipitated will be dissolved and converted to the complex.

The pH is usually adjusted before the addition of the fluoride in order to prevent or slow down the air oxidation of iodide ion which is rapid in acid solution. The pH adjustment is usually accomplished by the addition of ammonium hydroxide or of potassium biphthalate. If ammonia is used, it is added until a light blue precipitate of copper(II) hydroxide is formed or until the color of the solution begins to deepen due to the formation of the tetrammine copper(II) complex, $Cu(NH_3)_4^{2+}$. Addition of the ammonium bifluoride to such a solution lowers the pH to below 4 and prevents the interference of arsenic and antimony, which are often present in copper ores.

After the addition of the bifluoride it is necessary to work rapidly to prevent etching of the glassware by the hydrofluoric acid formed by the addition of the bifluoride. Potassium iodide is added and reacts with the copper(II) ion to form triiodide ion and copper(I) (cuprous) iodide, which is insoluble and separates as a brownish precipitate.

$$2Cu^{2+} + 5I^- \longrightarrow Cu_2I_2 + I_3^- \tag{24}$$

The triiodide is then titrated with thiosulfate using starch as the indicator. The starch should be added just before the loss of the color due to triiodide since the black starch complex formed when too large a concentration of triiodide is present is not completely reversible. The end point is the change from the starch-iodine color to colorless.

The cuprous iodide formed upon addition of iodide ion coprecipitates or occludes some of the triiodide ion. The extent of coprecipitation can be estimated from the color of the precipitate since pure cuprous iodide is white. The occluded triiodide ion is released by the addition of potassium or ammonium thiocyanate. Cuprous thiocyanate is less soluble (more insoluble) than cuprous iodide so that the cuprous iodide dissolves and is converted to cuprous thiocyanate.

$$Cu_2I_2 + 2KSCN \longrightarrow Cu_2(SCN)_2 + 2KI \tag{25}$$

As the precipitate disssolves, the occluded triiodide is released to the solution and produces a blue color with the starch. This color may be removed, usually by the addition of one or two drops of titrant.

The half reaction for copper in the determination is

$$2Cu^{2+} + 2I^- + 2e^- \longrightarrow Cu_2I_2 \qquad (26)$$

Since there is an electron change of 1 for each copper ion, the equivalent weight of copper equals the atomic weight in this reaction.

CHAPTER 12

Potentiometric Measurements

The potential of any given half-cell, as discussed in Chapter 10, depends upon the components of the half-cell and upon the active concentrations (activities) of the reactants and products of the half-cell reaction. The relationship is shown by the Nernst equation,

$$E = E^\circ + \frac{0.060}{n} \log \frac{(\text{Ox})}{(\text{Red})} \tag{1}$$

in which E is the actual potential produced by the half-cell, E° is the standard potential of the half-cell, n is the number of electrons involved in the balanced half-cell reaction, 0.060 is a constant which depends upon the temperature and has the value 0.060 at 30°C, and (Ox) and (Red) represent the active concentrations of the oxidized and reduced forms of the half-cell oxidation-reduction pair. In many calculations, molar concentrations may be substituted for active concentrations without major error.

The potential exerted by any given voltaic or galvanic cell is the difference between the actual potentials of the two half-cells of which it is composed, or

$$E_{cell} = E_1 - E_2 \tag{2}$$

in which E_1 and E_2 are the actual potentials of the two half-cells. Cells may be shown diagrammatically as in Fig. 10-1 or may be designated in type by using a single vertical line to represent the juncture between a solid and a liquid and a double vertical line to show the juncture between two liquids or solutions. Thus the cell composed of a zinc-zinc sulfate half-cell and a copper-copper sulfate half-cell connected by a saturated potassium chloride salt bridge and shown diagrammatically in Fig. 10-1, may be represented as

$$\mathrm{Zn}|\mathrm{ZnSO_4}_{\,(solution)}\|\text{saturated KCl}\|\mathrm{CuSO_4}_{\,(solution)}|\mathrm{Cu} \tag{3}$$

By convention the half-cell written on the left is designated as half-cell 1 and the one written on the right as half-cell 2. If the two cells are reversed, the numerical value of the potential will be the same but the sign will be changed. As written above the cell potential will be negative, but if written to show the copper half-cell on the left, the potential would be positive.

A. Concentration Cells

Since the half-cell potentials depend in part upon the active concentrations of the half-cell reactants, a potential will be developed between two half-cells which are identical except for the concentrations of the solutions. Such cells are known as concentration cells and may be used to determine the activity in a solution by measurement against a solution of known activity. For example, a galvanic cell composed of two silver-silver nitrate half-cells will produce a potential if the two solutions have different concentrations. The cell potential for the cell

$$\mathrm{Ag}|(\mathrm{AgNO_3})_s\|\text{saturated } \mathrm{KNO_3}\|(\mathrm{AgNO_3})_u|\mathrm{Ag} \tag{4}$$

where the subscripts s and u refer to the standard and unknown solutions, respectively, is given by

$$E_{cell} = E_{\mathrm{Ag}_s} - E_{\mathrm{Ag}_u} \tag{5}$$

or

$$E_{cell} = [E^\circ_{\mathrm{Ag}} + \frac{0.060}{1}\log(\mathrm{Ag}^+)_s] - [E^\circ_{\mathrm{Ag}} + \frac{0.060}{1}\log(\mathrm{Ag}^+)_u] \tag{6}$$

Equation (6) may be simplified by cancellation of E°_{Ag} and rearranged to give

$$E_{cell} = 0.060\,[\log(\mathrm{Ag}^+)_s - \log(\mathrm{Ag}^+)_u]$$

and

$$\log(\mathrm{Ag}^+)_u = \log(\mathrm{Ag}^+)_s - \frac{E_{cell}}{0.060} \tag{7}$$

As an example, calculate the active silver ion concentration in the unknown solution for the following cell:

$$Ag|AgNO_3(0.100A)||\text{saturated } KNO_3||AgNO_3 \text{ (unknown)}|Ag$$

for which the cell potential is 0.080 V. Note that it is not necessary to know the standard potential of the silver half-cell since the standard potential does not appear in eq. (7).

$$\log(Ag^+) = \log 0.100 - \frac{0.080}{0.060}$$
$$= -1.000 - 1.333 = -2.333$$
$$(Ag^+) = 10^{-2.333} = 10^{0.667} \times 10^{-3}$$
$$= 4.65 \times 10^{-3}\ A$$

If the molar instead of the active concentration of the standard is known, the active concentration may be determined by calculation of the activity coefficient (f) at the ionic strength of the solution as shown in Chapter 6. The activity is the product of the molar concentration and the activity coefficient.

B. Potentiometric Determination of *p*H

Even though *p*H or hydrogen ion activity can be determined using a concentration cell composed of two hydrogen electrodes, it is usually measured potentiometrically using a reference electrode and a hydrogen electrode or a substitute hydrogen electrode. A reference electrode shows the same potential irrespective of the composition (or changes in composition) of the solution in which it is immersed. Substitute hydrogen electrodes are electrode systems which respond quantitatively to variations in hydrogen ion activity, as does the hydrogen electrode.

1. Reference Electrodes

The most popular reference electrodes are the calomel electrodes of which the saturated calomel electrode is used more than any other. The saturated calomel electrode is constructed as indicated in Fig. 12-1. An inner chamber contains mercury and mercurous chloride (calomel) in contact with a solution that is saturated with mercurous chloride and potassium chloride. A conductor, usually platinum wire, connects the mercury to the

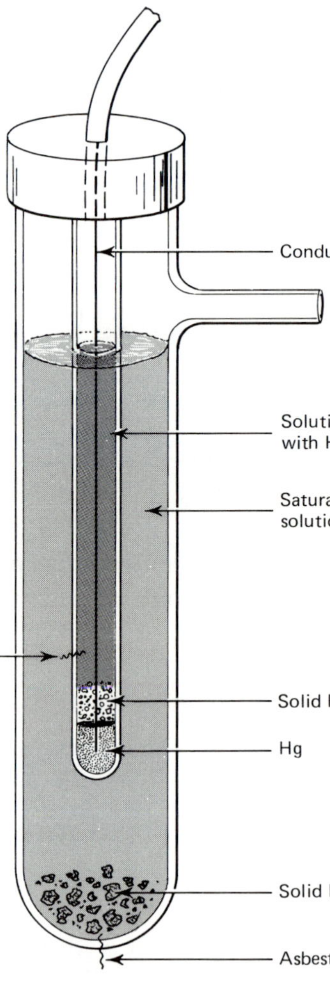

Figure 12.1. Saturated calomel reference electrode.

connecting wire for the electrode. A fiber connector is used to make contact between the inner electrode and the outer chamber which is filled with saturated potassium chloride. In some electrodes, a minute hole or a porous frit is used in place of the fiber. An asbestos fiber is used to make the connection between the outer chamber and the solution in which the electrode is immersed. Other calomel electrodes utilize molar or 0.1 *M* KCl solutions in place of the saturated KCl solution and are known as the molar (or normal) and tenth-molar calomel electrodes. Thus the descriptive adjective representing concentration refers to the concentration of the potassium chloride rather than the mercurous chloride. The potential of the electrode depends upon the KCl concentration, varying from 0.240 V for the saturated electrode to 0.334 V for the tenth-normal electrode at 30°C.

The reaction which takes place in the calomel electrode is

$$Hg_2Cl_2 + 2e^- \rightleftharpoons 2Hg + 2Cl^- \quad (8)$$

and the potential remains constant when only small amounts of current flow since the activity of the mercurous ion remains constant irrespective of the direction of flow. If the electrode accepts electrons and thus converts some mercurous ion to mercury, some of the solid mercurous chloride dissolves to resaturate the solution; if the electrode produces electrons, some mercury is converted to mercurous ion, but since the solution is saturated with mercurous chloride, the mercurous ions produced precipitate as mercurous chloride. Since there is a very high concentration of chloride ion, the small changes in chloride ion concentration do not affect the potential appreciably since the main effects of the chloride ion are to depress the solubility by the common ion effect and to maintain a constant ionic strength and thus a constant value for the activity coefficient.

2. The Hydrogen Electrode

Even though the hydrogen electrode is the standard electrode upon which the values of all standard potentials are based, it is actually used only occasionally due to the difficulty of operation. One form of the hydrogen electrode is shown in Fig. 12-2. A square of platinum black foil is arranged

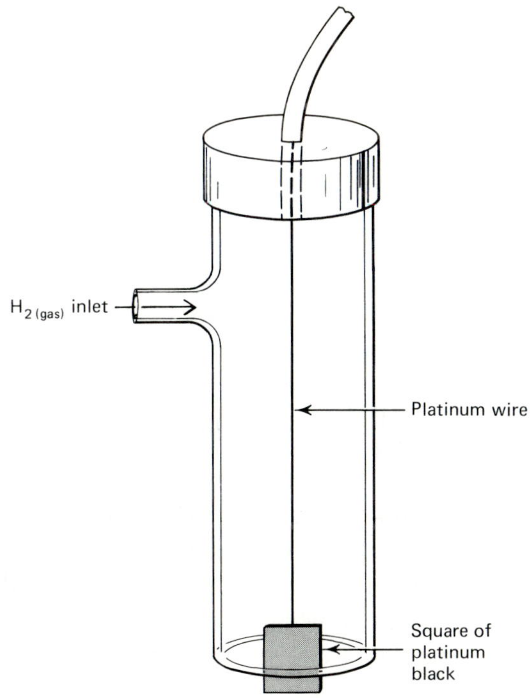

Figure 12.2. The hydrogen electrode.

so that half extends below the bottom of the glass tube. Hydrogen gas, under pressure, enters through the side arm and forces the liquid out of the tube so that the square is immersed half in the solution and half in the gaseous hydrogen. Under these conditions, the electrode acts like a metal-metal ion electrode and produces a potential which is dependent upon the hydrogen ion activity in the solution. The main objections to the hydrogen electrode are the difficulties encountered due to impure hydrogen, the fact that some ions may be reduced by the hydrogen and that hydrogen sweeps out any other acidic or basic gases dissolved in the solution. The actual *p*H measured is the *p*H after all such reactions take place and not the *p*H of the solution before immersion of the electrode.

3. Substitute Hydrogen Electrodes

a. THE QUINHYDRONE ELECTRODE. Quinhydrone ($C_6H_4O_2 \cdot H_2C_6H_4O_2$) is a 1:1 addition compound of quinone ($C_6H_4O_2$) and hydroquinone ($H_2C_6H_4O_2$) which is slightly soluble in water and which dissociates slightly into quinone and hydroquinone in solution. For convenience we shall use Q to represent quinone, H_2Q to represent hydroquinone, and QH_2Q to represent quinhydrone. If a solution is saturated with quinhydrone and a platinum wire is immersed in the solution, the potential of the platinum wire will be dependent upon the hydrogen ion concentration in the solution. Quinone and hydroquinone constitute an oxidation-reduction pair, as shown by

$$\mathrm{Q} + 2\mathrm{H}^+ + 2e^- \rightleftharpoons \mathrm{H_2Q} \tag{9}$$

The potential, as given by the Nernst equation, is

$$E_\mathrm{Q} = E^\circ_\mathrm{Q} + \frac{0.060}{2} \log \frac{(\mathrm{Q})(\mathrm{H}^+)^2}{(\mathrm{H_2Q})} \tag{10}$$

Substitution of the activity coefficient times the molar concentration for active concentration and rearrangement gives

$$E_\mathrm{Q} = E^\circ_\mathrm{Q} + 0.030 \log \frac{[\mathrm{Q}]}{[\mathrm{H_2Q}]} + 0.030 \log \frac{f_\mathrm{Q}}{f_\mathrm{H_2Q}} + \frac{0.060}{2} \log (\mathrm{H}^+)^2 \tag{11}$$

The second term on the right equals zero since the molar concentration of Q and H_2Q are equal because they are produced in equal numbers by dissociation of quinhydrone. The activity coefficients are almost equal to 1 since both quinone and hydroquinone are molecular so that the third term on the right also is essentially zero. As a consequence, eq. (11) reduces to

$$E_\mathrm{Q} = E^\circ_\mathrm{Q} + 0.060 \log(\mathrm{H}^+) \tag{12}$$

The quinhydrone electrode is easy to use since the only requirement is to saturate the solution with quinhydrone and to insert a platinum wire and a reference electrode. It has the disadvantage that it cannot be used above *p*H 9 since hydroquinone is a weak acid which ionizes at high *p*H so that Q is no longer equal to H_2Q. Also, some oxidizing and reducing agents will react with quinone or hydroquinone and upset the equilibrium. Measurements must be made without unnecessary delay since hydroquinone is oxidized slowly by air.

b. THE GLASS ELECTRODE. The most popular substitute electrode is the glass electrode, which has many advantages and only a few disadvantages when compared to the hydrogen electrode for the potentiometric determination of *p*H. The glass electrode is based upon the fact that when two solutions of different hydrogen ion activity are separated by a thin glass membrane a potential is developed across the membrane, the value of which is given by

$$E_M = E_A + 0.060 \log \frac{(\mathrm{H}^+)_o}{(\mathrm{H}^+)_i} \tag{13}$$

where E_M is the membrane potential, E_A is known as the asymmetry potential, and the subscripts *o* and *i* refer to the two solutions. The asymmetry potential will differ for each electrode; however, the value is small and usually decreases to zero as the membrane is aged. A diagram of a glass electrode is shown in Fig. 12-3. The barrel of the electrode is usually prepared from ordinary glass with a thin bulb of special glass sealed on the end. The solution inside the electrode is usually tenth-normal hydrochloric acid. A constant potential electrode must be enclosed in the glass electrode. This can be obtained by using a conductor of silver coated with silver chloride which will produce a silver-silver chloride-chloride ion half-cell of constant potential. In other electrodes, the hydrochloric acid is saturated with quinhydrone and the conductor is a platinum wire.

A complete explanation of a glass electrode is beyond the scope of this text. Evidently the potential across the membrane is a function of the collisions of hydrogen ions with the surface of the membrane, and the potential developed varies with differences in the rate of collisions on the two sides. The rate of collisions on the inner side is constant due to the constant concentration of the inner solution so that variations in potential are due to changes in the hydrogen ion activity in the solution in which the electrode is immersed.

The cell produced by immersing a glass electrode and a calomel electrode into a solution may be represented as

$$\text{Inner electrode}|\text{solution}_i|\text{glass}|\text{solution}_o\|\text{saturated KCl}|\mathrm{Hg_2Cl_2} + \mathrm{Hg} \tag{14}$$

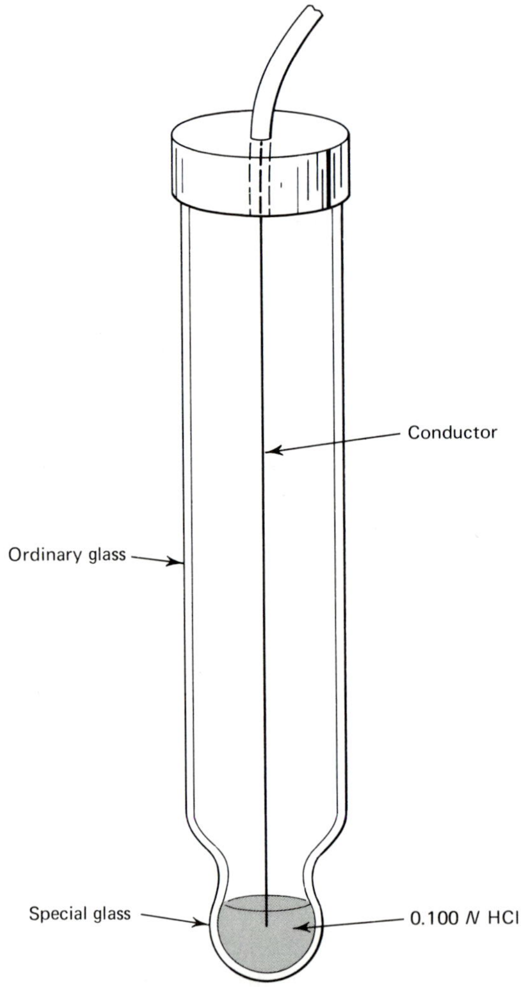

Figure 12.3. The glass electrode. The conductor may be silver coated with silver chloride to form a silver-silver chloride-chloride ion inner electrode or may be platinum immersed in an acid solution saturated with quinhydrone to form a quinhydrone inner electrode.

and the potential actually measured is that between the inner electrode and the calomel electrode. The subscripts *i* and *o* refer to the inner and outside solutions. Assuming that the inner electrode is a quinhydrone electrode, the cell potential is

$$E_{cell} = E_{\mathrm{Q}} + E_M - E_{\mathrm{Hg}} \tag{15}$$

where E_M is the potential across the glass membrane. The potential of the inner electrode is

$$E_{\mathrm{Q}} = E^{\circ}_{\mathrm{Q}} + 0.060 \log(\mathrm{H}^+)_i \tag{16}$$

Substituting this value and the value for E_M given in eq. (13) into eq. (15) gives

$$E_{cell} = E_Q^\circ + 0.060 \log(H^+)_i + E_A + 0.060 \log \frac{(H^+)_o}{(H^+)_i} - E_{Hg}$$

which, upon combination of the logarithmic terms and allowing E_g° to be equal to $E_Q^\circ + E_A$, becomes

$$E_{cell} = E_g^\circ + 0.060 \log(H^+)_o - E_{Hg} \tag{17}$$

To use the glass electrode for measurement of hydrogen ion activities, it is necessary to know the value of E_g°. This value cannot be calculated but must be measured against a hydrogen electrode for each electrode used. Since glass electrodes having the same type of inner electrode have essentially equal values of E_g°, *p*H meters are constructed so that the meter may be adjusted to offset differences in the E_g° values of different electrodes.

The main limitations on the use of the early glass electrodes dealt with the fact that they could not be used at high *p*H values (over approximately 10) or at low *p*H values (below approximately 2) without error and that they offered high resistance which precluded the use of a slide wire potentiometer. These limitations have been offset with present-day electrodes by the use of special glass to reduce the errors at high and low *p*H and by the use of vacuum tube voltmeters. The glass electrode is probably the closest approach to a perfect electrode for the measurement of *p*H potentiometrically.

4. Calculation of *p*H

a. HYDROGEN ELECTRODE AND CALOMEL ELECTRODE. The cell involved for the measurement of *p*H using a hydrogen electrode against a saturated calomel electrode as reference may be shown as

$$\text{Pt} + H_{2(gas)}|\text{solution}||\text{saturated KCl}|Hg_2Cl_2 + Hg \tag{18}$$

for which the cell potential is

$$E_{cell} = E_H - E_{Hg} \tag{19}$$

or

$$E_{cell} = E_H^\circ + 0.060 \log(H^+) - E_{Hg}$$

which may be rearranged to give, since $E_H^\circ = 0$,

$$-\log(H^+) = \frac{-E_{cell} - E_{Hg}}{0.060}$$

and, since by definition $-\log(H^+)$ equals pH,

$$pH = \frac{-E_{cell} - E_{Hg}}{0.060} \tag{20}$$

For example, calculate the pH of a solution if the cell potential between a hydrogen electrode and a saturated calomel electrode is -0.485 V. The E_{Hg} for saturated calomel electrode is 0.240 V at 30°C. (Note that this E_{Hg} differs from the value in the table of standard potentials which is the value at 25°C.)

$$pH = \frac{-(-0.485) - 0.240}{0.060} = \frac{0.245}{0.060} = 6.08$$

b. SUBSTITUTE ELECTRODES AND SATURATED CALOMEL ELECTRODE. For all substitute electrodes measured against a saturated calomel electrode, the cell potential is given by

$$E_{cell} = E_{sub} - E_{Hg} \tag{21}$$

and

$$E_{sub} = E^\circ_{sub} + 0.060 \log(H^+) \tag{22}$$

so that

$$E_{cell} = E^\circ_{sub} + 0.060 \log(H^+) - E_{Hg}$$

Rearrangement gives

$$-\log(H^+) = pH = \frac{E^\circ_{sub} - E_{cell} - E_{Hg}}{0.060} \tag{23}$$

For example, calculate the pH of a solution which produces a potential of 0.092 V between a glass electrode and a saturated calomel electrode. The E°_g is 0.692 V.

$$pH = \frac{0.692 - 0.092 - 0.240}{0.060} = \frac{0.360}{0.060} = 6.00$$

C. Potentiometric Titrations

In potentiometric titrations, the potential existing between two electrodes immersed in the solution is measured as the titrant is added. One of the electrodes must be a reference electrode and the other must be an indicating electrode, the potential of which will vary quantitatively with the activity of the measured ion in the solution. The most popular reference electrode is the saturated calomel electrode. The most popular indicating electrode for neutralization titrations is the glass electrode, and for oxidation-reduction titrations it is the platinum electrode. In any potentiometric titration there must be a sharp break in the measured potential at the equivalence point. Potentiometric titrations are capable of greater accuracy than visual indicator titrations and can be used in those cases where no satisfactory indicator is available or where substances present in the titrated solution interfere with the indicator action. Often two or more substances may be determined in the same solution by a single potentiometric titration. The main disadvantage to manual potentiometric titrations is the fact that they usually require more time and that the end point must be determined by graphical methods. Both these disadvantages are eliminated by the use of an automatic titrator of which several commercial models are available. Some automatic models use a recorder and plot the actual shape of the titration curve, while others titrate to a preset potential.

1. Graphical Methods of Selection of End Point

A graph of the *p*H versus the milliliters of titrant added will produce a curve similar to that in Fig. 9-1 in which there is a sharp break in *p*H (or potential) at the equivalence point. It can be shown mathematically that the equivalence point in any potentiometric titration coincides with the steepest point on the graph so that the problem becomes one of determining the location of the steepest point. Since the desired accuracy in estimating the volume of titrant necessary is in the range of 0.02 ml, the plot must be made so that the position of the steepest point can be estimated with this accuracy. This means that each small division of the paper must represent no more than 0.05–0.10 ml. Most titrations which involve a 50-ml burette are designed so that the volume necessary to reach the equivalence point is between 30 and 40 ml. A graph which shows the entire titration using 0.05 or 0.1 ml values for the smallest division of the paper will require a very large piece of paper. Selection of the equivalence point, however, requires only that portion of the graph in which the sharp break occurs. As a consequence, it is only necessary to plot the data for one or two ml on either side of the equivalence point. The steepest point of the curve usually is not obvious by inspection and must be located by some graphical method. Two such procedures are shown in Fig. 12-4. The straight line portions of the graph of *p*H versus milliliters (shown as a solid line) before and beyond the equivalence point break are extrapolated. Theoretically these two lines should be parallel, but in actual practice they seldom are. The perpendicular between these two extrapolations should be bisected by the curve at the steepest point on the curve. As a consequence, the distance between the two extrapolations is measured, and the perpendicular is drawn at a point at

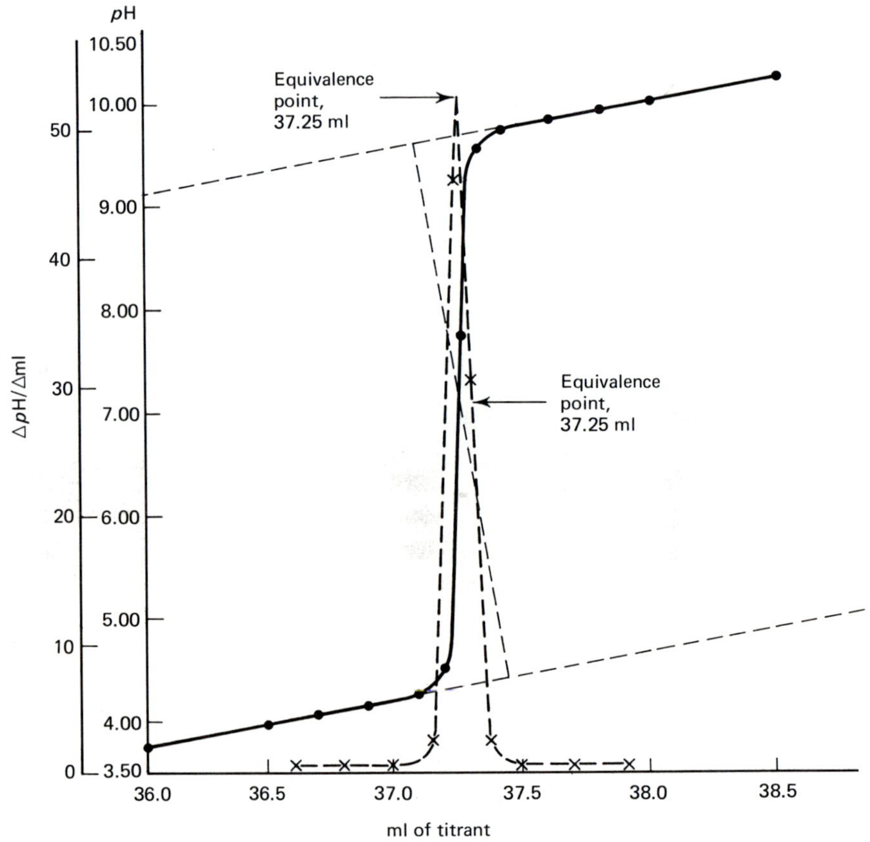

Figure 12.4. Expanded graph of the equivalence point region for a potentiometric titration showing graphical methods of selecting the end point.

which it will be bisected by the curve. The intersection of the curve and the perpendicular is the equivalence point.

The equivalence point can also be determined by the differential method. In this method the rate of change of *p*H with volume of titrant is plotted against the midpoint of each incremental addition of titrant. The difference between the *p*H values (Δ*p*H) for a given increment of titrant is divided by the titrant increment (Δml) obtain the average rate of change of *p*H with volume over that particular increment. These values are then plotted against the milliliters of titrant. The differential value should be the greatest at the steepest point on the curve, and the curve should show very low values of Δ*p*H/Δml until close to the equivalence point. In the equivalence point region the curve should rise sharply to a maximum and then decrease sharply. Such a graph is shown as the dashed curve in Fig. 12-4 for the same titration. Note that the values are plotted at the midpoint of the increments.

The equivalence point is the intersection of the lines obtained by extrapolating the best straight lines through the points on either side of the *p*H break. The values obtained by the two methods should agree.

D. Problems

1. Calculate the *p*H for each of the following cell potentials obtained using a hydrogen electrode and a saturated calomel electrode at 30°C. E_{Hg} at 30°C is 240 mV: (a) −0.666 V, (b) −0.446 V, (c) −0.303 V, (d) −0.999 V, and (e) −0.990 V.

   Ans. (b) 3.43
2. Calculate the *p*H for each of the following cell potentials obtained using a quinhydrone electrode and a tenth molar (normal) calomel electrode at 30°C $E^\circ_Q = 696$ mV and $E_{Hg} = 335$ mV at 30°C: (a) −61 mV, (b) 100 mV, (c) 301 mV. (d) −205 mV, and (e) 205 mV.

   Ans. (b) 4.35
3. Calculate the *p*H for each of the following cell potentials obtained using a glass electrode and a molar (normal) calomel electrode at 30°C. $E^\circ_g = 692$ mV and $E_{Hg} = 277$ mV at 30°C: (a) 0.325 V, (b) −0.385 V, (c) −0.015 V, (d) 115 mV, and (e) 256 mV.

   Ans. (b) 13.33
4. Calculate the silver ion activity in the following cell for the potentials given:

   $$Ag|0.0440A\ AgNO_3\,||\,\text{saturated } KNO_3\,||\,\text{unknown } AgNO_3|Ag$$

   (a) −0.085 V, (b) 0.085 V, (c) −0.010 V, (d) 0.025 V, and (e) −0.060 V.

   Ans. (*b*) = $1.69 \times 10^{-3}A$
5. The following data were taken during the potentiometric titration of 25.00 ml of HCl with 0.1067 *N* NaOH. (a) Plot the data in the equivalence point region as *p*H versus milliliters of titrant, (b) select the end point and calculate the normality of the hydrochloric acid, (c) calculate the Δ*p*H/Δml values and plot against milliliters of titrant in the equivalence point region, and (d) select the end point from this differential graph and calculate the normality of the acid:

Burette readings	*p*H	Burette readings	*p*H	Burette readings	*p*H	Burette readings	*p*H
0.55	1.70	25.70	3.45	26.00	7.50	26.40	10.52
24.50	3.00	25.80	3.50	26.05	9.95	26.50	10.56
25.00	3.17	25.85	3.60	26.10	10.20	27.00	10.74
25.50	3.37	25.90	3.75	26.20	10.35	27.50	10.92
25.60	3.41	25.95	5.00	26.30	10.47		

6. The following data were taken during the potentiometric titration of 25.00 ml of 0.1167 *N* HCl with NaOH. (a) Plot the data in the equivalence point region as *p*H versus milliliters of titrant, (b) select the end point from the graph and calculate the normality of the base, (c) calculate the Δ*p*H/Δml

values and plot against milliliters of titrant in the equivalence point region, and (d) select the end point from this differential graph and calculate the normality of the base:

Burette readings	pH	Burette readings	pH	Burette readings	pH	Burette readings	pH
0.25	1.60	25.40	3.35	25.70	7.40	26.10	10.42
24.20	2.90	25.50	3.40	25.75	9.85	26.20	10.46
24.70	3.07	25.55	3.50	25.80	10.10	26.70	10.64
25.20	3.27	25.60	3.65	25.90	10.25	27.20	10.82
25.30	3.31	25.65	4.90	26.00	10.37		

CHAPTER 13

Compleximetric Titrations

The formation of complex ions and neutral and ionic chelates plays an important role in many analytical processes. Many chelates are insoluble in water and form the basis of many rather specific precipitation reactions in gravimetry. Both chelates and coordination complexes may be used as masking agents to prevent an undesired reaction from interference with a desired reaction. Many chelates and complexes are highly colored and form the basis of colorimetric and photometric procedures. Furthermore, the solubility of neutral chelates in organic solvents allows the extraction of metals into the organic phase and thus is the basis of many important analytical separations. In this chapter we shall be interested primarily in the uses of complexes in analytical titrations.

The formation and properties of coordination complexes will not be discussed in detail. The student should review the principles involved in the formation and properties of these compounds which can be found in most recent general chemistry texts. Chelates, however, usually are not discussed in detail in general chemistry texts since chelates are derivatives of organic compounds which have at least two functional groups so arranged that the metal can form a ring structure with the compound. A detailed discussion of the principles of formation, properties, and uses of chelates is beyond the scope of this text. The student is referred to more advanced analytical texts for such information.

A. Stability and Instability Constants

Most complex ions contain more than one ligand coordinated to the metal, and the formation of the complex takes place in steps. Actually, practically all metal ions are hydrated in water, and the formation of the complex can be considered as the displacement of the coordinated water by a stronger ligand. Even though we often write the formula of the cupric ion in solution as Cu^{2+}, it is actually present as $Cu(H_2O)_4^{2+}$, and the blue color associated with cupric ions in solution is the color of the hydrated ion. The anhydrous ion is colorless, as is indicated by the fact that anhydrous cupric sulfate is white.

The formation of complexes in solution is illustrated in the first two equations below in which M is assumed to be a metal ion with coordination number of n and L represents a ligand other than water which can occupy one coordination position on the metal.

$$M(H_2O)_n + L \rightleftharpoons H_2O + M(H_2O)_{n-1}L \tag{1}$$

$$M(H_2O)_{n-1}L + L \rightleftharpoons H_2O + M(H_2O)_{n-2}L_2 \tag{2}$$

and the reaction continues in steps until the highest complex, ML_n, is formed. For simplicity in writing equations, the water of the aquo complexes is usually omitted. Furthermore, since the metal ion can be mono- or polyvalent, and since the ligands can be neutral or ionic, the ionic charges are omitted in the following general discussion.

Each stage in the formation of a complex is governed by the law of mass action and forms an equilibrium so that each step has its own formation or stability constant. Thus, for the formation of ML_n from M and L, the stepwise sequence is

$$M + L \rightleftharpoons ML \qquad k_1 = \frac{[ML]}{[M][L]} \tag{3}$$

$$ML + L \rightleftharpoons ML_2 \qquad k_2 = \frac{[ML_2]}{[ML][L]} \tag{4}$$

$$ML_2 + L \rightleftharpoons ML_3 \qquad k_3 = \frac{[ML_3]}{[ML_2][L]} \tag{5}$$

$$\vdots$$

$$ML_{n-1} + L \rightleftharpoons ML_n \qquad k_n = \frac{[ML_n]}{[ML_{n-1}][L]} \tag{6}$$

The overall formation constant for any given stage of complexation is given by the product of the successive k values; thus

$$M + 2L \rightleftharpoons ML_2 \qquad K_2 = k_1 k_2 = \frac{[ML_2]}{[M][L]^2} \tag{7}$$

$$M + 4L \rightleftharpoons ML_4 \qquad K_4 = k_1 k_2 k_3 k_4 = \frac{[ML_4]}{[M][L]^4} \tag{8}$$

The total molar concentration of the metal in a solution containing a metal ion and a complexing ligand is the sum of the equilibrium concentrations of each species present, or

$$[M]_t = [M] + [ML] + [ML_2] + [ML_3] + \cdots + [ML_n] \tag{9}$$

Another way of stating this is that all the possible species will be present even though some are present in such small concentrations that they are negligible compared to the predominant species.

In many cases successive constants are almost equal so that there can be rather wide ranges of ligand concentration in which adjacent species will be present in almost equal amounts. Furthermore, in the presence of high ligand concentration, the highest complex is usually favored, and the presence of the intermediate species may be neglected in many calculations.

The reverse reaction or the ionization of the complex also occurs in steps from the complex with the greatest number of ligands to the free metal ion. The ionization or instability constants for these reactions are the inverse of the constants for the formation reactions. Tables of constants for both complex formation and ionization are given in the literature and are also known as stability constants (K_{stab}) and instability constants (K_{instab}). The student should note carefully which type of table is being used to obtain information. In general, the easiest method of distinguishing is to note the values of the constants. In most cases the values for the stability constants will be large numbers (exponents will be positive) while the values for the instability constants will be very small numbers (exponents will be negative).

B. Titration of Chloride with Mercuric Nitrate

In reality, the titration of chloride with mercuric nitrate is an example of the formation of a slightly ionized compound rather than complex formation since mercuric chloride is only slightly ionized in water solution (less than 1% in 0.1 *N* solution). Since it is very unusual for a metallic chloride to be nonionic, this titration is usually discussed with complex formation titrations.

Several indicators are available for the titration, with the most popular being sodium nitroprusside [$Na_2Fe(CN)_5NO$] and diphenyl carbazone [$C_6H_5(NH)_2CON_2C_6H_5$]. The nitroprusside forms a precipitate (turbidity) with excess mercuric ion and should not be added until very close to the end point. The diphenylcarbazone forms a soluble violet-colored complex with mercuric ions and is the indicator of choice. The *p*H is very important for the proper functioning of the diphenylcarbazone indicator and must be maintained between 3.0 and 3.5 with nitric acid. For this reason, the indicator solution is often a mixture of diphenylcarbazone with bromphenol blue to aid in the adjustment of *p*H. The color of the bromphenol blue in this *p*H range also increases the sharpness of the end point.

The mercurimetric diphenylcarbazone method for chloride is suitable for titration in very dilute solutions in the range of 1 to 100 ppm and is used for determination of chloride in natural waters and in blood. The method is also used in the determination of bromide, thiocyanate, and cyanide and is remarkably free of interference.

Most directions call for the mercuric nitrate to be made up in concentrations of 0.025 *N* or less in dilute nitric acid and standardized against primary standard sodium chloride solutions. The sample should not contain more than 20 mg of chloride. Before titration, the *p*H is adjusted with 0.05 *N* HNO_3 to the yellow color of bromphenol blue followed by the addition of 1 ml excess of the acid. The end point is the appearance of a blue violet color which deepens as more titrant is added.

C. Titration of Cyanide with Silver Nitrate

The addition of a silver nitrate solution to a solution of a cyanide causes the formation of the soluble dicyanoargentate(I) ion, $Ag(CN)_2^-$. After sufficient silver has been added to react with all the cyanide by this reaction, the first excess will react to form insoluble silver dicyanoargentate(I), $Ag[Ag(CN)_2]$, which is often referred to and written as silver cyanide, AgCN. The ionic equations for the reactions are

$$Ag^+ + 2CN^- \rightleftharpoons Ag(CN)_2^- \tag{10}$$

$$Ag^+ + Ag(CN)_2^- \rightleftharpoons Ag[Ag(CN)_2] \tag{11}$$

As a consequence the appearance of a permanent precipitate indicates the quantitative completion of the first reaction and can be taken as the equivalence point. It should be noted that the equivalent weight of cyanide in the reaction is 52.06, or twice the radical weight since it requires two cyanides to react with each silver ion.

The end point is not sharp due to formation of solid silver dicyanoargentate(I) caused by local excess concentrations in the zone of mixing of the titrant and the solution. This precipitate is slow to redissolve in the equivalence point region. The end point can be sharpened by the addition of ammonia and iodide ion since the ammonia will prevent the precipitation of silver dicyanoargentate(I) but does not prevent the precipitation of silver iodide. As a consequence, the first excess of silver precipitates as silver iodide, which forms as a yellow opalescence in the solution and constitutes a very sharp end point. The silver iodide does not form before the equivalence point since it is kept in solution by the excess cyanide in which it is soluble. The concentration of ammonia used is very important since too little will cause the end point to appear early and too much will cause the end point to appear late. The concentration of iodide is also important for similar reasons. Most directions call for the solution to be approximately 0.3 M in ammonia and to contain approximately 0.2 g of potassium iodide per 100 ml.

Under the proper conditions, the reaction is accurate, and the end point is sufficiently sharp. The experiment is seldom carried out in beginning courses in analysis, however, due to the danger associated with the use of extremely poisonous cyanide solutions.

D. Titration with Ethylenediaminetetraacetic Acid (EDTA)

Ethylenediaminetetraacetic acid (also known as ethylenedinitrilotetraacetic acid and commonly referred to as EDTA) is the most important example of a class of reagents introduced in 1945 by Schwartzenbach which form soluble stable chelate complexes with many metals. This group of compounds is often called *complexones*, *chelons*, or *versenes* as a general group name and are usually aminopolycarboxylic acids. They can also be described as tertiary amines containing several carboxylic acid groups.

EDTA is a tetraprotic acid with the formula $(HOOCCH_2)_2N-CH_2-CH_2-N(CH_2COOH)_2$, which forms soluble 1:1 metal complexes of widely different stabilities with metals. The acid, usually written as H_4Y for convenience, is insoluble in water, and solutions are usually prepared with the disodium salt Na_2H_2Y, which is available in high purity as the dihydrate. The solution, however, is referred to as EDTA rather than as disodium EDTA. Since EDTA has both donor nitrogen and donor oxygen atoms, it can form as many as six five-membered chelate rings and consequently forms complexes with practically all metals except the alkali metals of Group I. However, the 1:1 mole reaction ratio makes equilibrium calculations simpler than for most complexes which are formed by stepwise equilibria.

EDTA is a weak acid which ionizes stepwise, as indicated in eqs. (12) through (15):

$$H_4Y \rightleftharpoons H^+ + H_3Y^- \qquad k_1 = \frac{[H^+][H_3Y^-]}{[H_4Y]} \tag{12}$$

$$H_3Y^- \rightleftharpoons H_2Y^{2-} + H^+ \qquad k_2 = \frac{[H^+][H_2Y^{2-}]}{[H_3Y^-]} \tag{13}$$

$$H_2Y^{2-} \rightleftharpoons H^+ + HY^{3-} \qquad k_3 = \frac{[H^+][HY^{3-}]}{[H_2Y^{2-}]} \tag{14}$$

$$HY^{3-} \rightleftharpoons H^+ + Y^{4-} \qquad k_4 = \frac{[H^+][Y^{4-}]}{[HY^{3-}]} \tag{15}$$

The pK values for the stepwise constants are $pK_1 = 2.00$, $pK_2 = 2.67$, $pK_3 = 6.16$, and $pK_4 = 10.26$, and the cumulative constant is $pK_{stab} = 21.09$. The secondary ion, H_2Y^{2-}, reacts with metals by displacement of hydrogen ions, as exemplified by

$$Mg^{2+} + H_2Y^{2-} \rightleftharpoons MgY^{2-} + 2H^+ \tag{16}$$

$$Al^{3+} + H_2Y^{2-} \rightleftharpoons AlY^- + 2H^+ \tag{17}$$

Equations (16) and (17) indicate that the smaller the stability constant, the higher the pH of the solution must be to form the complex. In general, complexes of divalent ions are stable in alkaline or slightly acidic solution, complexes of the trivalent ions are stable at pH 1 or 2, and complexes of the tetravalent cations are stable in strong acid solutions. A number of exceptions to these general statements can be found.

EDTA is used primarily as a titrant in volumetric analysis but is also utilized as a masking agent in various reactions and extractions and as a method of dissolving insoluble salts. Our discussion will deal only with its use as a titrant.

1. Titration Methods

Several titration methods are available because of the wide range of stabilities, the effect of pH, and the ability to use masking agents in the presence of the titrant. The most important of these are the direct and indirect titration procedures.

a. DIRECT TITRATION. Several metal ion indicators are available for use in direct titrations. In general, the solution of the metal ion is buffered to the proper pH, masking agents and the indicator are added, and the solution is titrated with EDTA to the color change at the end point.

b. INDIRECT OR BACK TITRATION. Metals which cannot be determined easily by direct titration may often be determined by the addition of an excess of a standard EDTA solution followed by the titration of the excess with a standard magnesium or zinc solution.

c. DISPLACEMENT OF A METAL ION. Addition of a solution of zinc or magnesium EDTA complex to a solution of a metal which forms a more stable complex with EDTA than zinc or magnesium causes the displacement of zinc or magnesium from the complex. As an example, when a solution of magnesium EDTA is added to a solution of ferric ion, the ferric ion replaces the magnesium in the complex.

$$Fe^{3+} + MgY^{2-} \rightleftharpoons FeY^{-} + Mg^{2+} \qquad (18)$$

The liberated Mg^{2+} is then titrated with standard EDTA and is a measure of the amount of Fe^{3+} in the solution.

d. TITRATION OF HYDROGEN LIBERATED. According to eqs. (16) and (17), the reaction of a metal ion with EDTA releases two hydrogen ions. The hydrogen ion can be titrated wth standard NaOH to determine the amount of cation indirectly.

e. TITRATION OF ANIONS. Anions which form insoluble compounds with metal ions which are titratable by EDTA can be determined by adding an excess of a standard solution of the cation. The excess cation is then titrated with standard EDTA. An example is the determination of sulfate by the addition of an excess of standard barium solution which is then titrated by EDTA.

$$SO_4^{2-} + Ba^{2+} \rightleftharpoons BaSO_4 + Ba^{2+}(XS) \qquad (19)$$

$$Ba^{2+}(XS) + H_2Y^{2-} \rightleftharpoons BaY^{2-} + 2H^{+} \qquad (20)$$

The end point has a tendency to fade since the $BaSO_4$ will dissolve in any excess EDTA added beyond the end point.

2. Indicators

A number of visual indicators are available which form stable highly colored complexes with metal ions. The metal-indicator complex ion must be less stable than the corresponding EDTA complex so that EDTA can remove the metal from the indicator complex. Using magnesium and Erio T (erio-chrome black T, also known as Indicator F241) as the example,

$$\text{Mg-Erio} + \text{EDTA} \rightleftharpoons \text{Mg-EDTA} + \text{Erio} \qquad (21)$$

The end point is the appearance of the color of the uncomplexed indicator.

Erio T, which was one of the first indicators to be used in EDTA titrations, works satisfactorily only in solutions in which the pH is approximately 8 to 10. At lower *p*H values, the end point is not sharp due to the fact that the dye has a purple red color similar to the color of the metal dye complexes, and at higher *p*H values it is converted to an orange form which does not function as an indicator. In the *p*H range of 8 to 10, Erio T can be used to determine the sum of all the alkaline earth metals present but cannot be used for the determination of a single alkaline earth in the presence of any of the others. It is also used for titrations of Zn, Cd, and Pb.

CalRed (also known as the indicator of Patton and Reeder) can be used in highly basic solutions (*p*H 12 to 14) for the determination of calcium. At this high *p*H, magnesium precipitates as the hydroxide and does not interfere. The color change for CalRed is from wine red to pure blue, which is the same as the Erio T change.

Other indicators used, together with the useful *p*H range and chemical type, are listed in Table 13-1.

TABLE 13.1. Indicators for EDTA Titrations

Indicator	Chemical type	Useful *p*H range	Typical metals determined
Erio T	Azo	8–10	Mg, Zn, Cd, Pb, (Mg + Ca)
Erio R	Azo	8–12	Mg, Zn (Mg + Ca)
CalRed	Azo	12–14	Ca
Calmagite	Azo	9–11	Mg, Zn, Cd (Mg + Ca)
PAN	Azo	2–11	Cu, Pb
Murexide	Purine	6–13	Ca, Cu, Co, Ni
Calcein Blue	Iminodiacetate	5–9, 13–14	Cu, Ca
Salicylic acid	—	2–3	Fe
Zincon	Hydrazo	5–10	Zn, Cu, Ca

3. Standardization

Disodium EDTA dihydrate can be purchased in high purity and can be used in some cases as a primary standard for the preparation of standard EDTA solutions. For most samples, however, it is preferable to standardize the solution against a standard solution of the metal to be determined. Many directions call for the standardization against a standard magnesium solution using Erio T in a solution buffered to a *p*H of 9 with an ammoniacal buffer. The end point is the change from a wine red or purple to a pure sky blue. Since solutions of Erio T are not stable for extended periods of time, indicator solutions should be prepared fresh each week. To offset the trouble of preparing fresh indicator periodically, the indicator can be coated on crystals of a soluble salt, which will not interfere with a titration such as potassium chloride, potassium sulfate, or sodium sulfate. The solid indicator is stable so that dispersions on solid crystals are also stable. Sufficient crystals are added to the solution being titrated to produce a satsfactory color.

Most standard solutions are 0.01 *M* (0.02 *N*) and are often also specified as having a calcium carbonate titer of 1.00 mg/ml, especially for solutions used to determine hardness of water.

4. Determination of Total Hardness in Water

One of the main uses of EDTA is the determination of total hardness in water. Total hardness is due to the presence of both calcium and magnesium ions so that total hardness (or the sum of calcium and magnesium) is determined readily using Erio T as the indicator. The stability constant for the calcium-EDTA complex ($\log K = 11.00$) is larger than the stability constant ($\log K = 8.69$) for the magnesium complex. As a result, the addition of EDTA first converts the calcium to the EDTA complex and converts the magnesium to the EDTA complex only after all the calcium is complexed. The stability constant of the magnesium-EDTA complex is larger than the stability constant for the magnesium-Erio T complex ($\log K = 7.00$), so that the end point occurs when all the magnesium is converted from the Erio T complex to the EDTA complex. The color of Erio T is a pure sky blue at *p*H values between 8 and 10, while the color of the magnesium complex is wine red. Since total hardness is usually expressed as parts per million (milligrams per liter) of calcium carbonate, irrespective of the ion causing the hardness, the calculation becomes simply the multiplication of the volume of titrant by the calcium carbonate titer, followed by conversion to the number of milligrams in 1 liter of the water.

5. Determination of Calcium and Magnesium in Limestone

Prior to the advent of EDTA as a titrant, calcium and magnesium were determined in limestone and other rocks by the precipitation of calcium as the oxalate followed by precipitation of magnesium as magnesium ammonium phosphate. The calcium oxalate was either converted to the oxide or carbonate and weighed or dissolved and titrated with standard permanganate. The magnesium ammonium phosphate was ignited to magnesium pyrophosphate and weighed. Early work with EDTA using Erio T as the indicator showed that accurate results of the sum of the magnesium and calcium present could be obtained in rock samples but there was no really satisfactory indicator for calcium or magnesium individually in the presence of each other. As a consequence, procedures to determine both calcium and magnesium depended upon the oxalate separation followed by titration of the magnesium in the filtrate with EDTA using Erio T as the indicator.

CalRed became available in 1956 as an indicator for the titration of calcium in the high *p*H range in which magnesium is quantitatively precipitated as the hydroxide and allowed the determination of calcium in a sample. Combination of the data for the calcium only titration with the data from the titration for the sum of the calcium and magnesium makes possible the calculation of both calcium and magnesium individually. Other indicators are now available for titration of individual alkaline earths.

Many directions for calcium and magnesium in limestone call for the dilution of the filtrate from the R_2O_3 precipitate to a definite volume and the titration of aliquots of this solution with EDTA using Erio T and CalRed as indicators for separate aliquots.

a. TITRATION FOR SUM OF CALCIUM AND MAGNESIUM. The aliquot to be titrated is buffered with ammoniacal buffer to a *p*H in the range of 9, and masking agents are added. Ordinarily, small amounts of both hydroxylamine and potassium cyanide solutions are added. The hydroxylamine reduces ferric iron to the ferrous form, which does not interfere as much as ferric ion. The cyanide is added to form complexes with iron and other heavy metals such as copper to prevent their interference with the indicator action. The indicator is added as a solution or dispersed on crystals, and the solution is titrated to a pure sky blue end point. The theory of the titration is the same as the determination of total hardness discussed above.

b. TITRATION FOR CALCIUM ONLY. A concentrated (8 *N*) solution of potassium or sodium hydroxide is added to the aliquot to be titrated for calcium to cause the quantitative precipitation of magnesium as the hydroxide. In some cases, the high *p*H will also cause the precipitation of some calcium hydroxide, and the end point will be indistinct since the calcium hydroxide dissolves slowly as the titrant is added. The calcium precipitation can be prevented by the addition of some of the standard EDTA before the addition of the base. The EDTA complexes part of the calcium and prevents its precipitation as the hydroxide but does not prevent precipitation of magnesium hydroxide. Care must be used to ensure that the amount added is not greater than the amount needed for the titration. Small amounts of hydroxylamine and potassium cyanide solutions are added to reduce and mask interfering metals. The CalRed indicator is added as a solution or dispersed on crystals and the solution titrated with standard EDTA to the end point, which is a change from wine red to pure sky blue.

c. CALCULATIONS. The amount of calcium present is obtained by direct calculation from the data of the CalRed titration for calcium. The amount of magnesium is determined by the difference between the volumes of EDTA used for identical aliquots in the Erio T and the CalRed titrations. The excess volume of EDTA required in the Erio T titration over that required in the CalRed titration is a measure of the amount of magnesium in the aliquot.

CHAPTER 14

Colorimetric and Photometric Methods of Analysis

Colorimetric and photometric methods of analysis constitute the most widely used analytical methods and have been developed for most elements and many organic compounds. All these methods measure in some way the intensity of color of a solution or the absorbance, transmittance, or reflectance of radiant energy by a solid, liquid, or gas. The majority of the measurements concern the absorbance of a solution of a compound or element. If the measurement is made with the eye the method is usually referred to as a visual colorimetric method, while methods which involve measurement using an instrument are known as photometric or spectrometric procedures. For a long time all such determinations were called colorimetric since light in the visible region was utilized. However, wavelengths outside the visible region (such as the x-ray, ultraviolet, and infrared regions) are used in many analyses so that colorimetry is really a misnomer. In general, present practice is to call all methods photometric, spectrometric, or spectrophotometric irrespective of the wavelength region except when the eye is used to estimate intensity of color.

Photometric methods are used most often for minor constituents, which are present in quantities from trace amounts (as low as 0.00001 per cent) up to about 2 per cent. However, some methods have been proposed which can determine the amount of a major constituent or even the per cent purity of a high-purity substance. Since such methods often call for special photometric measurements or extreme dilution, they are not as popular as the methods for minor amounts. The relative error in most photometric measurements is between 0.1 and 1.0 per cent, with 0.5 per cent being accepted as the general average. As a consequence, ordinary titrimetric and gravimetric procedures are often preferable for major constituents. Whenever colorimetric or photometric methods can be used for repetitive samples, they are usually the method of choice since they tend to be more rapid and require less manipulative procedures.

A. The Nature of Radiant Energy

Radiant energy may be defined as energy transmitted as electromagnetic radiation. It appears to have the characteristics or properties of both wave motion and particles. In diffraction and refraction, radiant energy shows the properties of waves, while in emission and absorption it has the properties of particles (called photons).

Radiant energy is characterized by the wavelength (λ) and the frequency (ν) of the radiation. The wavelength is the distance between two points which are in phase on adjacent waves measured along the line of propagation. It is often thought of as being the shortest distance between the maxima or the minima of two adjacent waves. The frequency is the number of cycles in a unit of time or is the number of maxima which pass a given point in a unit of time. The frequency increases as the wavelength decreases.

The energy of a photon is given by

$$E = h\nu \tag{1}$$

in which E is the energy in ergs, ν is the frequency in cycles per second, and h is Planck's constant, 6.62×10^{-27} erg-sec. As a consequence the energy is directly proportional to the frequency and inversely proportional to the wavelength. High energy is associated with short wavelengths and high frequencies, while long wavelengths and low frequencies have relatively lower energies.

The electromagnetic spectrum covers some 20 orders of magnitude and thus is difficult to show on a linear scale. Figure 14-1 shows the major regions of the spectrum together with the approximate wavelengths of the region boundaries and the type of energy change which is caused in a substance by absorption of that wavelength of radiant energy. It should be noted that the visible region covers a very small portion of the spectrum.

Energy change involved	Approximate wavelength	Region
Nuclear, inner-shell electrons	less than 10 nm	Gamma rays X rays
Ionization of atoms or molecules	10 nm to 200 nm	Vacuum Ultraviolet
Valence electrons	200 nm to 400 nm	Ultraviolet
Valence electrons	400 nm to 800 nm	Visible
Molecular vibration	800 nm to 25 μm	Infrared
Molecular rotation, Electro-magnetic spin	25 μm to 25 cm	Far Infrared Micro Waves Radio Waves

Wavelength	Color absorbed	Color observed
400 nm		
	Violet	Yellow-green
435 nm		
	Blue	Yellow
495 nm		
	Green	Purple
560 nm		
	Yellow	Blue
595 nm		
	Orange	Greenish-blue
650 nm		
	Red	Bluish-green
800 nm		

Figure 14.1. The electromagnetic spectrum. Designations of region boundaries are approximate. Wavelength units are related as follows: 1 m = 100 cm = 1000 mm = 1 million μm (micrometer) = 1 billion nm (nanometer) = 10 billion Å; or 1 Å = 10^{-1} nm = 10^{-4} μm = 10^{-7} mm = 10^{-8} cm = 10^{-10} m. The micrometer was formerly known as the micron (μ) and the nanometer as the millimicron (mμ).

B. The Absorption Process

When radiant energy impinges upon a substance it can be absorbed, transmitted, reflected, or refracted by the substance. For the energy to be absorbed it must correspond to the energy required to cause some change in an atom or molecule. As a consequence, some wavelengths of energy are absorbed and others are not absorbed by any given substance. The changes in the substance may be nuclear, electronic (changes in the energies of the electrons distributed about the atoms in the substance), vibrational (change in the average separation of the nuclei of two or more atoms), or rotational. The absorbed energy is usually dissipated as heat but may be emitted (as in fluorescence) or be utilized to initiate a chemical reaction (as in photochemistry).

As seen in Fig. 14-1, the highest energies and shortest wavelengths are associated with the nuclear changes caused by gamma rays. These energies are measured in the millions of electron volts (MeV). X-rays have less energy (in the range of several hundred to several thousand electron volts) and cause changes in the inner-shell electrons, which are firmly bound in the atom. All these changes are independent of the chemical combination of the atom or are the same irrespective of the chemical state of the atom involved.

Absorption in the ultraviolet and visible regions involves the transformation of the outer shell or valence electrons from their ground or lowest energy state to an excited state and thus is dependent upon the state of chemical combination. As an example, hydrated manganese(II) ion is pale pink, permanganate ion is dark purple, and manganate ion is dark green.

In the infrared region, absorption of the shorter wavelengths causes vibrational changes such as bending or stretching, while molecular rotation is associated with absorption of the longer wavelengths. Absorption of the very low-frequency, long-wavelength energy in the microwave and radio wave regions can cause changes in the electromagnetic spin characteristics.

When polychromatic light (light of many colors or of broad wavelength distribution) impinges upon a sample solution, light of some wavelengths is absorbed, while light of other wavelengths passes through the solution essentially unimpeded. With respect to the visible region, as shown in Fig. 14-1, a solution has the color of the light which is *not* absorbed or may be said to show the complementary color of the absorbed color. As an example, potassium dichromate solution is seen to be yellow-orange since it absorbs at a blue wavelength of approximately 450 nm.

C. Nomenclature

Much confusion exists in the use of terminology in the science of photometry, and several different symbols and names for the same concept or property can be found in the literature. To obtain some consistency in nomenclature, a Joint Committee on Nomenclature in Applied Spectroscopy was established by the society for Applied Spectroscopy (SAS) and the American Society for Testing Materials (ASTM) to recommend and define terms to be used by all writers. The definitions, symbols, and names used in this chapter are those recommended by this committee. A summary of the most common terms can be found in the December issue each year of *Analytical Chemistry*. As each new symbol or concept is introduced in the following discussion, some of the meanings or symbols previously used will be listed to aid the student in correlation between various sources of material. Some of the more common terms to be utilized are

Absorbance, A: [formerly known as optical density (OD), extinction (E), absorbancy]. The absorbance is the logarithm to the base 10 of the reciprocal of the transmittance, or, $A = \log 1/T$. Also defined as the logarithm to the base 10 of the ratio of the radiant power incident upon the sample to the radiant power transmitted by the sample, or $A = \log P_0/P$.

Absorptivity, a: (formerly known as the extinction coefficient, absorbancy index, specific extinction). The absorptivity equals the absorbance divided by the product of the concentration (grams per liter) and the sample path length (in centimeters), or, $a = A/bc$, where b is path length and c is concentration. The absorptivity is sometimes referred to as the *specific absorbance* since it is the absorbance per unit optical path length and per unit concentration. Both the path length and the concentration units should be specified clearly if they are not in centimeters or grams per liter.

Absorptivity, molar, ε: (formerly known as molar extinction coefficient, molar absorbancy index, molar absorption coefficient). The molar absorptivity is the product of the absorptivity and the molecular weight, or $\varepsilon = a\text{MW}$. This is the value obtained for the absorptivity if the concentration is expressed as molar or moles per liter instead of grams per liter.

Radiant Power, P: (formerly known as intensity, I). The radiant power is the rate at which energy is transported in a beam of radiant energy. It is the quantity measured by a detector such as the eye or a photocell. The radiant power incident upon a cell is given the symbol P_0 and that transmitted by a cell the symbol P.

Transmittance, T: (formerly known as transmittancy). The transmittance is the ratio of the radiant power transmitted by a sample to the radiant power of the beam incident upon the sample, or $T = P/P_0$. It is assumed that the measurements are made at the same wavelength with parallel radiation which is incident to the sample at right angles to the plane parallel surfaces of the sample.

D. Fundamental Laws of Absorbance

1. The Bougert or Lambert Law

Lambert, in 1760, expressed the relationship between the intensity of color of a solution and the depth of solution through which the light passed. Actually the same law had been proposed by Bougert in 1729, but this was not discovered until long after the law was known as the Lambert law. According to Lambert and Bougert, the intensity of light transmitted by a solution diminishes in geometric (exponential, logarithmic) progression as the depth of solution increases linearly. That is, the relationship is not arithmetical as one might suppose. This can be expressed as

$$T = a^{-b} \qquad (2)$$

where T is the transmittance, a is a constant characteristic of the solution and the wavelength, and b is the path length of the light through the solution. It may also be expressed logarithmically as

$$-\log T = -\log P/P_0 = ab \qquad (3)$$

The relationship can be understood by reference to Fig. 14-2, which shows the value of the incident light and the transmitted light for a series of cells of the same path length containing the same solution. For a transmittance of 0.5, the radiant power of the transmitted light after one cell is $0.5P_0$, after the second cell is $0.25P_0$, after the third is $0.125P_0$, etc., so that the radiant power of the transmitted beam after n cells will be T^nP_0, or 0.5^nP_0 for the example in Fig. 14-2.

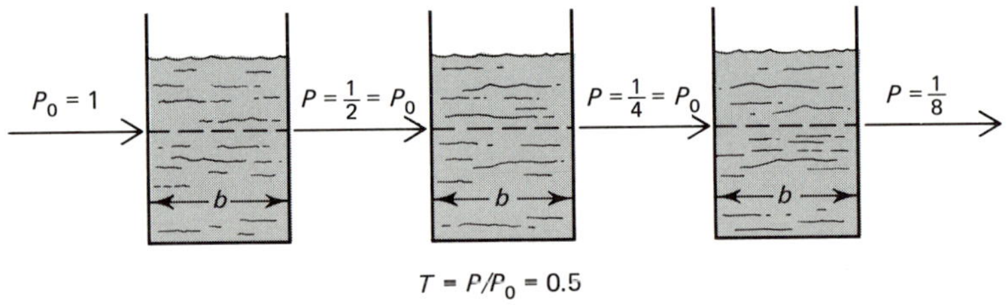

Figure 14.2. Diagrammatic representation of the Bougert-Lambert law showing the exponential effect of increasing the depth of the solution through which the light passes.

2. The Beer Law

Beer enunciated the law relating transmittance to concentration in 1852. This law states that the intensity of light transmitted by a solution decreases geometrically (exponentially, logarithmically) as the concentration increases linearly, or

$$T = a^{-c} \qquad (4)$$

where T is the transmittance, a is a constant depending upon the wavelength and the color absorbing species, and c is the concentration. This may also be expressed logarithmically as

$$-\log T = -\log P/P_0 = ac \qquad (5)$$

The relationship between concentration and transmittance can be understood by referring to Fig. 14-3, which shows the value of the incident and transmitted light for a series of identical cells which contain solutions of different concentration. For a transmittance of 0.5, the radiant power of the

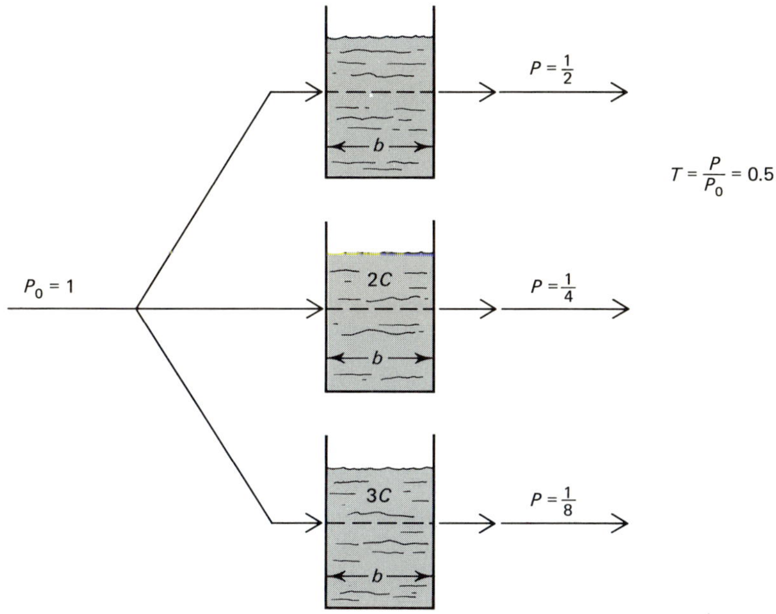

Figure 14.3. Diagrammatic representation of the Beer law showing the exponential effect of increases in concentration.

transmitted beam for a solution of concentration c is $0.5P_0$, for a concentration of $2c$ is $0.25P_0$, and for a concentration of $3c$ is $0.125P_0$, so that the radiant power transmitted through a cell containing a solution of concentration nc would be T^nP_0, or, for $T = 0.5$, 0.5^nP_0.

3. The Combined Bougert-Beer Law

The two laws just discussed are usually combined and used together in practically all absorption analysis. The combined law is often referred to simply as the Beer law, and is stated in the following ways:

$$A = abc \tag{6}$$

$$A = -\log T = -\log P/P_0 = \log P_0/P = abc \tag{7}$$

where A is the absorbance, T is the transmittance, P_0 is the incident radiant power, P is the transmitted radiant power, a is the absorptivity, b is the optical path length in centimeters, and c is the concentration in grams per liter. The value of a depends upon the units of b and c. As a consequence, the path length and the concentration must be expressed clearly whenever they are not expressed in centimeters or grams per liter. Since concentrations in analytical procedures are often expressed as moles per liter (molar), the expression may be written as

$$A = \varepsilon bc \tag{8}$$

where ε is the molar absorptivity, b is the depth in centimeters, and c is the molar concentration.

E. Visual Colorimetric Methods

All visual colorimetric methods use the eye to estimate the intensity of color transmitted by a solution compared to the intensity of color of a standard solution of the substance being determined. Since it is more practical to vary some factor until the two colors are exactly the same than it is to estimate the relative difference between two colors, all visual methods also depend upon the matching of colors by the operator. There are several methods which may be used, and the selection of any one depends upon several factors, including the number of determinations to be made. When only one or a few determinations are to be made, the duplication (titration) method, the dilution method, or the method of balance (variable depth) are the methods of choice. If a large number of samples are to be estimated, the standard series method or the method of balance are usually used. In all these methods, except the method of balance, Nessler tubes are utilized. A Nessler tube is a glass cylinder of uniform diameter with a flat, clear bottom so that colors may be viewed either horizontally or vertically. The tubes are calibrated to some volume and should be used in matched sets of equal diameter so that the depth of solution viewed will be the same for all tubes whether the measurement is made vertically or horizontally. The various methods are summarized below.

1. The Duplication (Titration) Method

In this method the color is developed in the sample and diluted to volume in a Nessler tube. All the color development reagents are then added to a second tube, and a standard solution of the substance being determined is added from a burette until the colors almost match. The standard is then diluted almost to volume, and the color is matched by addition of a few drops of the standard solution. When the colors match, the concentrations of the unknown and the standard are identical since the absorptivity and depth of solution are the same. The tubes must be viewed under equal lighting conditions and may be viewed either vertically or horizontally. The concentration of the unknown is calculated from the volume and concentration of the standard which had to be added to match the color. This is a good method for one or two samples if a suitable standard solution is available. Uniform bore test tubes may be substituted for Nessler tubes if the solutions are viewed horizontally.

2. The Dilution Method

In this method, the sample and a standard of almost equal concentration are placed in Nessler or other tubes of equal diameter and viewed horizontally. The darkest solution is then diluted until the two colors match and the volumes of the two solutions are measured. Since the absorbance, absorptivity, and path length are the same for the two solutions, the concentrations are equal, and the unknown can be calculated from the known concentration of the standard and the volume ratio. This is probably the simplest of the visual colorimetric methods and may be used when only one or two samples are to be determined or when a rapid estimate of concentration is necessary. Since the concentrations are equal and the volumes known,

$$\frac{g_u}{V_u} = \frac{g_s}{V_s} \tag{9}$$

where g is grams, V is volume, and the subscripts u and s refer to unknown and standard, respectively.

3. The Method of Balance or Variable Depth

A Duboscq colorimeter (Fig. 14-4) is used in this method, which involves varying the depth of two solutions through which light of equal incident intensity is passing until the colors match. One of the solutions is the standard and the other is the unknown. When the colors are matched, the absorbances are equal and

$$ab_u c_u = ab_s c_s \tag{10}$$

and, since the absorptivities are the same,

$$b_u c_u = b_s c_s \tag{11}$$

or

$$c_u = \frac{b_s c_s}{b_u} \tag{12}$$

The Duboscq colorimeter is constructed so that optical plungers may be inserted into special cells with plane glass bottoms. The cells are placed upon movable platforms which have scales attached which indicate the distance between the platform and the bottom of the optical plunger. In operation,

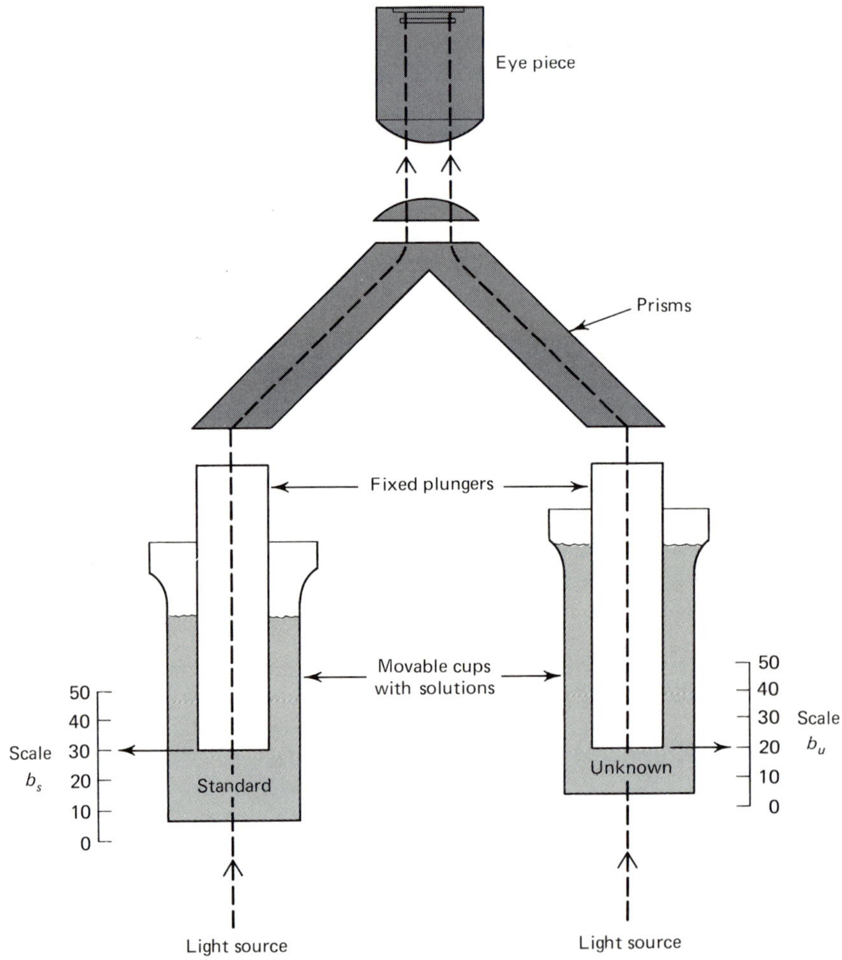

Figure 14.4. Simplified diagram of a Duboscq colorimeter. Not drawn to scale. The readings shown are $b_u = 20$ mm and $b_s = 30$ mm.

the standard is usually set at some particular path length, and the unknown is varied until the colors match in the eyepiece of the instrument. It is usually best to take the average of several readings of which approximately half are obtained by approaching the match by darkening the variable color and the other half by lightening the color. This was the method of choice in colorimetry before the advent of photometers in which a light sensitive device is substituted for the eye as the measuring instrument. It has the disadvantage of causing eye fatigue and using ordinary white light (polychromatic) instead of monochromatic light upon which the Beer law is based. It is an excellent method of demonstrating the Beer law by showing that under conditions showing the same absorptivity, the transmittance or absorbance of a solution is a function of the product of the concentration and the optical path length.

4. The Standard Series Method

In this method, the color of the unknown is compared to the colors of a series of standards. The color of the unknown will match one of the standard colors exactly or else will fall between two of the colors. As a consequence, the concentration cannot be estimated closer than one-half the concentration increment between two standards. The procedure is slow due to the necessary preparation of several standard solutions with a constant concentration increment between successive standards. As a consequence it is seldom used except when a large number of samples is to be run or when a series of permanent standards is available. Several comparitors are on the market which furnish a series of permanent standards, either glass or solutions, so that comparisons can be made with little or no preparation. Such an instrument is beneficial in field work where access to electrical current is limited.

The standard series method is the basis of all calibration curves used in photometry. These calibration procedures measure the absorbance or transmittance of a series of standard solutions of the proper concentration sequence and then graph the results so that future measurements of absorbance can be related directly to the concentration from the graph.

F. Photometric Instruments

To simplify measurements and to offset some of the errors associated with visual measurements, instruments have been designed which substitute a photosensitive device for the eye as the method of measurement. These instruments are called photometers, spectrophotometers, or spectrometers and are classified in several different ways. One classification is upon the basis of the type of wavelength selector used (filter and spectrophotometers); another is based upon whether one or two light beams are used (single or double beam instruments); and a third is based upon whether the result is automatically plotted by a recorder or whether the reading must be made visually from a meter of some sort. A further classification is based upon the region of the spectra which is being measured (ultraviolet, visual, infrared). No matter what type of photometer is being used, all have the same basic components, which are illustrated in the simplified diagram of a single-beam photometer, shown in Fig. 14-5.

Light from some source passes through an entrance slit into the instrument and is focused by use of mirrors and lenses through a wavelength selector, the sample cell, and an exit slit onto a photosensitive device, the output of which is fed through the proper circuitry to some sort of a readout. Each of these main features will be discussed separately.

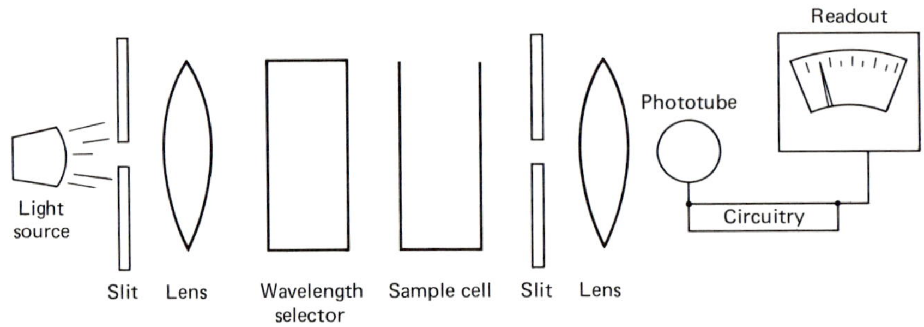

Figure 14.5. Simplified schematic diagram of a single-beam photometer.

1. Light Sources

Light sources vary with the region of the spectra being utilized. In the ultraviolet region, the light source is usually a hydrogen discharge lamp; in the visible region, a tungsten lamp is used; and in the infrared a globar or Nernst glower furnishes the radiant energy. The intensity of the light emitted by the source is a function of the potential applied across the source. As a consequence, the potential must be controlled carefully between narrow limits for reproducible results. In the visible region, the tungsten lamp is operated at either 6 or 12 V dc and most instruments incorporate a constant voltage transformer in the system to help maintain a constant potential. A variation of more than a few thousandths of a volt may cause a variation in the photocurrent from the detector of more than 0.2%.

2. The Optical System

The optical system consists of the slits, lenses, mirrors, cell, and wavelength selectors, but the wavelength selectors will be discussed separately in the next section. The entrance slit on all photometers is variable in width and is used to control the amount of light entering the photometer. It is utilized to set the 100 per cent transmittance or zero absorbance for the instrument at the beginning of a set of measurements.

The exit slit helps to govern the band width of the light entering the detector. It is fixed in less expensive instruments and is variable in more expensive instruments. When fixed, it usually allows a band width of 20–40 nm to pass; when variable, the slit may be controlled in many instruments to allow band widths as small as 5 nm or less.

The lenses are used to focus the light and mirrors may be used to extend the path length inside the instrument or to change the direction of the light beam. The sample cells, cuvettes, are made of optical glass, plastics, or quartz. The path length through the cell must be controlled carefully for reproducible results between several cells. The outside surfaces must not reflect or diffract too much light and must always be kept scrupulously clean. Most cuvettes are square cylinders with a path length of 1.00 cm, but round or rectangular cuvettes can be used. In IR photometers, the optics and cells are often made of sodium chloride or other soluble salts.

3. The Wavelength Selectors

The wavelength selectors are used to produce some degree of monochromatic light. Since the Beer and Lambert laws are based upon the use of monochromatic light, the use of polychromatic light will often decrease the concentration range in which the laws will hold. The efficiency of the wavelength selector is measured by the maximum percentage of incident light which will be transmitted and upon the half-width of the wavelength band that is allowed to pass. The half-width is the width of the band at one-half the height of the maximum transmittance. Wavelength selectors are classified as filters or monochromators.

a. FILTERS. Filters may be made of colored glass, gelatine, paper, solutions, polyethylene, thin layers of metals, or other similar materials. A comparison of the efficiency of three types of filters is given in Fig. 14-6. Filters made of colored glass usually have the lowest maximum per cent transmittance and the widest band pass and consequently are the poorest filters. Combination filters made of two or more interference filters show

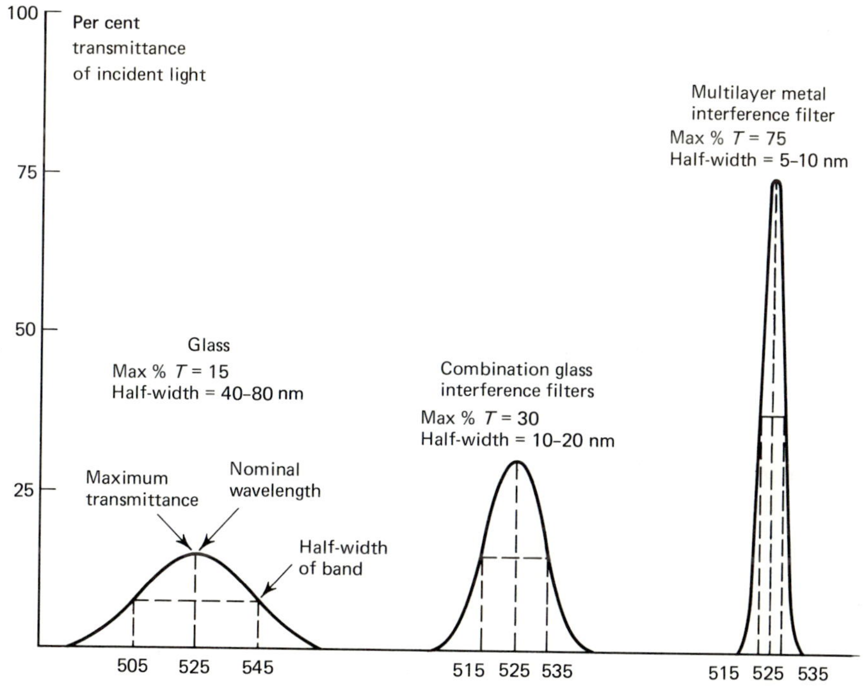

Figure 14.6. Efficiency of filters.

almost twice the efficiency of glass. An interference filter is a filter which goes from zero transmittance to maximum transmittance over a short wavelength span so that the combination of a filter which shows increasing transmittance with one which shows decreasing transmittance in the same wavelength region is more efficient than colored glass alone. Multilayer metal interference filters are the most efficient filters, both as to peak transmittance and band pass. The simplest example of this type of filter consists of two highly reflecting but partially transmitting films of silver separated by a spacer film of nonabsorbing material. This combination is deposited on a glass plate by high vacuum methods and a cover plate is then cemented on for protection. Light incident upon the filter is reflected back and forth between the metal films so that only light which has a wavelength of twice the depth of the spacer will pass through the filter. More complex filters are made by alternating the spacer and metal films until as many as 25 layers are obtained. Such filters allow the passage of some second and higher orders of wavelengths, but these may be screened out by glass blocking filters if necessary.

b. MONOCHROMATORS. The two types of monochromators used in photometers are prisms and diffraction gratings. Prisms may be made of glass, quartz, or other materials and may have various shapes. Most prisms are triangular in shape and are able to separate polychromatic light into individual wavelengths since light is bent upon passing from one medium to another and the extent of bending is a function of the wavelength. Rotation of the prism in the light path allows the focusing of any particular wavelength through the sample and the exit slit. Diffraction gratings consist of a large number of closely spaced parallel lines ruled upon a surface. The surface can be flat or curved, and gratings are classified as plane gratings or convex gratings depending upon the shape. When light strikes a diffraction grating it is diffracted and the extent of diffraction depends upon the wavelength. As a consequence rotation of a grating in the light path also allows light of a particular wavelength to be focused through the sample and the exit slit.

Photometers which use filters are classified as filter photometers, and those using a prism or grating are classified as spectrophotometers. Filter photometers can operate only at the wavelengths of the filters available, but spectrophotometers may operate at any wavelength simply by rotation of the prism or grating to the proper position. Original gratings are very expensive and are not used except in the most expensive instruments. Several companies, however, use replica gratings in comparatively inexpensive spectrophotometers satisfactorily. Many of the moderately expensive instruments which formerly used prisms are now using high-grade replica gratings.

4. Light-Sensitive Devices

Some light-sensitive devices used in the visible region may also be used in the ultraviolet region but cannot be used in the infrared region since they do not respond to the longer wavelengths of light found in the infrared region. Only those which can be utilized in the visible and ultraviolet regions will be discussed.

a. VACUUM TUBE PHOTOCELLS (PHOTOEMISSIVE CELLS). The vacuum phototube contains a semicylindrical cathode or negative electrode and a wire anode sealed inside an evacuated glass tube. The surface of the cathode is coated with a metal such as cesium which will emit electrons when struck by light. The anode is maintained at a positive potential by means of an external circuit so that electrons emitted by the cathode will move to the anode and produce a flow of electrons or a current in the circuit. This current is amplified and fed to the readout device, which is usually a meter calibrated in per cent transmittance or absorbance. The flow of electrons from the cathode must be proportional to the intensity of the light striking the cathode so that measurements of the current will be related to the radiant power of the light.

b. BARRIER-LAYER CELLS (PHOTOVOLTAIC CELLS). A barrier-layer cell is composed of a thin sheet of a conducting metal upon which a thin layer of a semiconductor has been deposited. The semiconductor is then covered with a grid or screen of conducting metal such as silver which is connected through the external circuit to the backing plate. The semiconductor may be selenium or cuprous oxide deposited on iron or copper as the backing plate. The grid acts as the collector electrode. External power is not required since the cell produces a potential.

When light passes through the grid and strikes the semiconductor, electrons flow from the semiconductor to the grid. The current thus produced flows through the external circuit to the backing plate. The term *barrier layer* comes from the fact that electrons will not flow in the opposite direction from the grid to the semiconductor. The current is small but can be measured with a galvanometer or microammeter if the external resistance is not too great. Due to the low internal resistance, the current cannot be amplified readily. Under proper conditions, the amount of current is proportional to the intensity of the light striking the cell and thus can be used to measure the intensity of the incident light. Barrier-layer cells suffer from fatigue and should not be irradiated for extended periods of time. Proper design of the circuit and selection of conditions for measurement can eliminate this difficulty in most cases. Barrier-layer cells are normally used only in the visible region in instruments which furnish fairly high levels of illumination. The response of the selenium cell is approximately the same as the human eye, having the greatest sensitivity in the 500- to 600-nm region.

c. PHOTOMULTIPLIER TUBES. Photomultiplier tubes have been devised which have several dynodes at increasing potential differences so that the electrons from the photocathode can be increased by each dynode. The dynodes are made of materials which, because of the applied potential, will emit several electrons for each electron that strikes the surface, and, as a consequence, the number of electrons striking the photoanode is increased. The amplification can be as great as 10^6. Photomultiplier tubes can be used for low light intensities to increase the sensitivity. Since they are limited to a current output of a few milliamperes, they can be used only with very low-intensity light. Essentially, photomultipliers are vacuum-tube photocells which provide their own internal amplification.

G. Photometric Methods

1. Absorption Curves

Each substance capable of absorbing light absorbs only those wavelengths of light which have energies which can cause some transformation in the atom or molecule absorbing the light. These changes can be nuclear, electronic, or molecular, as discussed in Section C. The wavelengths at which any given substance will absorb are determined by measuring the absorbance or transmittance of a solution as the wavelength is varied. The graph of absorbance versus wavelength is known as the absorption curve or the absorption spectra of the substance. Absorption curves for permanganate ion and the orthophenanthroline complex of ferrous iron are shown in Fig. 14-7. Curves A and B of this figure represent the absorption curves of two solutions of potassium permanganate measured under the same conditions and differing only in concentration, with B being more concentrated than A. It should be noted that the shape of the curves is identical but that the greatest distance between the curves occurs at the peak at 525 nm. This peak represents the maximum absorbance for permanganate and is known as the wavelength of maximum absorbance or minimum transmittance. There is a greater change in absorbance per unit change of concentration at this wavelength than at any other wavelength. As a consequence the measurement of concentrations is more sensitive at the wavelength of maximum absorbance than at any other wavelength.

The greater sensitivity at the wavelength of minimum transmittance limits the concentrations which can be measured within the desired range of 20–80 per cent transmittance (0.70–0.10 absorbance). If a wider range of concentrations needs to be determined, the measurements can be made at a wavelength of 550 nm, at which the sensitivity is less. Still wider ranges (with still less sensitivity) can be measured at wavelengths of 510 and 590 nm at which "humps" appear on the spectra. Measurements should not be made at wavelengths on the steepest part of the curve since minor errors in wavelength setting could cause major errors in the measured absorbance of the solution.

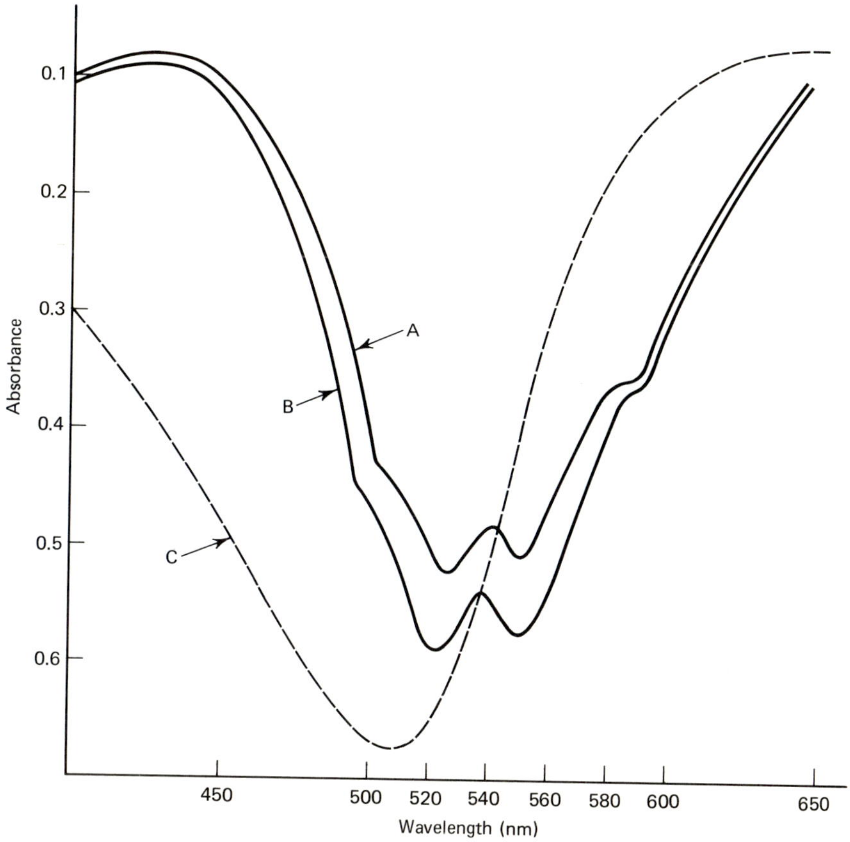

Figure 14.7. Absorbance curves. **A** and **B**, Potassium permanganate of different concentrations. **C**, Iron(II) orthophenanthroline complex.

2. Determination of Concentration from Absorbance

a. BY CALCULATION. The concentration for any given unknown solution can be calculated from the measured absorbance of the solution, and the absorbance of a standard solution determined under identical conditions

$$A = abc \tag{6}$$

so that

$$A_u = a_u b_u c_u$$

and

$$A_s = a_s b_s c_s$$

where the subscripts u and s refer to the unknown and standard solutions, respectively. Since the solutions are measured under the same conditions in cells of equal light path, $a_u = a_s$ and $b_u = b_s$, and the ratio of absorbances becomes

$$\frac{A_u}{A_s} = \frac{a_u b_u c_u}{a_s b_s c_s} = \frac{c_u}{c_s} \tag{13}$$

or

$$c_u = \frac{A_u c_s}{A_s} \tag{14}$$

It should be remembered that eq. (14) does not hold for transmittance and that readings in T or per cent T must be converted to A ($-\log T = A$) before the calculation can be made.

b. CALIBRATION CURVES. When a large number of repetitive samples are being run, it is preferable to construct a calibration or standard curve so that concentrations may be read directly from the graph. Standard curves are obtained by measuring the transmittance or absorbance of a series of standard solutions in the range of the solutions being measured in the determination. Each standard solution should be prepared in identically the same way as the solutions being measured; that is, all the reagents added to the sample should be added to the standard and the standard should be taken through all the steps involved in the sample determination. The transmittance or absorbance values are then plotted against concentration. Absorbance values may be plotted on regular rectilinear graph paper, but transmittance or per cent transmittance must be plotted on semilogarithm paper. The use of the logarithmic scale for transmittance obviates the necessity of converting transmittance to absorbance before using the graph. A representative calibration is shown in Fig. 14-8 plotted on semilog paper as percent T versus concentration and on regular paper as A versus concentration. It should be noted that the two calibrations will superimpose on each other when equivalent units are used for A, T, and concentration.

The meters or readout devices of most photometers are designed so that either per cent transmittance or absorbance can be read. In most cases, fewer mistakes are made if the per cent T scale is used since it reads from 0 on the left to 100 on the right in equal divisions. The absorbance scale reads from 2 on the left to 0 on the right and the divisions vary in size throughout the scale. Since the photometers are set on 100 per cent T or 0A using the solvent or a blank solution, the zero concentration-zero absorbance (100 % T) point should be considered as a measured point in constructing the graph. Ordinarily, the best straight line is drawn through all the points,

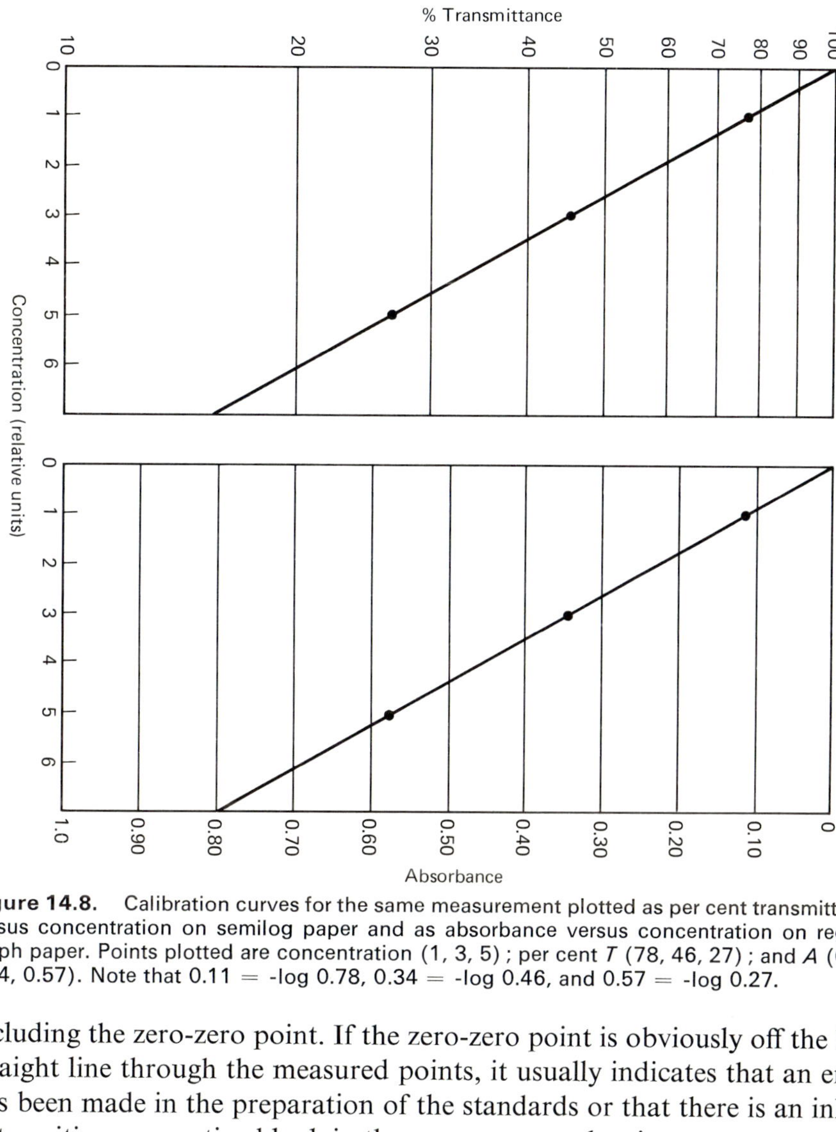

Figure 14.8. Calibration curves for the same measurement plotted as per cent transmittance versus concentration on semilog paper and as absorbance versus concentration on regular graph paper. Points plotted are concentration (1, 3, 5); per cent *T* (78, 46, 27); and *A* (0.11, 0.34, 0.57). Note that 0.11 = -log 0.78, 0.34 = -log 0.46, and 0.57 = -log 0.27.

including the zero-zero point. If the zero-zero point is obviously off the best straight line through the measured points, it usually indicates that an error has been made in the preparation of the standards or that there is an inherent positive or negative blank in the measurement that is not accounted for by the blank used to set the instrument.

3. Representative Procedures

There are such a large number of photometric determinations being performed in various types of laboratories that a complete listing is not possible in this text. Two procedures will be discussed as representative examples: the determination of manganese in steel or cast iron and the determination of iron in limestone.

a. DETERMINATION OF MANGANESE IN CAST IRON OR STEEL. Most directions call for the sample of iron or steel to be dissolved in nitric acid and any insoluble residue to be removed by filtration. A mild oxidizing agent such as ammonium persulfate is then added to oxidize carbon compounds which might interfere with the photometric measurement. The manganese present is then converted to the permanganate by oxidation with a strong oxidizing agent such as potassium meta-periodate at elevated temperatures. Phosphoric acid is added to the cooled solution, which is then diluted to a specified volume. The phosphoric acid converts the iron to a colorless complex which does not interfere with the measurement. The absorbance or transmittance of this solution is measured at 525 nm, and the concentration or manganese is read from a previously prepared calibration graph. Alternatively, a standard prepared by diluting standard permanganate or by using a standard iron or steel sample can be measured simultaneously and the concentration calculated. Chloride ion must be absent since it will decolorize the permanganate due to reaction to form elementary chlorine. The measurement can also be made visually by the standard series method or by the method of balance using a Duboscq colorimeter.

b. DETERMINATION OF IRON IN LIMESTONE. One of the most sensitive methods for the photometric determination of iron uses the intense red color developed by the iron(II) (ferrous) ion reaction with 1,10-phenanthroline (orthophenanthroline) to form the 1:3 molar ratio complex which has the formula

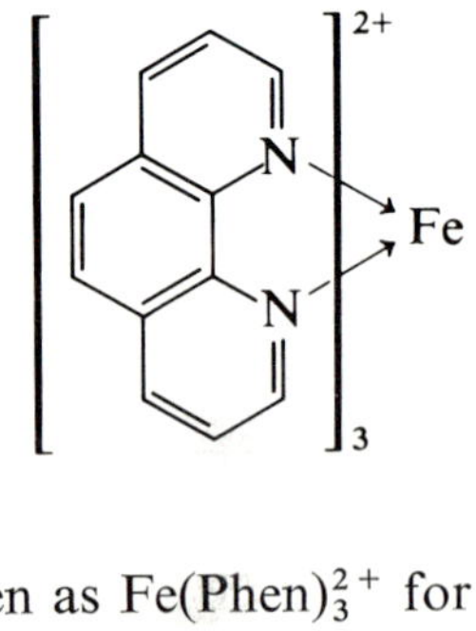

and which is usually written as $Fe(Phen)_3^{2+}$ for convenience. This complex has a molar absorptivity at 510 nm of approximately 11,000 so that it can be used for small quantities (1–10 mg/liter) of iron when using 1-cm cuvettes. 1,10-Phenanthroline also forms a complex with iron(III) (ferric) ions, which has a pale blue color. As a consequence, all the iron must be reduced to the ferrous state to ensure proper results. The *p*H is also important since full color development will occur only in the proper *p*H range. Most directions call for the use of an acetate buffer for *p*H adjustment.

The amount of iron in limestone varies considerably, and any procedure for the determination must be able to account for widely different iron concentrations. In many limestone analyses, the iron and aluminum are precipitated with ammonia after the removal of the acid-insoluble (silica) material.

This precipitate can be ignited and weighed and reported as combined aluminum and iron oxides, or it can be dissolved and the iron determined by any satisfactory titrimetric or photometric procedure.

Photometric procedures usually call for dissolving the precipitate of iron and aluminum oxides in dilute hydrochloric acid and diluting to a specified volume. Various sized aliquots of this solution can then be utilized for the development of color so that the absorbance is in the proper range (0.1–0.7) for satisfactory photometric determination. The iron in the aliquot is reduced by the addition of an excess of hydroxylamine hydrochloride followed by addition of an excess of orthophenanthroline and a concentrated acetate buffer. After dilution to volume, the absorbance or transmittance is read at 510 nm and the amount in the aliquot is determined from a previously prepared calibration curve. The amount of iron in the sample is then found by multiplication by the proper dilution factor.

H. Problems

1. Convert the following transmittance measurements to the corresponding absorbance: (a) 0.355*T*, (b) 35.5%*T*, (c) 0.368*T*, (d) 70%*T*, and (e) 0.200*T*.
   Ans. (b) 0.450
2. Convert the following absorbance readings to the corresponding per cent transmittance values: (a) 0.699, (b) 0.109, (c) 0.500, (d) 0.434, and (e) 0.222.
   Ans. (b) 77.8%
3. Calculate the molar absorptivities of 0.0050 *M* solutions which were measured in 1-cm cells and which had the absorbance values listed in problem 2.
   Ans. (b) 21.8
4. Calculate the absorptivity in liters per gram centimeter, for 0.01% (w/v) solutions which were measured in 5.00-cm cells and which showed the transmittance values listed in problem 1.
   Ans. (b) 0.900
5. Calculate the range of molar concentrations which can be used to produce transmittance readings between 20.0 and 60.0% if 1.50-cm cells are to be used and the compounds have the molar absorptivities listed: (a) 11,100, (b) 5500, (c) 25, (d) 365, and (e) 20,200.
   Ans. (b) 2.69×10^{-5} to 8.47×10^{-5}
6. (a) In the colorimetric determination of Mn in steel, the following data were obtained: weight of unknown, 0.65 g; weight of standard steel containing 0.40% Mn, 0.96 g. Both solutions were diluted to 250 ml before comparison in a Duboscq colorimeter. The standard was set at 37.3 mm and the following readings were obtained for the unknown: 34.7, 34.5, 34.3, 34.7, 34.4, and 34.4. Calculate the per cent Mn in the unknown.
   (b) The standard and unknown solutions in part (a) were then measured photometrically in identical cells and the transmittance of the standard was 31.0% while that of the unknown was 28.1%. Calculate the per cent Mn in the unknown.
   Ans. 0.639

7. Repeat problem 6(a), assuming that the standard contained 0.35% Mn and that 0.85 g was used.
8. Repeat problem 6(a), assuming that the standard contained 0.54% Mn and that 0.59 g was used. Also assume that the standard was set at 25.0 and that the readings on the unknown were 26.2, 26.4, 25.8, 26.0, 26.6, and 26.2.
9. After development of the $KMnO_4$ color and dilution to 250 ml, the absorbance of a 1.00-g sample of cast iron at 525 nm was 0.484 using a 1.00-cm cell. Using the same cell and wavelength, the absorbance of a solution prepared by diluting 5.00 ml of 0.0185 *M* $KMnO_4$ to 250 ml was 0.400. Calculate the per cent Mn in the cast iron.

   Ans. 0.614
10. Repeat problem 9, assuming that 5.00 ml of 0.0205 *M* $KMnO_4$ was used.
11. Repeat problem 9, assuming that the absorbance of the sample was 0.434 and that of the standard was 0.415.
12. A spectrophotometer was calibrated with standard solutions of glucose for which the concentrations were adjusted so that they had the same concentrations as protein-free filtrates from blood samples and were expressed as milligram per cent in the original blood sample. The following data were obtained:

Concentration (mg %)	% *T*	Concentration (mg %)	% *T*
50	84.0	150	59.0
75	76.8	200	49.0
100	70.2	250	41.0

A glucose tolerance test was run on a patient and the original blood sample before administration of glucose gave a reading of 68.4% *T*. A later sample, after dilution 1:1 with water before precipitation, gave a reading of 45.0% *T* Determine the milligram per cent glucose in each blood sample.

Ans. 108, 448

13. The transmittances of a series of $Cr(NO_3)_3$ solutions were determined using 1-cm cells at 550 nm and used to prepare a calibration curve for chromium solutions. The following data were recorded:

Concentration (*M*)	Transmittance	Concentration (*M*)	Transmittance
Blank	1.000	0.0300	0.387
0.0100	0.725	0.0400	0.281
0.0200	0.530	0.0500	0.207

(a) Graph the calibration curve as *T* versus *M* on semilog paper.
(b) Convert *T* to *A* and plot *A* versus *M* on rectilinear paper.
(c) A 1.234-g sample of chromite ore ($FeO{\cdot}Cr_2O_3$ + impurities) is dissolved and after proper treatment is diluted to 100.0 ml. The transmittance of this solution was 0.399. Read the concentration of chromium from each graph. Using the transmittance curve, calculate the percentage of chromite in the ore. Using the absorbance curve value, calculate the percentage of chromium in the ore.

14. In the determination of acetone in biological fluids, the following calibration curve between absorbance and concentration was obtained:

Acetone standards (mg/100 ml)	*A*	Acetone standards (mg/100 ml)	*A*
Reagent blank	0.035	4.0	0.127
0.5	0.047	6.0	0.172
1.0	0.059	8.0	0.219
2.0	0.082	10.0	0.264

Samples of blood and urine from a normal subject and from a ketotic patient were analyzed with these results: normal blood, $A = 0.068$, ketotic blood, $A = 0.189$; normal urine, $A = 0.097$, ketotic urine, $A = 0.198$ (1 to 25 dilution). Calculate the acetone concentration in each sample.

15. A standard chromium (III) solution was prepared by reduction of 5.00 ml of 1.200 *N* $K_2Cr_2O_7$ followed by dilution to 100 ml. This solution showed an absorbance of 0.310 when measured at 440 nm in a 1.00-cm cell. A 1.000-g sample of a chromium ore was dissolved, treated to ensure that all the chromium was in the trivalent state, and diluted to 100 ml. This solution showed an absorbance of 0.336 when measured in the same cell under the same conditions. Calculate the per cent Cr in the ore.

CHAPTER 15

Chromatography and Ion Exchange

A. Chromatography

Under favorable circumstances, a solution containing a mixture of two or more components can be poured through a column of a finely divided solid and the substances will be adsorbed on the solid at the very top of the column. Continued washing of the column with a liquid may cause the components to move down the column at different rates and to be separated into individual zones if the column is long enough. Such separations are called chromatographic separations and the process is known as chromatography. Complete and partial separation of components on a column is illustrated in Fig. 15-1. Column 1 of Fig. 15-1 represents the adsorption of the components of the mixture at the top of the column, column 2 shows the partial separation of the components as they move down the column, and column 3 indicates the complete separation into individual zones moving at their own characteristic rate down the column as more liquid is added.

Chromatographic separations are based on physical properties such as adsorption rather than chemical changes such as precipitation, complex formation, neutralization, or oxidation-reduction, and, as a consequence, the original substances are available for testing after the separation. All chromatographic methods involve a mobile phase and an immobile phase and are classified as being differential migration methods. The mobile phase usually provides the driving force which moves the components to

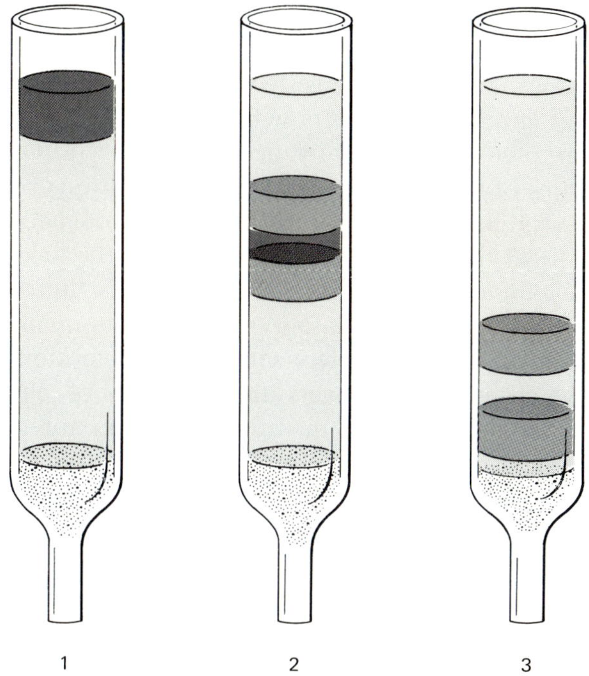

Figure 15.1. Column chromatography. **1**, Adsorption of mixture at top of column. **2**, Partial resolution of components. **3**, Complete resolution of components.

be separated through the medium, and the immobile phase usually provides the selective retarding force which causes each component to move through the medium at its own rate. The mobile phase can be liquid or gas, and the immobile phase is usually a finely divided solid or a finely divided solid coated with a thin layer of a liquid. The driving force is the flow of the liquid or gas, and the retarding forces are due either to adsorption on a solid or to partition between two fluids (gas–liquid or liquid–liquid).

Chromatographic methods can be classified as to the mobile phase (gas or liquid chromatography), as to the retardation mechanism (adsorption or partition), and as to the equipment used (column, thin-layer, paper).

The first reported chromatographic separation was the separation of plant pigments by the Russian botanist, Tswett, in 1906. Tswett allowed a petroleum ether extract of plant leaves to percolate through a column of finely divided calcium carbonate and noticed that the pigments were strongly adsorbed near the top of the column and were separated into yellow and green bands. Continued washing of the column with the petroleum ether solvent caused movement of the bands down the column with still further separation or resolution.

Tswett called the process chromatography since he considered it to be the separation of colored materials by selective adsorption. It is now known that colorless materials can also be separated by the process; in fact, chromatography is used to separate more colorless materials than colored. Several other names have been suggested from time to time but none has replaced chromatography, and it is now firmly fixed in the literature.

Chromatography was not really developed as a separation method until after the work of Kuhn and Lederer in the separation of plant carotenes in 1931. There was gradual acceptance of the method between 1931 and 1943, and since 1943 the number of publications in the field has increased rapidly with constant acceleration so that now literally thousands of papers which include some type of chromatography appear annually.

There is probably no completely satisfactory definition of chromatography. One which probably covers the majority of chromatographic methods is that chromatography is the resolution of mixtures by differential migration from a narrow zone in porous media, the migration being caused by flow of liquid or gas. This definition does not include electrophoresis in which the migration is caused by electrical potential. However, it is generally applicable since there is no mention of the retarding mechanism which causes the differential movement.

A complete discussion of all phases of chromatography is beyond the scope of this book. The student is referred to various advanced analytical texts and to reference works on the general subject and on particular aspects of the various types of chromatographic separation. We shall consider only the simpler aspects of column chromatography, paper chromatography, and gas chromatography.

1. Column Chromatography

a. SEPARATION MECHANISM. The following discussion is based upon the premises that the retarding force is selective adsorption upon a solid, that the driving force is the flow of a liquid, and that the column can be considered to be composed of a series of "theoretical plates" in which equilibrium is attained between the adsorbed substance and the dissolved substance.

In any adsorption process, equilibrium is attained between the dissolved and adsorbed solute, and the relation between the two is given by the typical adsorption isotherm, shown in Fig. 15-2. From the isotherm it is apparent that the extent of adsorption increases with concentration and that there is a greater rate of change of extent of adsorption per unit change of concentration at low concentrations.

If a solution of concentration c is equilibrated with a given amount of solid adsorbent, the concentration in the solution will be decreased to c_1 due to adsorption of m_1 of the solute upon the adsorbent. If the solution is removed by filtration and again equilibrated with fresh solid adsorbent, the concentration will be reduced to c_2 due to adsorption of m_2 of solute

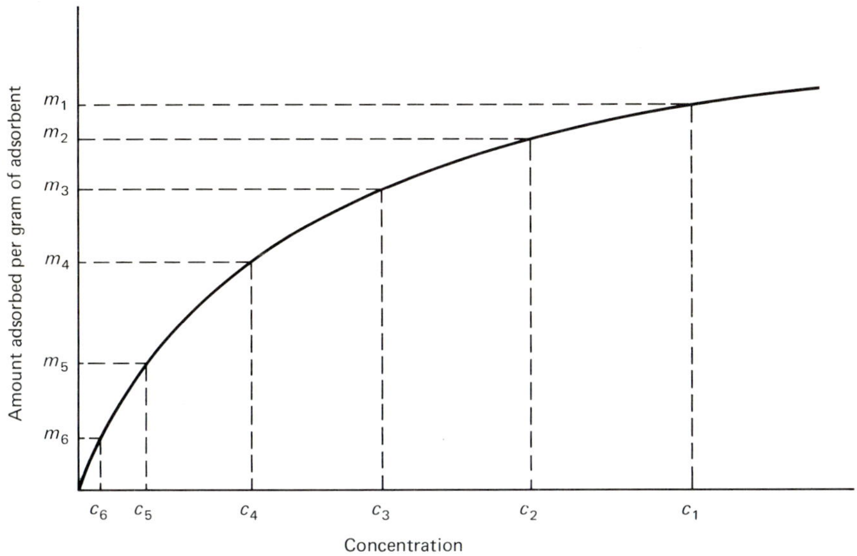

Figure 15.2. Removal of solute from solution by a series of batch adsorptions using fresh adsorbent for each step. The solution and adsorbent are separated by filtration after each stage. The concentration of the solution is decreased in each step until only a negligible amount of solute remains in the solution.

on the adsorbent. If this process is continued, the concentration in the solution will be reduced to a negligible value after a sufficient number of equilibrations have been made.

Assume that the solution is poured down a column of solid adsorbent instead of being equilibrated successively with different batches of solid adsorbent. Also, assume that the column is composed of a number of thin layers or plates, each of which acts as one batch in successive batch separations. That is, assume equilibrium is attained between the dissolved solute and adsorbed solute in each layer. As the solution moves through each layer, some solute is removed by adsorption until a negligible amount remains, as indicated in Fig. 15-3A.

If pure solvent is now poured through the column, some of the solute will be desorbed from the first layer to establish equilibrium between the dissolved solute and the adsorbed solute. This will continue as the solvent (now a solution) passes through the successive plates until the concentration leaving one plate exceeds the equilibrium concentration in the next plate (note plate 5 of Fig. 15-3B). In this plate some of the dissolved solute will be adsorbed, and the concentration will be decreased to the proper equilibrium value. As a consequence, the solute will be moved down the column by decreasing the amount adsorbed at the top of the zone and increasing the amount adsorbed toward the bottom of the zone, as indicated by Fig. 15-3B.

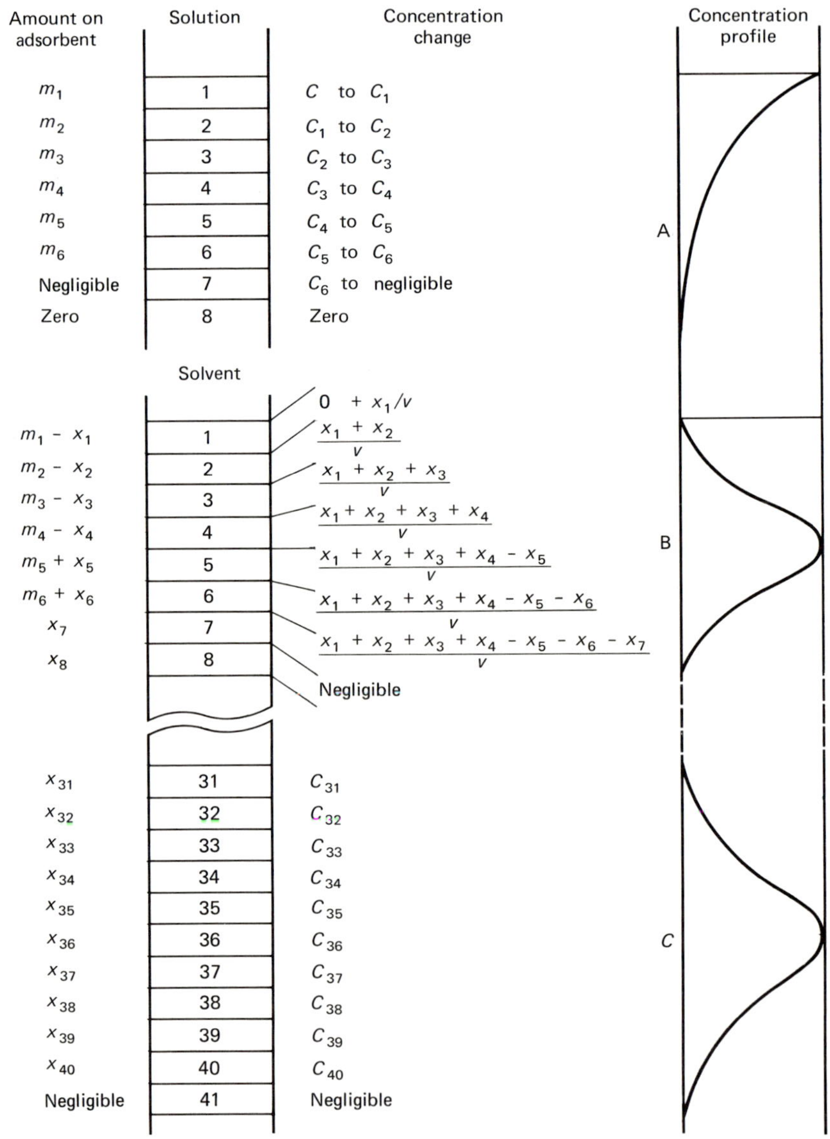

Figure 15.3. **A**, Adsorption on column. **B**, Effect of solvent. **C**, Distribution as zone moves down column.

As more solvent is poured through the column the substance is moved down the column and is distributed between the solid and the liquid, as shown in Fig. 15-3C. It should be noted that the concentration distribution in the zone does not follow the normal distribution curve since the distribution curves in Fig. 15-3 indicate definite tailing. The tailing is caused primarily by the lateral diffusion in the solution and by failure to attain equilibrium in each plate.

For separations to be effective, the rate of movement of the individual components must be sufficiently different and the column must be long enough. The rate of movement of a substance down a column may be stated in several different ways. One method is the *R* factor, which is defined as the ratio between the rate of movement of a zone and the rate of movement of the solvent. To be uniform, the leading edges are usually measured and stated as the R_f factor, or

$$R_f = \frac{\text{rate of movement of leading edge of solute zone}}{\text{rate of movement of leading edge of solvent}} \tag{1}$$

Since rate equals distance divided by time and since the time is the same for both the zone and the solvent, distance can be substituted for rate in eq. (1) to obtain

$$R_f = \frac{\text{distance moved by leading edge of solute zone}}{\text{distance moved by leading edge of solvent}} \tag{2}$$

The R_f factors are characteristic of the solutes and the conditions under which the separation is performed and can be used to identify substances if conditions are reproduced.

b. METHODS OF OPERATING CHROMATOGRAPHIC COLUMNS. Methods of operation of chromatographic columns are usually based upon the liquid poured through the column after the mixture to be separated has been adsorbed on the column. Use of the original solution is known as frontal analysis, use of a different solvent is referred to as displacement analysis, while use of the same solvent is called elution analysis. Each of these will be discussed separately.

(1) *Frontal analysis.* When the sample solution is continuously poured through the column, the component which is least strongly adsorbed moves through the column faster than the other components, appears first in the eluate, and is present in the eluate from that time on. After a period in which the eluate contains only the fastest-moving component, the second will also appear and be continuously present in further eluates. A graph of the determination of the components in the eluate (known as a chromatogram) is shown in Fig. 15-4a. Frontal analysis produces a small portion of pure A but cannot be used as a quantitative separation method since the eluate has the same composition as the original solution after a certain volume has passed through the column. However, it does give information as to the minimum number of components which can be present in the solution.

(2) *Displacement analysis.* In displacement analysis, the mixture to be separated is adsorbed in a narrow zone at the top of the column.

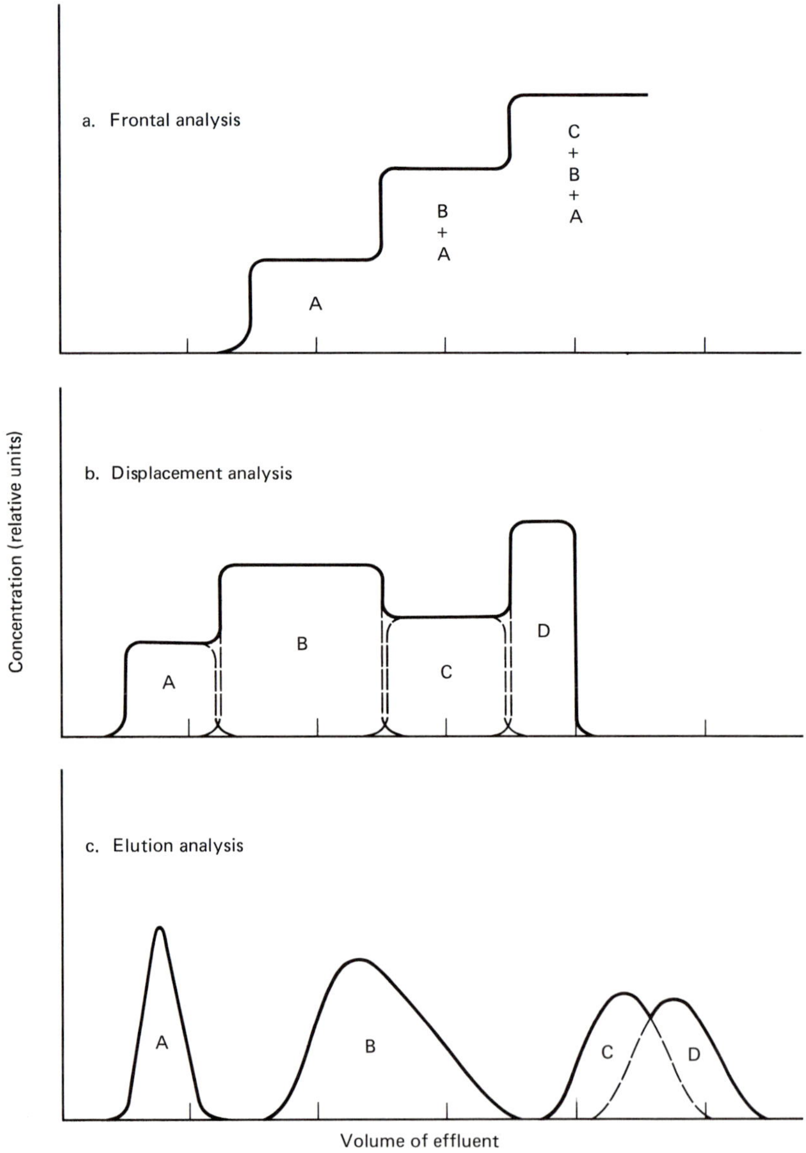

Figure 15.4. Typical chromatograms from different types of column operation.

A solvent (or a solution) which is more strongly adsorbed by the adsorbent than any of the components is then poured through the column. All the components of the mixture are displaced by the developing solvent or solution and in turn displace each other as they are moved down the column. The least strongly adsorbed moves the fastest since the other components displace it. As a result, the components are separated into adjacent zones which move through the column ahead of the developing liquid. Theoretically each component should move through the column as a zone of pure material and thus appear as pure material in the eluate. In practice there is a slight overlapping of the zones so that the method cannot be used as a quantitative separation measure. It does give information as to the minimum number of components and can be used to determine substances quantitatively since the length of the zone (and thus the volume of eluate which contains it) is a measure of the amount of the substance present in the mixture. A typical chromatogram is shown in Fig. 15-4b.

(3) *Elution analysis.* Elution analysis is the type of chromatography introduced by Tswett in that pure solvent is poured through the column after the sample is adsorbed in a narrow zone at the top of the column. Each component moves down the column at a rate which is a measure of the strength with which it is adsorbed. Elution analysis can be used to resolve very complex mixtures if the column is sufficiently long and the flow rate is slow enough that equilibrium can be established in each theoretical plate. Each solute is distributed in a zone which moves down the column at a characteristic rate and can be determined in the eluate as it comes out the bottom of the column. A typical chromatogram is shown in Fig. 15-4c, which shows two components appearing as completely separated zones and two as incompletely separated zones. The longer the column or the smaller the R_f factor, the more diffuse is the zone.

Elution analysis can be used to separate and determine the components of a mixture since the area under the curve in the chromatogram is a measure of the amount of the substance present in the mixture. Also, since the R_f factors are characteristic and reproducible under controlled conditions, each component can be determined in the eluate by analyzing that portion of the eluate in which it is contained. Elution analysis is by far the most popular type of column chromatography.

(4) *Partition analysis.* Partition chromatography is similar to elution chromatography but differs in the mechanism of selective retardation. In adsorption the retarding force is the selective adsorption of the solutes on the solid, while in partition the retarding force is the distribution of the solutes between two liquids. As a consequence, partition uses two immiscible solvents instead of a solid and a solvent. One solvent is coated on an inert solid and is used as the immobile phase, while the other is used as the mobile phase.

A solute which is soluble in two immiscible liquids will be distributed between the two liquids when the liquids are in contact. The ratio of concentrations in the two liquids is known as the distribution or partition coefficient and is a constant at a given temperature.

$$K_d = \frac{c_1}{c_2} \tag{3}$$

The symbols P, D, and α are often used for the distribution coefficient. The ratio of concentrations is often approximately the same as the ratio of the solubilities of the solute in the two solvents, but there are other factors in addition to solubility which affect the value of the distribution coefficient.

As the solutes are washed through the column, they are continuously partitioned between the immobile liquid phase and the moving solvent phase and consequently move through the column at their own individual rates, which are controlled in part by the partition coefficients. A differential migration is thus set up which allows separations to be effected.

The relationship between the rate of movement and the partition coefficient is given by

$$R_f = \frac{A_m}{A_m + PA_s} \tag{4}$$

where

$$R_f = \frac{\text{distance solute zone migrates}}{\text{distance solvent front migrates}}$$

$$A_m = \text{cross-sectional area of mobile phase}$$

$$A_s = \text{cross-sectional area of immobile phase}$$

$$P = \frac{\text{solute concentration in immobile phase}}{\text{solute concentration in mobile phase}}$$

This relationship can be used to calculate R_f factors from known partition coefficients, and the calculated values agree well with measured values.

For column partition separations to be effective, the two solvents must be essentially insoluble in each other, the flowing solvent must be saturated with the immobile liquid, and the substances to be separated must be soluble in both liquids and must have different distribution ratios.

2. Thin-Layer Chromatography (TLC)

Thin-layer chromatography is similar to column adsorption chromatography with the exception that it is performed using a thin layer of adsorbent (0.025 μm or less) spread evenly over a glass plate. The sample is deposited as a small spot and the solvent is allowed to move upward over the spot, usually by capillary action. Since the thin layer is used, sample sizes must be much smaller than with columns, and the time required for separations is much shorter. The general principles which apply to separations on a column also apply to thin-layer chromatography so that thin-layer experiments are often used to determine proper column conditions before the column separation is run. Thin-layer separations are especially beneficial when the sample size is limited. The size of the spot made by a component can be used to estimate the amount present if a series of standards is run on the same plate. Replication from plate to plate is not precise since it is difficult to prepare plates which have identical thicknesses of adsorbent layer.

Plates are usually prepared by spreading a thin layer of a slurry over a glass plate. The layer is then smoothed out by some device which ensures approximately equal thickness of the desired depth over the entire plate, and the plate is dried at room or elevated temperatures. Various devices are on the market which are of aid in the preparation of uniform plates and several companies market prepared plates of constant thickness and composition.

The chromatograms must be run in a closed system to prevent evaporation of the solvent as it flows through the adsorbent. The closed system can be a simple wide-mouth jar with a screw top, a device which allows a cover plate to be used above the thin layer in a sandwich arrangement with a small air space between the adsorbent layer and the cover, or a thermostated closed container which may cost up to several thousand dollars.

3. Paper Chromatography

The principles of paper chromatography are similar to the general principles of column chromatography with the exception that paper is used as the solid immobile phase in place of a finely divided solid adsorbent. The sample is spotted on the paper as with thin-layer chromatography, and the solvent or mobile phase is allowed to flow over the spot to cause the separation.

The exact retarding mechanism which causes the differential migration is not known. In some cases it appears that the main mechanism is adsorption on the surface of the cellulose fibers; in others it is thought to be due primarily to partition between the flowing solvent and the water which is adsorbed on the paper; while in still others ion exchange seems to play an important role. Since it is difficult to differentiate between which of these possible retarding mechanisms is the actual cause of the differential migration, it is usually assumed that all three play a part in all separations.

The partition viewpoint is substantiated by the fact that some R_f factors can be predicted satisfactorily from known partition coefficients. However, this is not true for a large number of cases. Carboxyl groups are formed on the cellulose during processing or manufacture, and the number can be increased by certain treatments. These carboxyl groups can furnish sites for ion exchange of metal or organic ions and thus play a part in the retardation mechanism. The flow of solvent through the paper is by capillary action through the interspaces between fibers and in the capillary openings in the fibers themselves. Adsorption probably occurs in the fiber capillaries and also upon the surface of long fibers.

a. GENERAL METHODS. Chromatographic paper is usually available in large sheets (18 × 22 in.) or in 1-in. wide strips in rolls of several hundred feet. Since the "grain" of the paper has some effect upon the flow rates, the long direction of the paper which is the "machine" or "grain" direction should be used.

(1) *Ascending chromatography.* The sample is spotted about one-half inch from the end of the paper strip, and the strip is allowed to hang down with the lower one-fourth inch immersed in a liquid. The liquid flows upward by capillary action. The chromatographic chamber must be closed to ensure equilibrium between the liquid and vapor in the chamber to prevent evaporation of the solvent during the development of the chromatogram. The chamber can be a test tube, a milk jar, a specimen flask, or a commercially available chromatographic chamber, which can be very expensive. The effectiveness of the separation is limited by the fact that the solvents will rise only to a limited height by capillary action. The procedure is also known as countergravity separation.

(2) *Descending chromatography.* The sample is spotted several inches from one end of the strip, which is folded between the spot and the short end of the strip. The strip is suspended at the fold, and the short end is allowed to dip into a tank of the solvent being used. The solvent rises to the fold and then moves down the paper over the spot by capillary action. The strip must be protected from siphon action by a dry rod or container edge. This technique permits longer strips, which is the equivalent of lengthening the column in column chromatography. The chromatogram must be developed in a closed chamber to prevent evaporation of the solvent. Simple equipment, such as a graduate covered by a beaker, up to commercially available equipment which can run into several thousand dollars can be used.

(3) *Two-dimensional chromatography.* In two-dimensional chromatography a large sheet of chromatographic paper is used together with two solvents. The sample is spotted in one corner and run in one direction by one solvent. The chromatogram is then dried, turned 90°, and developed with the second solvent. The advantage of two-dimensional over single-dimension chromatography is that several substances which are not separated by the first solvent may be separated by the second so that the efficiency of separation is increased tremendously. Two-dimensional paper chromatography is used to separate complex mixtures of amino acids produced by hydrolysis of protein. Many complex mixtures which cannot be separated by any other method can be resolved effectively by two-dimensional chromatography by proper selection of the two solvents used. As in the other methods a closed chamber must be used.

b. EVALUATION OF CHROMATOGRAMS. The location of the various zones or spots which indicate the positions of the components of the mixture can be determined readily if all the components are colored. For colorless components, the separated components must be converted into colored compounds before location is attempted. This can be done by spraying the chromatogram with a color-producing reagent such as ninhydrin for amino acids. Other procedures involve streaking or dipping the chromatogram. In some cases radioactive tracers may be utilized, and the locations may be determined by covering the chromatogram with photographic paper followed by later development of the exposed photographic film.

The amounts of the various components can be evaluated by the size of the spot, the intensity of the color, or the amount of darkening of the photographic film. Qualitative identification is made by the R_f factor.

The R factor for paper chromatography is defined as the ratio of the rate of movement of the solute zone to the rate of movement of the solvent. The leading edge, the center of the spot, or the trailing edge can be used to make such ratio measurements. For uniformity, and since it is usually easier to measure and is more reproducible, the leading edge is usually used and the ratio is called the R_f factor. The R_f factor is defined as

$$R_f = \frac{\text{rate of movement of leading edge of zone}}{\text{rate of movement of leading edge of solvent}} \tag{1}$$

and since the time involved for both solute and solvent zones is the same, distances may be substituted for rates to give

$$R_f = \frac{\text{distance moved by leading edge of solute zone}}{\text{distance moved by leading edge of solvent}} \tag{2}$$

The R_f factors are reproducible as long as all conditions which affect them are held constant or are reproduced. They are also characteristic of the individual compound and thus can be used to identify substances as long as conditions are reproduced. Conditions which affect R_f factors include the solvent used, the paper used, the method of development, vapor equilibrium with the solvent, and the temperature.

Since two substances may have almost equal or identical R_f factors with a given solvent under a given set of conditions, reproduction of an R_f factor under one set of conditions is not considered absolute identification. However, if the substance has the same R_f factor as a known standard under several sets of conditions using different solvents, different types of paper, different temperatures, and different methods, it is essentially sure proof that the substances are identical.

4. Gas Chromatography

Gas chromatography is similar to column chromatography except that a gas is used as the mobile phase in place of a liquid. Gas chromatographic separations are based upon selective adsorption on a solid (called gas-solid chromatography or GSC) or upon partition between the gas and an immobile liquid phase (called gas-liquid chromatography or GLC). The latter is by far the most popular and has many more applications than gas-solid chromatography. Credit for discovery of column partition chromatography is usually credited to A. J. P. Martin in 1941. In the same paper, Martin described the possibilities of GLC, but essentially no work was done until after Martin published the first successful gas-liquid chromatographic separation in 1951. Since that time, the field has grown by leaps and bounds, and today there are several journals devoted solely to gas chromatography, as well as literally hundreds of papers each month in other periodicals which deal at least in part with GSC or GLC. In the early days, progress was hindered by slow development of the necessary instrumentation, but as the great possibilities of GLC as a separation method were realized and more and more applications were proposed, tremendous strides have been made in the development of immobile phase liquids, sensitive detectors, fast recorders, and proper circuitry for stable operation.

a. INSTRUMENTATION. All gas chromatographs, whether GSC or GLC, have the same basic components, which are shown in Fig. 15-5. The carrier gas from a tank is fed through a reducing gauge to the inlet or reference side of the detector. The sample is injected into the gas stream after it leaves the reference side of the detector and before it enters the column. After passing through the column, the gas, with the separated components of the sample, passes through the exit or sample side of the detector and then through a flow meter and the exit vent. The detector is connected to a recorder.

The gas is the driving force for the movement of the zones through the column, and the solid (for GSC) or liquid (for GLC) furnishes the selective retarding force. The detector may be considered as the brain of the chromatograph and the column as the heart.

Hydrogen, helium, nitrogen, and air are the most widely used gases. Hydrogen probably has more advantages than the others but is dangerous to use; helium is second best but is expensive; nitrogen is inexpensive but gives reduced sensitivity; and air is used only when the oxygen in the air is beneficial to the detector or separation.

The most popular detector used is the differential thermal conductivity dectector, which is based upon the fact that the temperature and thus the resistance of a wire through which a current is flowing is dependent upon

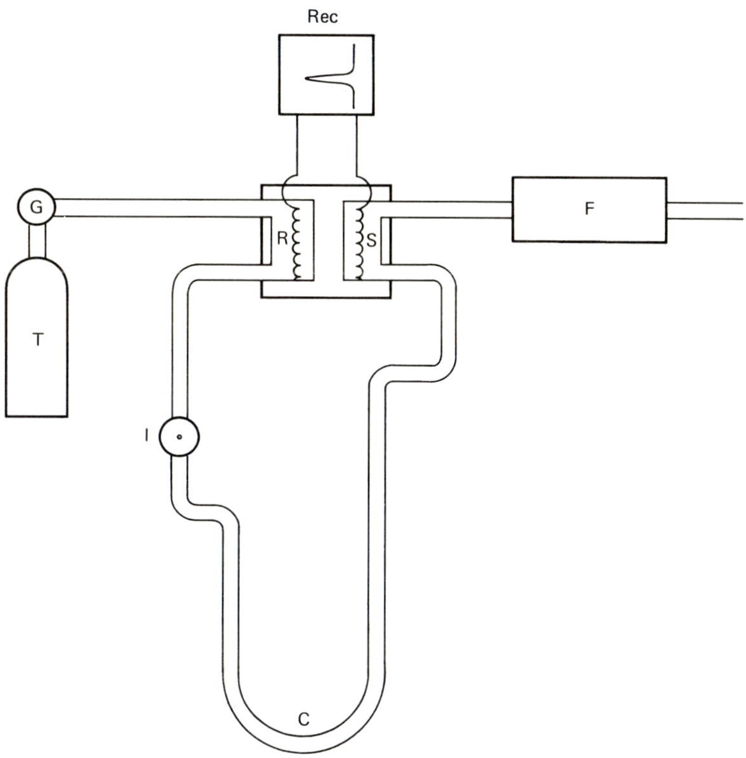

Figure 15.5. Typical components of a gas chromatograph. T, carrier gas tank; G, reducing valve and gauge; R, reference side of detector; I, sample injection port; C, column; S, sample side of detector; F, flow meter; and Rec, recorder.

the thermal conductivity of the gas in which it is immersed. The thermal conductivity of a gas is a function of its composition. The detector has two sides (known as the reference and sample sides) in which identical wires are used. These wires are connected to a Wheatstone Bridge and heated by the passage of a current. The carrier gas flows through the reference side before the sample is injected and thus the reference side always contains gas of the same composition. The gas flows through the sample side as it exits from the column and thus contains the separated components as well as the carrier gas. When there is no component exiting from the column, the gas in the two sides is identical, and the Wheatstone Bridge remains in balance since the wires have the same temperature. As soon as a sample component enters the sample side, the composition of the gas in the sample side differs from the reference side. This causes a change in temperature of the wire and an imbalance in the bridge. The circuit is designed to measure the extent of the imbalance and to feed this signal to a recorder. Since the degree of imbalance is a function of the concentration of the component in the carrier gas, the recorder draws the elution curve for the chromatographic separation.

The column can be constructed of glass or metal tubing and, for analytical work, is usually 4–8-mm i.d. It can be coiled, bent, or straight. The column is packed with the immobile phase—a solid adsorbent for GSC or a solid coated with a high boiling liquid for GLC. Adsorbents for GSC include silica, alumina, and various grades of diatomaceous earth. The solid supports used for GLC include firebrick and various grades of Chromosorb. Mesh sizes from 30/50 to 80/100 are utilized, with 30/60 and 60/80 being the most popular.

The solid support is coated with a high-boiling liquid called the substrate, which acts as the immobile phase in GLC. There are literally hundreds of substrates commercially available, and a complete listing is beyond the scope of this book. The general requirements of the liquid phase are (1) good solvent properties for the components, (2) differential partitioning of sample components, (3) high thermal stability, and (4) a low vapor pressure at the column temperature. Substrates used include paraffin oil, squalane, silicone oils and greases, high-molecular weight esters such as dinonylphthalate, polyglycols, etc.

The column is usually enclosed in a thermally insulated chamber which can be maintained at a constant temperature or which can be programmed for increasing the temperature at a definite rate. As a consequence, the vapor pressure of the liquid substrate must be high enough to prevent "bleeding" of the substrate off the column during use at elevated temperatures.

b. EVALUATION. The efficiency of a column is expressed by the number (N) of theoretical plates in the column or by the height equivalent of a theoretical plate (HETP). The larger the number of theoretical plates or the smaller the HETP, the more efficient the column is for separations. A theoretical plate is that distance on the column in which equilibrium is attained between the solute in the gas phase and the solute in the liquid phase and is equivalent to one equilibrium stage in a distillation.

The number of theoretical plates in any column can be calculated from the elution curve, as shown in Fig. 15-6. The distance (d) from the point of injection to the peak maximum can be measured in units of length, time, or volume and is usually called the retention time (t) or retention volume (V_r). In Fig. 15-6, H is the height of the peak maximum, W is the width of the peak measured as the distance between the intersections of the tangents to the inflection points with the base line, β is the width of the peak at a height of H/e, t is the time lapse between the injection and the emergence of the peak maximum, t_i and t_f are the time lapses between injection and the initial rise of the peak and the return of the peak to the base line, respectively, and the area is calculated as the height multiplied by the width of the peak at one-half the height.

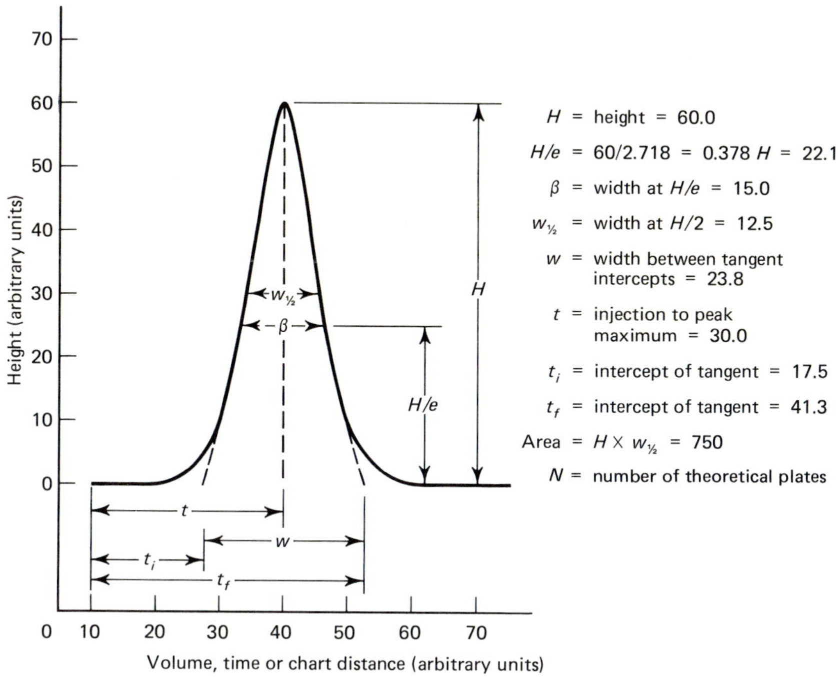

Glueckauf equation:

$$N = 8\left(\frac{t}{\beta}\right)^2 = 8\left(\frac{30}{15}\right)^2 = 8 \times 4 = 32$$

Van Deemter or Martin equation:

$$N = 16\left(\frac{t}{w}\right)^2 = 16\left(\frac{30}{23.8}\right)^2 = 25.4$$

Area equation:

$$N = 2\pi\left(\frac{Ht}{A}\right)^2 = 2 \times 3.14\left(\frac{60 \times 30}{750}\right)^2 = 6.28\,(2.40)^2 = 6.28 \times 5.76 = 36.2$$

Equations for use with unsymmetrical peaks:

$$N = \left(\frac{4t_i}{w} + 2\right)^2 = 24.4 \qquad N = \left(\frac{4t_f}{w} - 2\right)^2 = 24.4 \qquad \text{Avg } N = 28.5$$

Figure 15.6. Calculation of number of theoretical plates.

The Glueckauf equation is probably the simplest to use since the width is measured at *H/e* instead of the intersections of the tangents with the base line. The values by the Glueckauf equation and the Van Deemter or Martin and Synge equation agree satisfactorily. The other equations shown are utilized when the peak is not symmetrical.

The number of theoretical plates is a good measure of the efficiency of a column but does not give any information concerning the effect of various parameters upon the efficiency. The Van Deemter equation, which is based upon the rate theory, gives the relationship of the HETP to the flow rate and includes the effect of many operational parameters in the various terms of the equation. The Van Deemter equation may be written in simplified form as

$$\text{HETP} = A + \frac{B}{u} + Cu \tag{5}$$

where u is the gas velocity or flow rate. A is a constant which involves the packing effect in the column and the particle diameter and is known as the eddy diffusion term. B is a constant which includes the effect of diffusion in the gas phase and the correction for tortuosity of path and is called the molecular diffusion or nonequilibrium term. C is a constant which reflects the resistance to mass transfer between the gas and liquid and includes such factors as the thickness of the film, diffusion in the liquid phase, and the fraction of the solute in each phase.

The graph of a typical Van Deemter study, in Fig. 15-7, shows the effects of each term of the equation on the relationship between the flow

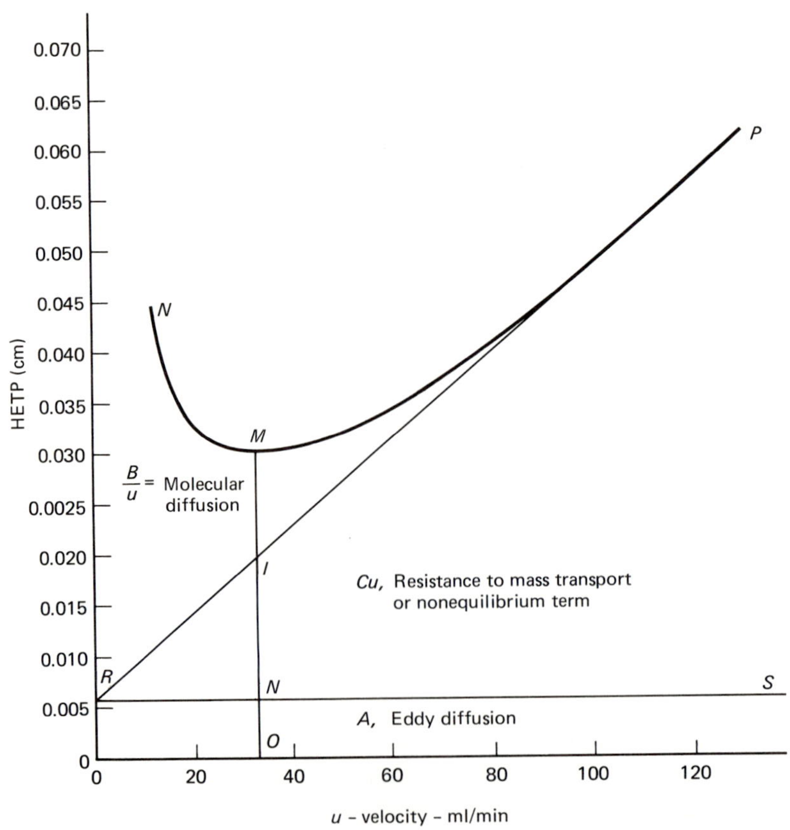

Figure 15.7. Typical Van Deemter graph. Optimum flow, 33; minimum HETP, 0.030, Values of constants as read from graph: *A*, 0.006; *B*, 0.033; and *C*, 0.0004.

rate and the HETP. Since A is constant and does not include the flow rate, it contributes the same amount to the HETP irrespective of the flow rate. The second term increases as flow rate decreases and contributes the largest effect at low flow rates, while the third term contributes its greatest effect at high flow rates since the effect increases as the flow rate increases. Due to the opposing effects of the second and third terms, there is an optimum flow rate at which there will be a minimum HETP. In Fig. 15-7, the optimum flow rate is 33 ml/min, which gives a minimum HETP of 0.030. Since the HETP increases sharply at flow rates below the optimum, it is preferable to operate at the optimum or slightly higher flow rates. The values of the various constants can be determined from the graph by calculating the contribution of each term to the optimum flow rate. The value of A is obtained by extrapolating the tangent to the curve to zero flow rate (line PR in Fig. 15-7). The contribution of the second term is given by line MI and the contribution of the third term by the line IN. B is calculated by multiplying the value of MI by the optimum flow rate and C by dividing the value of IN by the optimum flow rate.

B. Ion Exchange

Ion exchange is the exchange of ions of like sign between a solution and a solid-insoluble body in contact with it. For such an exchange to take place, the solid must be permeable to the solution and must contain ions of its own. Many naturally occurring substances such as clays, the humic acids in the soil, and some manmade resins have this property of exchanging ions. The most popular exchange materials in use today are the synthetic organic ion exchangers, the majority of which are derived from cross-linked polystyrene. These polymers are prepared by copolymerization of styrene and divinyl benzene and are relatively simple. They are more stable toward strong acids and bases than the naturally occurring siliceous materials and have high capacities.

1. Synthetic Ion-Exchange Materials

a. STRUCTURE. Typical structures of polystyrene resins are shown in Fig. 15-8. The molecular weight is very high, and the material is therefore quite insoluble but permeable to water and solutions. The solid granules swell when wet, but the amount of expansion is limited or controlled by the cross linking which joins the long chains together, so that the greater the cross linkage, the less the swelling. The structure is similar to a sponge or a three-dimensional fish net with the sulfonic acid groups attached firmly to the skeletal framework. Since the sulfonic acid group is a strong acid and thus essentially completely ionized, the result is a sequence of negatively

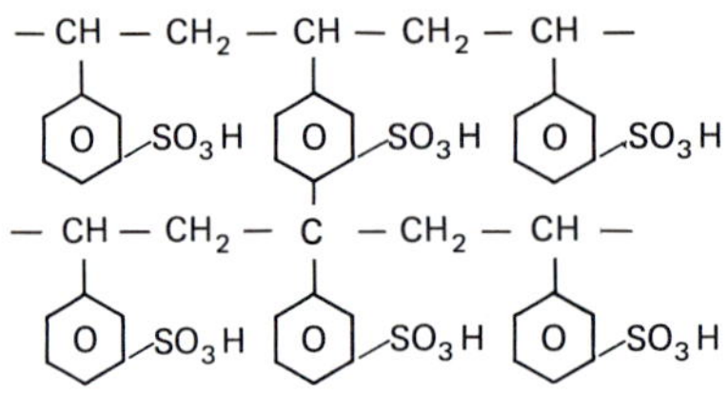

Cation resin

For an anion resin a basic group such as $-CH_2-N(CH_3)_3OH$ is substituted for the acidic $-SO_3H$ group in the cation resin.

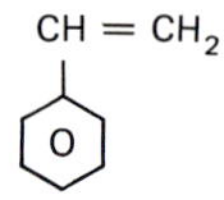

Styrene

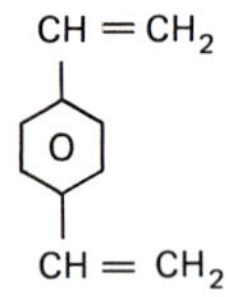

Divinyl benzene

Figure 15.8. Structure of polystyrene ion-exchange resins.

charged ionic groups which are fixed in position surrounded by hydrogen or other positive ions which are free to move. Even though the cations can move freely about inside the resin, they cannot leave the resin particle unless replaced by another ion of equal charge, since electrical neutrality must be maintained. The inside of the particle is very similar to a concentrated solution of a strong acid except that the negative charges are held firmly in position instead of being free to move as they are in solution.

b. MECHANISM OF EXCHANGE. When a synthetic or naturally occurring ion exchanger is placed in a solution, ions may enter the solid particle and be exchanged for the ions of like charge which are free to move. Ions of the same charge as the fixed exchange positions will be excluded from the particle. For a cation exchanger in the hydrogen form in contact with a solution of sodium chloride, the exchange may be represented as

$$Na^+ + H^+R^- \rightleftharpoons Na^+R^- + H^+$$

in which R^- represents the fixed negative sites on the polymer structure. The exchange is a reversible reaction which will attain equilibrium and which follows the law of mass action to a first approximation. If a solution containing a relatively small amount of sodium ion is percolated through a relatively large amount of resin, the sodium will be quantitatively removed from the solution and replaced with an equivalent amount of hydrogen ions. Similarly, if a concentrated solution of an acid is passed through a cation resin in the sodium form, the sodium will be essentially completely removed from the resin, which will be converted to the hydrogen form. Such regeneration of the resin may be repeated a large number of times without apparent damage to the resin.

c. SELECTIVITY RULES. The usual synthetic organic ion exchangers exhibit only slight selectivity for different ions and are not specific for any ion. There are, however, some general rules which may be used to predict the behavior of resins in contact with ionic solutions. For the strong cation exchangers such as sulfonated polystyrene, the higher the charge on an ion, the more firmly it is held by the resin. A trivalent cation such as aluminum will be more firmly bound to the resin than a divalent ion such as calcium and calcium more than a monovalent ion such as sodium. In exchange reactions, trivalent ions will tend to displace divalent and monovalent ions from the resins, and divalent ions will displace monovalent ions.

There is also some selectivity among ions of the same valence group. In these cases it is apparently the size of the hydrated ion which is the principal factor. The smaller the hydrated ion, the more strongly the ion is held by the resin. It should be noted that it is the diameter of the hydrated ion and not the bare ion as it exists in crystalline solids that is effective. In any periodic table group, the elements with lower atomic number bind more water molecules than the elements with higher atomic number and thus actually are larger in the hydrated state than those with higher atomic number. In the alkali metals (Group IA), the ${}_{3}Li(H_2O)_x^+$ ion is the largest and the ${}_{55}Cs(H_2O)_x^+$ ion is the smallest, and ${}_{39}K(H_2O)_x^+$ would be held more firmly than ${}_{23}Na(H_2O)_x^+$.

The extent of cross linkage also affects the selectivity of a resin to some extent, especially with large ions. The greater the amount of cross linkage, the less the swelling and the smaller the size of the pores through which water and ions enter the resin particle. In some cases, highly cross-linked resins will actually exclude large organic ions.

While the above generalizations apply to the selectivity of all forms of strong cation exchangers and to all forms of strong anion exchangers, other considerations will affect other types of exchange resins. The weak acid or carboxyl type of cation exchanger binds hydrogen much more strongly than other cations, and weak base anion exchangers bind hydroxyl more strongly than other anions. The newer chelating resins which have iminodiacetate or other chelating groups for the functional groups attached to the resin structure show greater specificity than the ordinary cation and anion resins. The iminodiacetate resins are essentially specific for divalent ions, especially the transition elements which tend to form coordination complexes readily.

d. DISTRIBUTION COEFFICIENTS. One measure of the strength with which a metal ion is held by a resin is the distribution coefficient, which is defined as

$$D = \frac{\text{mmoles of metal/g of dry resin}}{\text{mmoles of metal/ml of solution}}$$

The larger the distribution coefficient, the more strongly the metal is held by the resin. Distribution coefficients are measured by a batch process in which a given weight of resin is equilibrated with a known volume of a solution of known concentration. The concentration in the solution is determined after the equilibration, and the amount of metal in the resin is calculated from the values before and after equilibration.

2. Applications

a. DEIONIZATION OF WATER. If natural or domestic water is passed first through a cation resin in the hydrogen form and then through an anion resin in the hydroxide form, the resulting water is said to be demineralized or deionized. Deionized water can be used as a substitute for distilled water in many applications and often is much less expensive than distilled water. When the resins become loaded with ions after continued use, they can be regenerated into the hydrogen and hydroxyl forms by use of inexpensive, commercial grades of acid and base. Cation exchangers (either naturally occurring or synthetic) in the sodium form are used to soften water by exchanging sodium ions for calcium and magnesium ions and can be regenerated with a concentrated sodium chloride solution. Mixed-bed deionization (using a cation and anion resin mixed together instead of in separate columns) is now being used in many deionization installations. The resins are separated by a flotation process before regeneration and mixed again after regeneration. Since the exchange of ions follows the laws of equivalency, a cation exchanger in the hydrogen form can be used to determine the total salt content of natural or domestic waters. The *p*H is adjusted to some value before passing the water through a column of the exchanger and is then titrated back to the same *p*H after passage through the column. This procedure can also be utilized with an anion exchanger, but the results are not as satisfactory.

b. SEPARATION OF IONS OF OPPOSITE CHARGE. In many analytical methods, an ion interferes with the determination of an ion of opposite charge, and often these interfering ions may be removed by passage through a column of an appropriate exchanger. As an example, phosphate ores, which usually contain calcium phosphate or calcium apatites, are dissolved in acid for analysis. The phosphate can be titrated from the $H_2PO_4^-$ form to the HPO_4^{2-} form readily, but calcium causes precipitation as the *p*H is raised. Passage of the acid solution through a cation exchanger in the hydrogen form prior to titration removes the calcium ions and allows the titration to be completed satisfactorily. There are many hundreds of examples of this type of separation in the literature using either anion or cation resins.

c. CONCENTRATION OF TRACE CONSTITUENTS. One very important use of ion-exchange resins is the concentration of trace constituents. Large volumes of liquids containing small amounts of ionic materials can be passed through a column of an appropriate exchanger. The ionic materials which have been held by the column can then be eluted with a small volume of solution for analysis. An excellent example of this type of use is the development by the Public Health Service of a simple field method for obtaining samples of milk which may be contaminated with radioactive fallout. Radioactive iodine 131 (^{131}I) is the ion of greatest public health interest and is separated from raw milk by passing a gallon or more through a plastic cartridge containing the appropriate resin. The cartridge is then sent to the Public Health Service for determination of the amount of radioactivity. Similar methods are available for determination of radioactive cations such as strontium 90 (^{90}Sr) and cesium 137 (^{137}Cs).

d. CHROMATOGRAPHIC SEPARATIONS OF SIMILAR IONS. Ion-exchange resins may also be used in the chromatographic separation of similar ions which are difficult to separate by chemical means. If a mixture of relatively small amounts of two cations (or anions) is poured onto a column of exchanger, both ions will be removed from the solution in the top portion of the resin. If one cation (A) is held more strongly by the resin than the other cation (B), use of a satisfactory eluting agent will cause B to move down the column faster than A, and under favorable circumstances all of B will flow out the bottom of the column before A appears in the effluent. By proper choice of eluting agents, ion-exchange resin, column length, flow rate, particle size, and other variables, separations of complex mixtures may be effected by this procedure. Examples are the separation of sodium from potassium, calcium from magnesium, and chloride from bromide. Perhaps one of the most spectacular successes has been the separation of the rare earth elements which has allowed the preparation of very pure samples of many of these metals.

e. USE OF CHELATING RESINS. Ordinary strong acid or weak acid cation-exchange resins cannot be used to remove divalent metals from brine solutions such as ocean water and oil well brines since the high concentration of sodium ions effectively displaces them from the resin. This is true for the removal of any trace ion from a concentrated solution. Use of a chelating resin such as an iminodiacetate exchanger allows the separation of trace divalent metals from brines and other concentrated solutions of monovalent cations since monovalent cations are not bound to the resin and thus will not displace the divalent metals. Examples include the separation of manganese and other divalent ions from oil well brines, of polyvalent ions from sugar solutions, of calcium from concentrated caustic solutions, and of copper from water.

f. ION-EXCHANGE MEMBRANES. Ion-exchange membranes are now prepared by casting the resin as a thin film or by incorporating finely ground resin particles in a plastic but inert matrix. Membranes prepared from cation-exchange resins are permeable to cations but not to anions since the high concentration of fixed negative charges in the resin matrix offers a high resistance to the passage of anions. Similarly, membranes of anion-exchange materials readily pass anions but almost completely exclude cations. These membranes are now used in one method of desalinization of water by electrodialysis and increase the efficiency from about 18 to 80 per cent over ordinary nonselective membranes. In a multicompartment electrodialysis cell formed by alternating anion and cation membranes, the salt tends to concentrate in every other cell, and thus desalting occurs in every other cell.

CHAPTER 16

Reliability of Measurements

A. Accuracy, Precision, and Error

The concepts of accuracy and precision are often confused by beginning students in quantitative analysis, even though these terms have substantially different meanings. The accuracy is the closeness with which the results obtained agree with the accepted correct value, while the precision of a set of results expresses the reproducibility of the measurement and is defined as the amount of deviation within a series of measurements. Thus accuracy is related to how close the observed result is to the true value, and precision is related to how close the measurements are to each other. Any given set of measurements may be precise and also be inaccurate, but normally an accurate set of results is precise also.

Errors may be expressed as absolute errors or as relative errors. The absolute error (E) is the difference between the observed (O) and the true value (T) and is always expressed in the units of the measurement. The relative error (RE) expresses the relationship of the error to the true value and is usually expressed as percentage error or as parts per thousand. Thus

$$E = O - T \quad \text{(in units of measurement)}$$

$$RE = \frac{E}{T} \times 1000 \quad \text{(in ppt)}$$

For example, assume that two values are measured with the same absolute error, as

Measurement (g)	True value (g)	Absolute error (g)	Relative error (ppt)
10.545	10.544	0.001	0.094
0.105	0.104	0.001	9.6

While the absolute errors are equal, the relative errors show that the first measurement is relatively more accurate than the latter.

B. Classification of Errors

The two main classes of errors are the determinate and the indeterminate errors. Determinate errors (sometimes called systematic errors) are those which can be discovered and corrected for or else can be eliminated in a repetition of the determination. The errors can be variable or constant and can be either positive or negative. Variable errors will affect the precision but constant errors do not. Indeterminate errors (sometimes called random or accidental) are errors which cannot be discovered, corrected for, or eliminated in a careful rerun of the measurement. This type of error is usually due to the inability of the observer to control conditions, to correct for variation in external conditions, or to observe more carefully. Indeterminate errors follow the law of normal distribution or probability. The normal distribution curve is shown in Fig. 16-1 and indicates that small errors are more probable than large errors and that positive and negative errors are equally probable.

A constant error which occurs in each and every measurement does not affect the precision of the set but does affect the magnitude of the average of the set. Constant errors of this type are called a bias, and, if the extent of the bias can be determined, the results can be corrected for the bias by addition or subtraction of the constant error from each and every result. The effect of a bias on a determination is shown by the dashed curve in Fig. 16-1.

The shape of the distribution curve for a set of measurements is affected by the precision, as shown in Fig. 16-2. A narrow distribution curve indicates good precision (small standard deviation), while a broad distribution is caused by poor precision (large standard deviation).

1. Types of Determinate Errors

Determinate errors are classified as personal, instrumental, or method errors.

a. Personal errors are those due to the carelessness of the observer and include such things as misreading a burette (29.38 instead of the correct

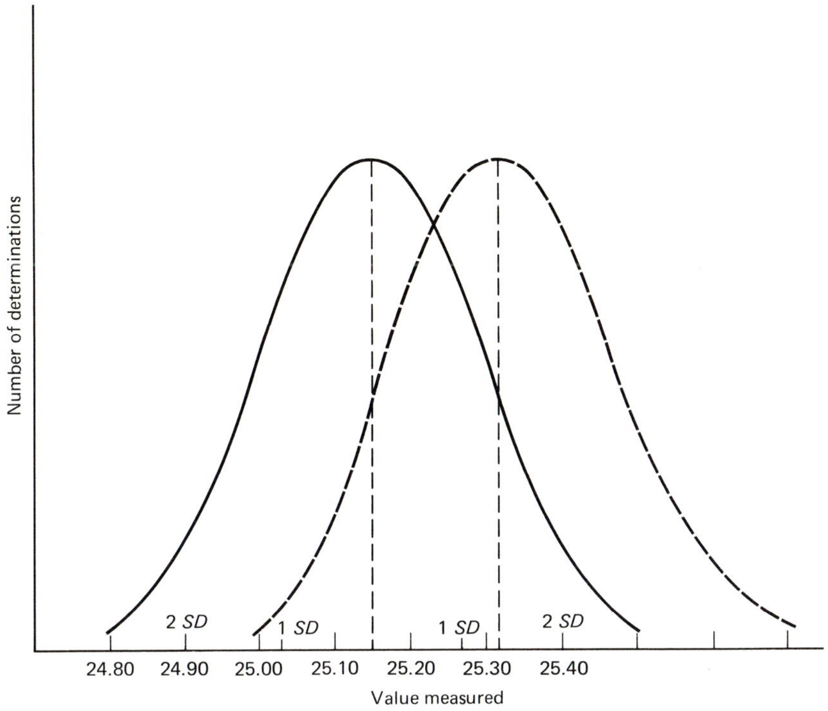

Figure 16.1. Normal distribution of errors. The solid curve represents measurements for which the average value is 25.15 with a standard deviation of 0.125. The dashed curve represents a measurement with the same degree of precision but with a bias of 0.17 units.

value of 28.62), listing weights incorrectly (listing a 2-g weight when a 1-g weight was used), misinterpreting directions (adding concentrated acid instead of 1:1 or adding 10 ml when directions call for 1.0 ml), overshooting an end point in a titration, and use of dirty equipment. It should be noted that these errors probably would be eliminated in a careful rerun of the determination.

b. Instrumental errors are those due to the instruments themselves and include the use of uncalibrated equipment (weights, pipettes, burettes, etc.), construction errors in the instruments (balance arms not equal), and damaged equipment (chipped knife edge).

c. Method errors are those which are inherent in the method itself which normally would not be eliminated in a rerun but which often may be discovered and corrected for by calculation or procedural changes. These include impurities in reagents (correct by using a blank determination), solubility of precipitates in precipitation medium or wash solution (determine and correct mathematically or revise procedure to decrease solubility), contamination in or on precipitates (dissolve and reprecipitate or remove

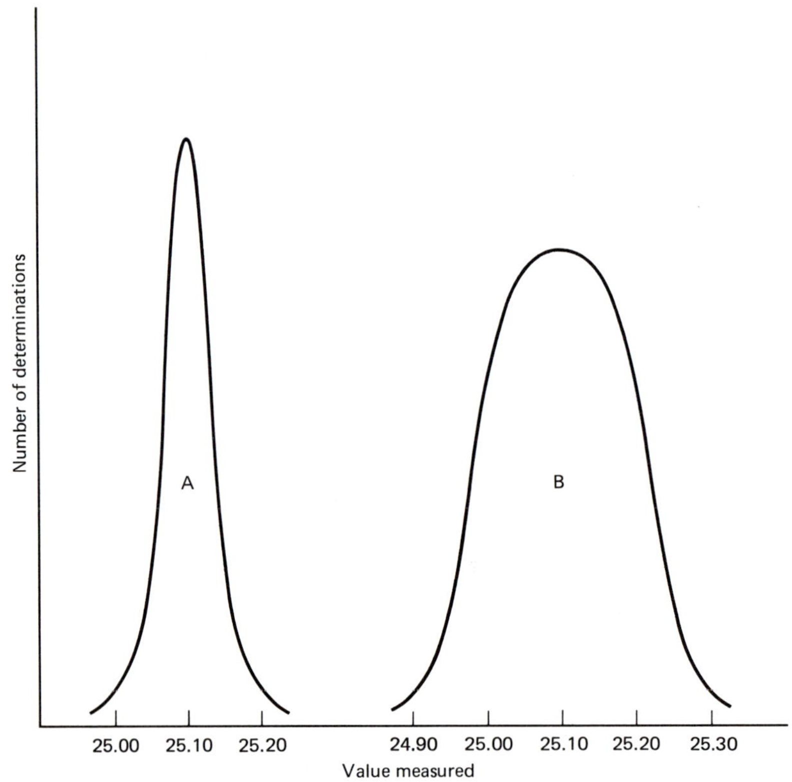

Figure 16.2. Effect of precision upon shape of distribution curve. **A**, Good precision with a standard deviation of approximately 0.03. **B**, Poorer precision with a standard deviation of approximately 0.10 even though the averages are identical.

contaminating substances by prior treatment), indicator errors (run blank), and reaction not stoichiometric (can often be corrected by careful standardizations).

2. Types of Indeterminate or Random Errors

Indeterminate errors are classified as personal, as instrumental, or as being due to variations in external conditions. Note that for all of the random errors listed below, essentially none can be discovered or corrected for and few, if any, would be eliminated in a repeat of the determination. This type of error is just as likely to occur during the second run as in the first.

a. Personal errors are due to the inability of the observer to distinguish between slightly different measurements or colors. Examples are misestimation of the position of the miniscus (estimating 25.56 instead of the correct value of 25.57), assuming two colors are identical when there is a slight difference in hue, misestimation of balance rest point, and misestimation of the position of a meter needle.

b. Indeterminate instrumental errors are due to random variations in the response of instruments, variations in line voltage, wandering rest points during a weighing, irreversible reactions caused by momentary nonhomogeneity of solutions, etc.

c. Variations of external conditions over which the observer has no control also cause random errors. Included here would be errors due to variations of barometric pressure, variations of humidity, slight variations of temperature, and momentary air currents.

C. Methods of Expressing Precision of a Set of Results

Precision may be expressed as the average deviation (*AD*), the relative average deviation (*RAD*), or as the standard deviation (*SD* or *s*). The square of the *SD* is known as the variance. Even though the relative average deviation is used more often in the laboratory for a first course in quantitative analysis, the standard deviation is a better estimate of the precision of the results.

1. Average Deviation and Relative Average Deviation

The absolute deviation (*D*) and the average deviation are expressed in the units of the measurement, while the relative deviation (*RD*) and the relative average deviation are expressed as parts per thousand or per cent. The deviations are calculated as the difference between individual measurements and the average or mean of the set and may be averaged to obtain the average deviation. The relative deviations are calculated by dividing the individual deviations by the average of the set and multiplying by 1000. These are averaged to obtain the relative average deviation. An example of these calculations is given below:

Measurements of % of Cl in unknown	*D*	*RD* (ppt)
45.23	0.05	$\frac{0.05 \times 1000}{45.28} = 1.1$
45.31	0.03	$\frac{0.03 \times 1000}{45.28} = 0.7$
45.30	0.02	$\frac{0.02 \times 1000}{45.28} = 0.4$
Avg = 45.28	*AD* = 0.03	*RAD* = 0.7

Note that AD is expressed to the same number of decimal places as the measurements and that RD and RAD are expressed to the closest 0.1 of 1 ppt.

Since accuracy and precision are both relative values, it is necessary to define the conditions under which a set of results will be considered accurate and/or precise. For the purpose of this text, a set of three results will be considered to be accurate if the RAD from the *true value* does not exceed 2.0 ppt, and the set will be considered to be precise if the RAD from the *average of the set* does not exceed 2.0 ppt and if any individual RD does not exceed 3.0 ppt.

2. Methods of Calculating Standard Deviations

a. SUM OF SQUARES METHOD. In this method, the following operations are performed:

1. Square each individual measurement and sum these squares. Do not round off.
2. Sum the individual measurements and square this sum. Do not round off.
3. Divide the square of the sum (2) by the number of measurements in the series, and then subtract from the sum of the squares of the individual measurements (1) to obtain the sum of the squares.
4. Divide the sum of squares (3) by the degrees of freedom (one less than the number of measurements) to obtain the variance.
5. Extract the square root of the variance to obtain the standard deviation.

Examples (columns R and S represent values obtained by different students running the same unknown for which the true value is 59.55):

	R		*S*	
	59.51	3,541.4401	59.09	3,491.6281
	59.75	3,570.0625	59.17	3,501.0889
	59.61	3,553.3521	59.27	3,512.9329
	59.60	3,552.1600	59.13	3,496.3569
			59.10	3,492.8100
			59.14	3,497.5396
Totals	238.47	14,217.0147	354.90	20,992.3564
Average	59.62		59.15	
Total^2/n	$\frac{(238.47)^2}{4}$	14,216.9852	$\frac{(354.90)^2}{6}$	20992.3350
Difference or sum of squares		0.0295		0.0214
Degrees of freedom		3		5
Variance		0.00983		0.00428
SD		0.0991		0.0654

b. DEVIATIONS FROM AVERAGE METHOD. In this method the following operations are performed:

1. Sum the individual measurements and divide by the number of measurements to obtain the average.
2. Subtract the average from each individual measurement to obtain the deviation of each.
3. Square each of these deviations and then sum the squares.
4. Divide the sum of the squares by the degrees of freedom to obtain the variance.
5. Extract the square root of the variance to obtain the *SD*.

	R			*S*		
	59.51	0.11	0.0121	59.09	0.06	0.0036
	59.75	0.13	0.0169	59.17	0.02	0.0004
	59.61	0.01	0.0001	59.27	0.12	0.0144
	59.60	0.02	0.0004	59.13	0.02	0.0004
				59.10	0.05	0.0025
				59.14	0.01	0.0001
Total	238.47	0.27	0.0295	354.90	0.28	0.0214
Average	59.62	0.07		59.15	0.05	
Variance	$\frac{0.0295}{3} = 0.00983$			$\frac{0.0214}{5} = 0.00428$		
SD	0.0991			0.0654		

c. FROM THE RANGE.[1] In this method, the range of a set of measurements is determined and multiplied by a deviation factor to obtain the standard deviation. The range is the difference between the smallest and largest observation, and the deviation factor may be obtained from the Dean and Dixon table, which is reproduced in Table 16-1. The table includes values for sets which contain from 2 to 10 measurements. The standard deviation calculated from the range is usually larger than the values obtained by the better known methods already given but is considered to be a better estimate of the actual *SD* of the set when small numbers of measurements are made. The range method also has the advantage of simplicity of calculation. Applied to the data of *R* and *S*,

	R	*S*
Range	$59.75 - 59.51 = 0.24$	$59.27 - 59.09 = 0.18$
SD	$0.24 \times 0.49 = 0.1176$	$0.18 \times 0.40 = 0.0720$

[1] See R. B. Dean and W. J. Dixon, "Simplified Statistics for Small Numbers of Observations," *Anal. Chem.*, **23**, 636 (1951).

TABLE 16.1 Values for Statistical Evaluation of Sets of 2 to 10 Observations

Number of observations (n)	Deviation (K_w)	Confidence factors $t_{w0.95}$	$t_{w0.99}$	Rejection quotient ($Q_{0.90}$)
2	0.89	6.4	31.83	
3	0.59	1.3	3.01	0.94
4	0.49	0.72	1.32	0.76
5	0.43	0.51	0.84	0.64
6	0.40	0.40	0.63	0.56
7	0.37	0.33	0.51	0.51
8	0.35	0.29	0.43	0.47
9	0.34	0.26	0.37	0.44
10	0.33	0.23	0.33	0.41

From Dean and Dixon, *Anal. Chem.*, **23**, 636 (1951).

D. Confidence Limits

Confidence limits are defined as the upper and lower values between which the true central value would lie if based on a large number of measurements. By confidence interval, we mean the distance on either side of the average of a small set of measurements in which we would expect to find, with a given probability, the true central value. For example, the true average would be expected to be within the 95 per cent confidence limits in 95 out of 100 such sets of values.

Since the confidence limits may be calculated from small sets of data and then used to predict the interval between which the true value would lie within a given probability, the chance of making an erroneous prediction based upon small sets of measurements can be reduced to any level desired by the investigator.

There are two methods of calculating confidence intervals for a given set of determinations. One of these ultilizes the t value of Student and the standard deviation of the average (the standard deviation of the individual measurements divided by the square root of the number of measurements). The other method utilizes the range and the t_w value given by Dean and Dixon for sets containing 10 or fewer measurements. Both of these methods will be illustrated using the data of R and S. (Student's t values depend upon the number of measurements and the confidence interval desired and are listed in tables in most books concerning statistics.)

1. Method Using Student's *t* Value

	R	*S*
Degrees of freedom	3	5
t for 95% limits	3.182	2.571
SD of individual determinations	0.0991	0.0654
Standard deviation of average ($SD/\sqrt{n}$)	0.0496	0.0267
95% limits ($t \times SD$ of average)	0.1562	0.0686

Thus, the true average value for R should be between 59.46 and 59.78 and that for S should be between 59.08 and 59.22.

2. Method Using Range and t_w

	R	*S*
Range	0.24	0.18
t_w for 95% for number of measurements	0.72	0.40
95% limits (range $\times\ t_w$)	0.174	0.072

Thus, the true average value for R should be between 59.45 and 59.79 and that for S should be between 59.08 and 59.22. It should be noted that these values agree closely with those calculated by the first method.

E. Method for Discarding Deviant Results in Sets of 10 or Less

This method entails calculating the difference between the deviant figure and the closest value to it when the results are arranged in increasing or decreasing order. This difference is then divided by the range to obtain the Q values. The calculated Q value is compared to Q values for 90% confidence as given by Dean and Dixon. The questionable value may be discarded if the calculated value of Q exceeds the value given in the table. This discarding of a deviant result will be correct 9 times out of 10.

Examples for data of *R* and *S*	*R*	*S*
Number of measurements	4	6
Range	0.24	0.18
Difference	0.14	0.10
Calculated Q value	0.583	0.556
Q_{90} from table	0.76	0.56

Since the calculated Q value is definitely below the table value for the R data, this value (59.75) should be retained. Since the calculated Q value for the data of S is very close to the table value, this value (59.27) can be either retained or discarded. In such cases, it is usually best to use some other criterion for discarding a given value.

F. Rules for Significant Figures

Since the precision of the measurement is denoted by the number of significant figures used to express the result, it is important that the correct number of significant figures be used in specifying any measurement. A significant figure is a digit which denotes the quantity in the place in which it stands. To determine the number of significant figures in any number, start counting from a digit other than zero to the left and count through the last digit, including zero, to the right. Note that the position of the decimal has nothing to do with the number of significant figures. Thus the number 102 has three significant figures, the number 0.000102 has three significant figures, and the number 0.01020 has four significant figures.

To express the correct precision of any measurement, only one insignificant digit is included, and it is understood that the variation in the last figure is plus or minus 1 unless a different precision is indicated. Thus the number 12.567 represents a measurement which is known to be between 12.566 and 12.568. If the measurement is such that the true value is known to be only between 12.563 and 12.571, the value should be listed as 12.567 ± 0.004.

In rejecting figures beyond the first uncertain figure, increase the last figure retained by 1 if the following rejected figure is 5 or more; thus 1.25015 rounds off to 1.2502, while 1.25014 rounds off to 1.2501. For addition and subtraction, retain in each number only as many figures to the right of the decimal as there are in the number having the least number of figures to the right of the decimal. For addition of the numbers 516.3, 34.36, and 1.349, the numbers actually added would be 516.3, 34.4, and 1.3 to obtain 552.0 as the correct sum. Note that each figure is rounded off before addition. In multiplication and division, retain as many significant figures in each factor and in the results as are necessary to provide the same degree of precision as in the least precise factor. The simplest way to offset errors in the estimation of equivalent precision of several numbers is to retain in each factor one more significant figure than is to be used to express the answer. If the answer is to be expressed in four significant figures, round off each number to five significant figures before use in multiplication or divison.

G. Problems

1. How many significant figures are there in each of the following measurements?
   (a) 123.45 (d) 1.69×10^{-5} (g) 987
   (b) 0.01120 (e) 0.62×10^{10} (h) 3,000,058
   (c) 30.30 (f) 9.00×10^{3} (i) 1.6295
   Ans. (b) 4
2. Add or subtract after rewriting the following additions and subtractions in vertical columns. Express each number (including the difference or the sum) in the proper number of significant figures.
   (a) 12.67 + 357.8 + 1.349 (e) 35.6982 − 13.4397
   (b) $0.0025 + 2.5 \times 10^{-3} + 0.1025$ (f) 1.634 − 0.29
   (c) 0.543 + 5.43 + 54.3 + 543 (g) 107.97 − 23
   (d) $1.2 \times 10^{3} + 5605 + 630.0$ (h) $6.53 \times 10^{-2} - 6.53 \times 10^{-3}$
   Ans. (b) 0.1075, (f) 1.34
3. Perform the indicated multiplications and divisions expressing each number, including the result, in the proper number of significant figures.
   (a) $\dfrac{3.50 \times 0.1563}{35.07 \times 0.562}$ (d) $\dfrac{3.5 \times 10^{-6} \times 7.66 \times 10^{-6}}{0.253 \times 35}$
   (b) $\dfrac{5.735 \times 0.565}{27.40 \times 6.8164}$ (e) $\dfrac{5.43 \times 543}{0.543 \times 54.3}$
   (c) $\dfrac{25.67 \times 0.1123}{1.023 \times 0.5000}$ *Ans.* (b) 0.01735
4. For each of the following sets of results, calculate (1) the average, (2) the deviations, (3) the average deviation, (4) the relative deviations in parts per thousand, (5) the relative average deviation in parts per thousand, (6) the standard deviation by the deviation from average method, (7) the standard deviation from the range, and (8) the 95% confidence limits.
   (a) 35.47, 35.49, 35.42, 35.46
   (b) 25.10, 25.20, 25.00
   (c) 63.92, 63.75, 63.90, 63.86, 63.84
   (d) 6.050, 6.048, 6.068, 6.054, 6.056
   (e) 50.00, 49.96, 49.92, 50.15
   Ans. (b) (1) 25.10, (3) 0.07, (5) 2.7, (6) 0.10, (7), 0.118, (8) ±0.26
5. For each set in problem 4, calculate the Q value for the deviant result and state whether or not the deviant value should be discarded.
   Ans. (b) 0.50; no.
6. For each set in problem 4 for which a deviant value can be discarded, recalculate the values requested in problem 4 after the deviant value has been discarded.
7. Assuming the true values are (a) 35.53, (b) 25.06, (c) 63.80, (d) 6.064, and (e) 49.80, state whether the sets in problem 4 are (1) both accurate and precise, (2) accurate but not precise, (3) precise but not accurate, or (4) neither accurate nor precise.
   Ans. (b) (4)
8. Classify the following errors as being (1) determinate-personal, (2) determinate-instrumental, (3) determinate-method, (4) indeterminate-personal, (5) indeterminate-instrumental, or (6) indeterminate-conditions:
   (a) End point and equivalence point do not coincide

(b) Reading a burette as 15.63 instead of correct value of 15.65
(c) Reading a balance weight as 13.5692 instead of the correct weight of 13.6592
(d) Misestimation of equilibrium point of balance
(e) Reading a burette as 15.63 instead of the correct value of 14.37
(f) Use of an uncalibrated pipette
(g) Variation of humidity during weighing of a hygroscopic solid
(h) Variation of temperature during a titration
(i) Solubility of a precipitate in the precipitation medium
(j) Using 1:10 HCl instead of the correct 1:1 HCl
(k) Use of nitric acid in place of sulfuric acid
(l) Failure to distinguish between two shades of orange
(m) Overshooting an end point by 0.03 ml
(n) Overshooting an end point by 10 drops
(o) Variation of the equilibrium point of a balance during a weighing
(p) Weighing a warm crucible
(q) The 5-g weight does not weigh exactly 5.000 g

CHAPTER 17

Laboratory Techniques

A. General

The importance of learning and using good laboratory technique cannot be overemphasized since good technique is the prime requisite for accurate analysis. Most students are able to develop the manual dexterity necessary for the proper handling of analytical equipment, but constant practice and care are necessary until the correct operations become automatic through repeated performance. In learning techniques, the student should first understand the steps required for correct usage of the equipment and should then practice the actual measurements using movements which are comfortable for him. Repeated practice teaches the manual dexterity necessary, and going slowly at the beginning will help in the learning of the movements. After the muscles have been trained to perform the technique, the student can increase the speed of the operation so that he becomes both accurate and quick. One of the main aids in learning and using good laboratory practice is to remember to be relaxed when making measurements since tension usuallly causes poor muscular control and consequent errors.

The techniques given here are general techniques used in most of the analyses performed by an analyst. Some of the more refined or special techniques which are not used repeatedly will be described in the specific procedure in which they are needed.

B. Transfer of Liquids from Beakers

In most cases, transfer of a liquid from a beaker is performed with the equipment at room temperature, but there are occasions when hot solutions must be transferred. As a consequence, the student should learn how to use

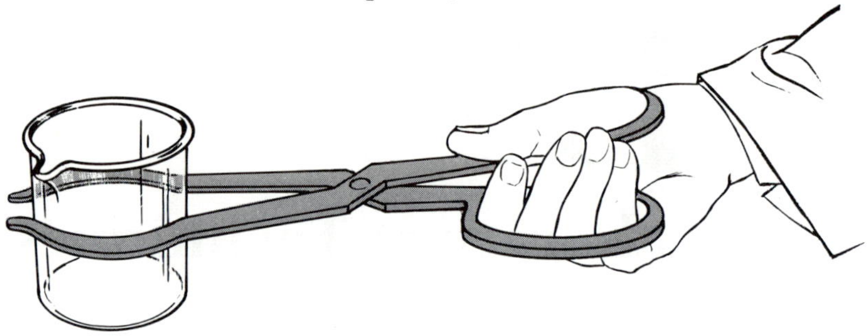

Figure 17.1. Use of beaker tongs.

Figure 17.2. Technique for pouring from a beaker.

the beaker tongs to hold the beaker in the proper manner so that the transfer is easy to control. The student's hands and arms should be in comfortable and normal positions which allow relaxed and efficient handling during the pouring of the liquid. The best way to do this is to have the tongs encircle the sides of the beaker with the handles at right angles to the lip of the beaker and held so that the wrist and palm of the hand is upward, as shown in Fig. 17-1. This position allows proper flexing of the wrist for adequate control of the beaker during the pouring operation.

A stirring rod should always be used when pouring from a beaker. The rod should be placed on the lip of the beaker and the liquid poured down the rod, as shown in Fig. 17-2. This ensures that all the liquid goes down the rod into the receiver. When the rod is not used, some of the liquid will flow down the outside of the beaker and be lost to the transfer. It should be remembered that in order for a transfer to be quantitative, all the liquid must end up in the receiving vessel, not on the outside of the original beaker.

C. Filtration

The purpose of a filtration is to remove a solid from contact with a solution. The solid is usually a precipitate which has been formed in the liquid, and the liquid is often referred to as the mother liquor. The liquid which passes through the filtration device is known as the filtrate. Various devices are used in the filtration process, the most common being paper, fritted glass crucibles, and Gooch crucibles. The crucibles are ordinarily used for suction filtration and paper for gravity filtration.

1. Gravity Filtration

Filter paper circles should be folded into half and then quarter circles, as shown in Fig. 17-3. One corner at the open end of the fold should be torn off and the paper should be seated in the funnel with the torn corner to the outside next to the glass. While holding the paper in the funnel, fill it with solvent (usually water) until the liquid level is above the top of the paper. Press down the folds and the top of the paper while the liquid is running out in order to be sure that no air can leak through the paper into the stem of the funnel. This will ensure a full column of water in the stem, which is necessary for adequate and rapid filtration. Fig. 17-3 shows a good funnel (stem is full of water) and a poor funnel (stem is filled partly with air).

2. Suction Filtration

Suction filtration is preferred when possible since it is much faster than ordinary gravity filtration. Gooch crucibles and fritted glass crucibles are used for suction filtration since the power of the suction will break most

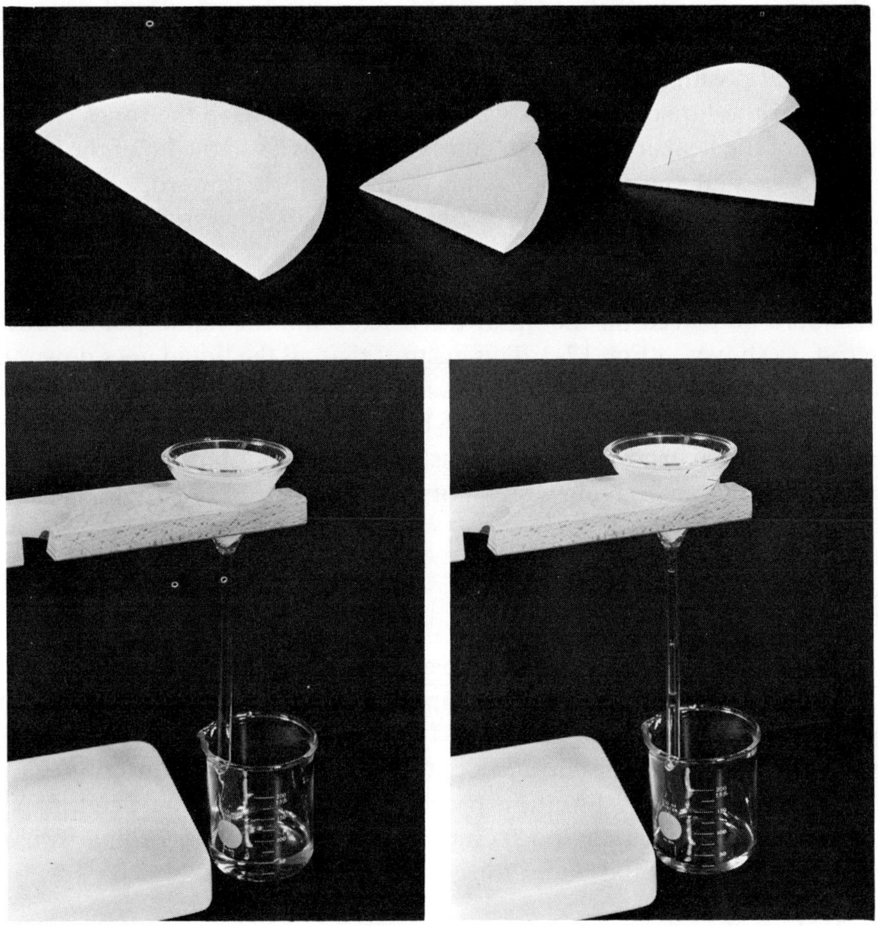

Figure 17.3. Preparation of funnel for gravity filtration.

ordinary filter paper. Paper can be used if the bottom of the paper is supported by a platinum or porcelain filter cone, but this is not a popular procedure. The crucible is supported on a rubber ring in a funnel in a suction flask, as shown in Fig. 17-4. Some of the other methods of support which may be utilized are shown in the same figure. The suction flask is always separated from the aspirator with a trap to protect the liquid filtrate from contamination with tap water, as shown in Fig. 17-5. Whenever there is a change in water pressure on the water line, the suction pulled by the aspirator changes, and when pressure falls it is possible for the water to flow from the aspirator to the flask if the trap is not in the line. When water pressure increases, the water will be pulled from the trap if placed in the line properly.

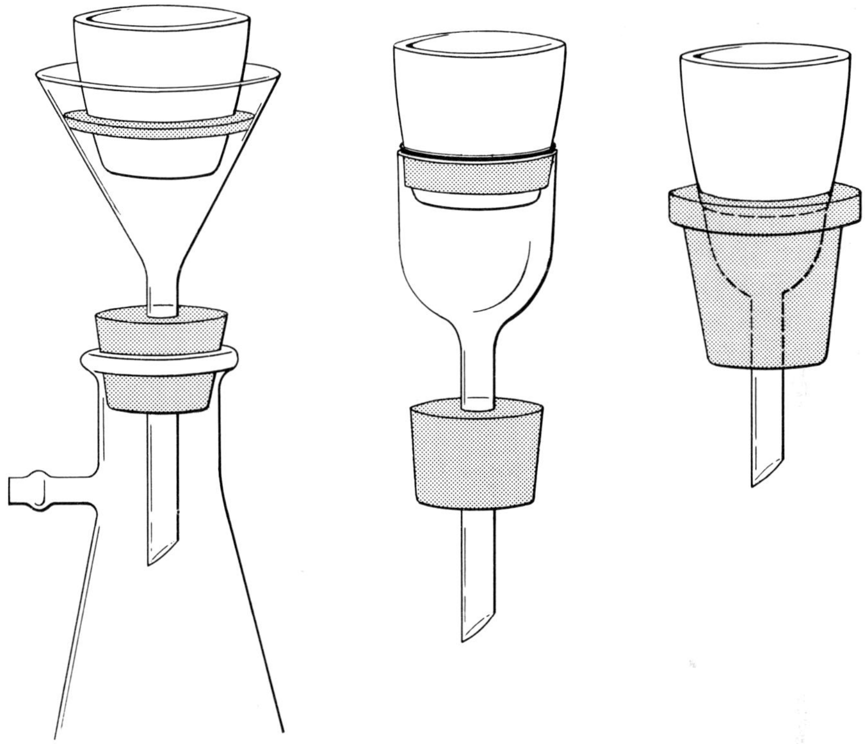

Figure 17.4. Methods of support for crucibles during suction filtration.

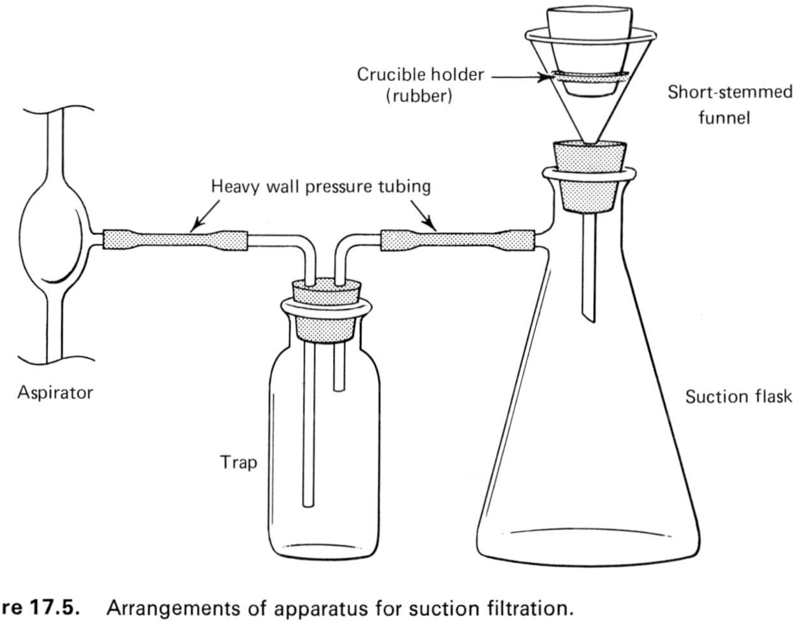

Figure 17.5. Arrangements of apparatus for suction filtration.

D. Washing

A precipitate is often washed by decantation before being physically transferred to the paper or crucible. In this procedure, which is shown in Fig. 17-6, the mother liquor is poured from the beaker into the filtration device until no more can be poured without the precipitate also leaving the beaker. The precipitate in the beaker is covered with some volume of wash solution, agitated, and allowed to settle. This wash liquid is then decanted in the same way as the mother solution, and the operation is continued as many times as desired. Washing by decantation is desirable when possible since it is more efficient in removing impurities and since it is much faster than washing in the filtration device after transfer of the precipitate. The

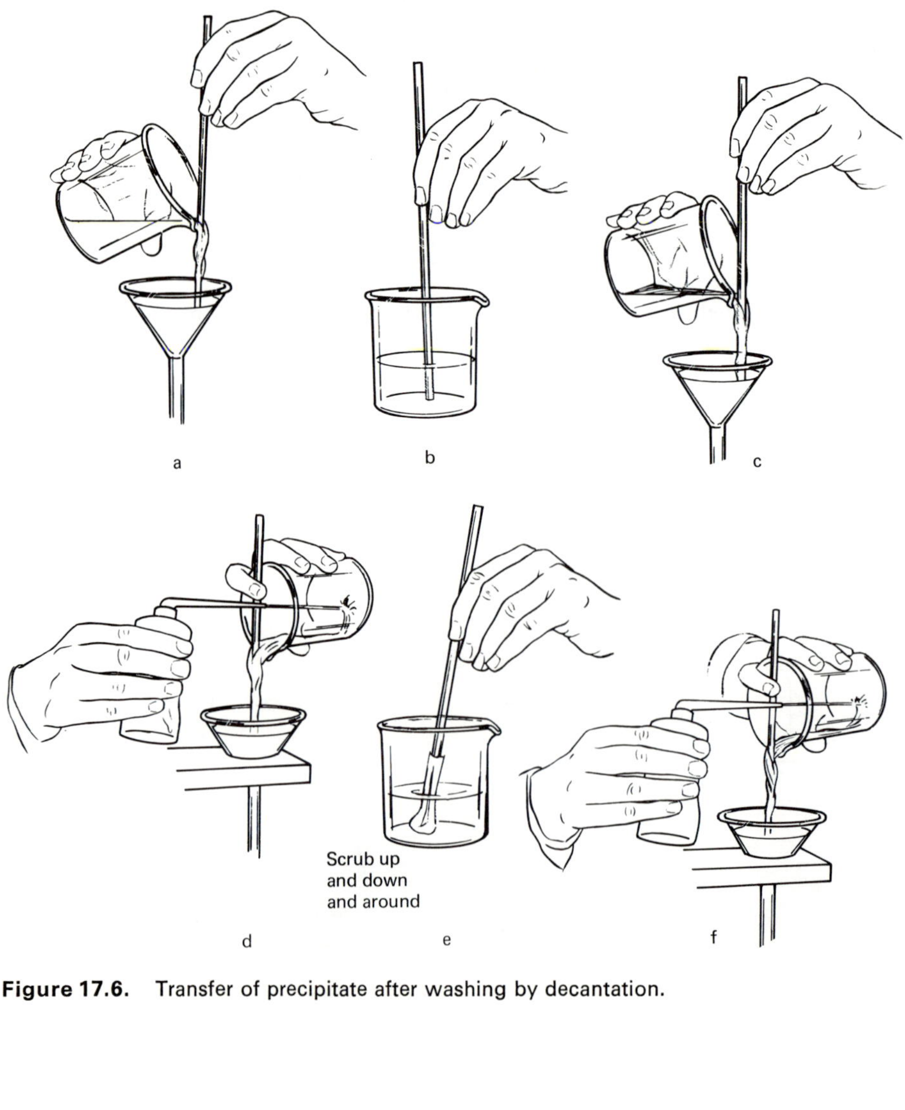

Figure 17.6. Transfer of precipitate after washing by decantation.

rate of filtration decreases significantly after the precipitate has been transferred.

During the washing by decantation, the liquid should be poured down a rod, as shown in Fig. 17-6a. When the precipitate itself is to be transferred, it must be washed from the beaker with a jet of water from the plastic wash bottle and allowed to flow down the rod in the same manner as shown in Fig. 17-6d. Be careful during this part of the transfer to keep the rod centered over the filtration device so that the solid will not be lost outside the device.

After as much of the solid has been transferred as possible, wash down the rod and return it to the beaker. Then scrub the beaker and rod with a policeman to loosen any solid adhering to the walls, as shown in Fig. 17-6e. Wash any loosened solid into the filtration device and continue washing the precipitate as directed for the particular procedure being run.

There are several shapes of rubber policemen on the market for attachment to the end of a stirring rod. Irrespective of the shape, a policeman must be used only for scrubbing glass containers to remove adherent particles. The rubber portion of the policeman rod should never be used as a stirring

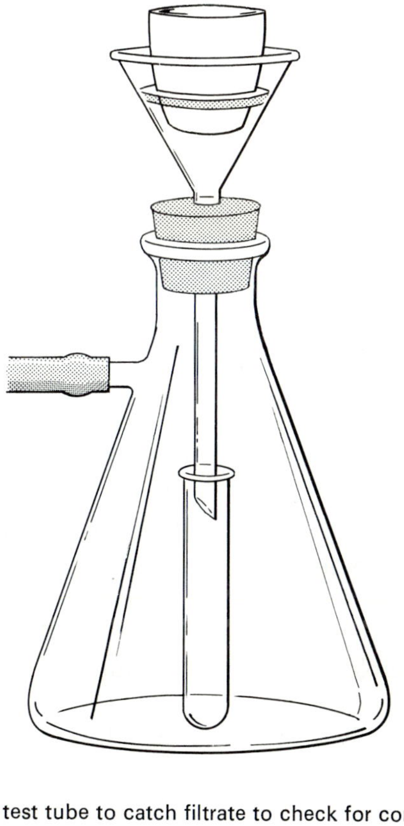

Figure 17.7. Use of test tube to catch filtrate to check for completeness of washing.

rod since the rubber will deteriorate, and there is grave danger of contamination of the solution from material on the policeman.

After washing is completed, it is usually necessary to check the filtrate for the presence or absence of some ion, which will indicate whether or not the washing is satisfactorily complete. For gravity filtration a few drops of the filtrate are caught on a watch glass. The proper reagents are then added to the watch glass to complete the test. For suction filtration, a few drops or milliters of the filtrate are caught in a test tube inserted in the flask, as shown in Fig. 17-7. The test is either made in the test tube or a small amount is transferred to a watch glass for the test.

E. Preparation of Gooch Crucibles

1. General

The Gooch crucible is a filtration device. The holes in the bottom must be covered with some material which will retain small particles of the precipitate and also allow free flow of liquid when suction is applied. Ignited and washed asbestos fibers are used to prepare the retaining pad. A properly prepared crucible mat will retain the smallest particles of the precipitate being filtered and will also allow rapid flow of liquid when suction is applied. One which fails in either of these two conditions is not satisfactory for use. To prepare a proper mat, the asbestos is applied in three steps: coarse fibers, medium fibers, and fine fibers.

2. Preliminary Steps

a. Clean a 400-ml beaker, the short-stemmed funnel, the suction flask, and a stirring rod.

b. Clean the Gooch crucibles with detergent and water and be sure they are marked so that each may be distinguished from each other. If marking is insufficient, call the instructor, who will show you how to mark the crucible permanently. *Do not mark with the wax glass marking pencil.*

c. Set up the suction arrangement as shown in Fig. 17-5. Connect the trap to the aspirator and to the suction flask with heavy-walled pressure tubing. Be sure the glass tubing which extends to the bottom of the trap is connected to the aspirator. Insert the short-stemmed funnel in a proper-sized one-hole rubber stopper and place it in the mouth of the suction flask. Support the rubber crucible holder in the funnel and place the Gooch crucible in the holder.

d. Obtain approximately 15 ml of concentrated asbestos suspension from the side shelf in a 400-ml beaker. Dilute to approximately 350 ml and stir.

3. Preparation of Asbestos Mat in Crucible

a. Do *not* turn on suction.

b. Agitate asbestos suspension vigorously, and completely fill the crucible with it to form the coarse part of the mat. Place the beaker containing the asbestos suspension on the desk.

c. Allow the liquid in the crucible to settle for 2 or 3 min.

d. Turn the suction on a slight amount until all water has been pulled through the crucible.

e. Tamp the asbestos mat gently with a well-rounded stirring rod.

f. Without reagitating the asbestos suspension in the beaker, again pour the crucible full of the suspension to make the medium fiber portion of the mat.

g. Allow the suspension in the crucible to settle for about 1 or 2 min and then increase the suction to moderate.

h. When water has been completely pulled through the Gooch, tamp pad lightly with the stirring rod and increase suction to full force. The pad now contains the coarse and medium layers.

i. Without reagitating the asbestos suspension, pour small amounts of the fine suspension down the stirring rod onto the pad until it is coated with fine particles of asbestos.

j. Wash 10–15 times with small portions (3–5 ml) of water by directing the stream from the wash bottle onto a stirring rod held vertically over the crucible. Do not pour water directly into the crucible as this may ruin the pad.

k. Cut off suction, remove the Gooch, and look through the bottom at a source of bright light. Only a faint amount of light should pass through the pad and there should be no bright spots.

l. Place the crucible in a marked beaker and dry it in the oven for at least 1 hr. Remove from the furnace, cool, store in the desiccator, and weigh.

m. Whenever the crucible is used for filtration, be sure to turn the suction on full before pouring any liquid into the crucible. Pouring liquid into the crucible with no suction usually disturbs the asbestos mat and ruins the crucible.

F. Ignition of Precipitates

Precipitates which have been caught in paper must have the paper removed before the precipitate can be dried or ignited for weighing since it is not possible to dry the paper itself reproducibly or with sufficient accuracy for most quantitative procedures. The precipitate in the paper is transferred to an ignition crucible, and the paper is removed by drying and charring

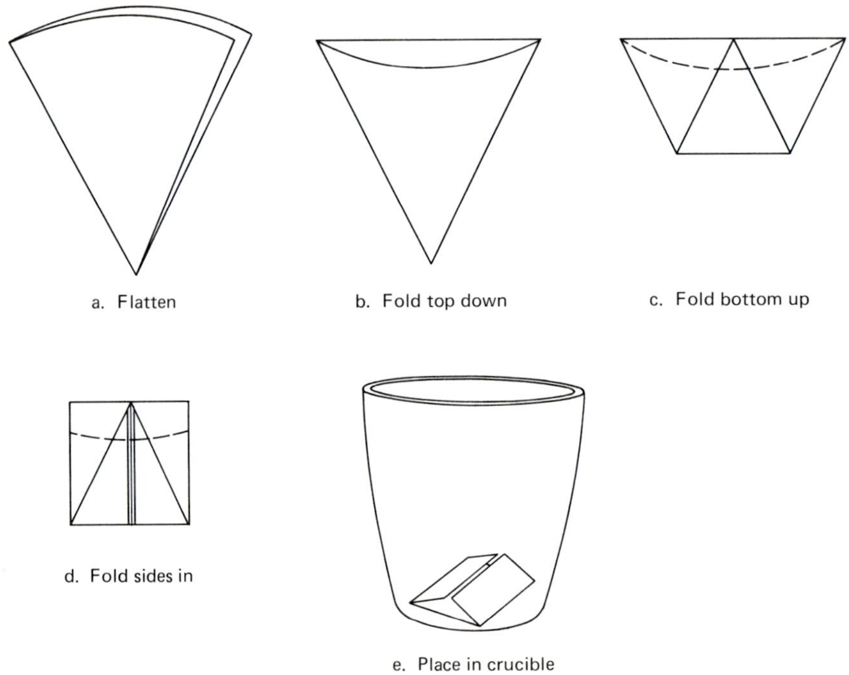

Figure 17.8. Procedure for folding filter paper containing a precipitate prior to ignition.

at low temperature followed by high-temperature ignition to convert the carbon to carbon dioxide.

The proper method of removing and folding the wet filter paper with the precipitate is shown in Fig. 17-8. After flattening, the paper is folded sequentially down from the top, up from the bottom, and finally in from the sides before insertion in the crucible. The purpose of the folding steps is to ensure that the precipitate is separated from the crucible by several thicknesses of paper.

The paper is dried and charred at low temperature in an upright position with the top at a slight angle to allow a small opening, as shown in Fig. 17-9a. During this part of the procedure, the temperature should be low enough that the bottom of the crucible never becomes red hot. High temperatures during the drying often cause spattering of the precipitate onto the sides and top, where it is easily lost in handling. After the paper is dried, the temperature should be raised slightly to cause charring to start. As the smoke from the charring decreases in volume, the temperature is raised again and this is continued until no more smoke appears even though a relatively hot flame is being used. *Do not allow the smoke or the paper to catch on fire during this part of the procedure.* After the paper is completely charred, the crucible is placed at an angle, and the top is adjusted as shown in Fig. 17-9b. The

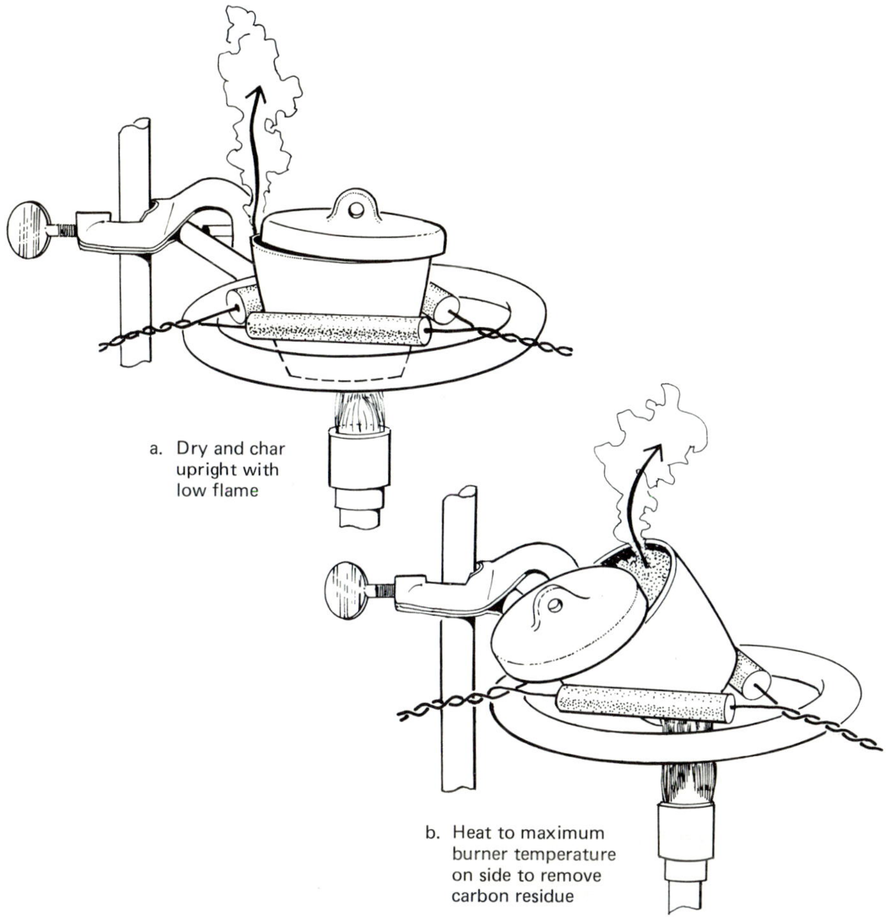

Figure 17.9. Drying, charring, and ignition of filter paper containing a precipitate.

flame from the burner should be increased to its maximum and placed at the bottom of the crucible so that the flame gases do not enter the top of the crucible. This position allows air to flow over the precipitate and paper and to burn the carbon more rapidly. Flame gases are reducing and often cause reduction of the precipitate to an undesirable form if allowed to enter the crucible while the precipitate is hot. The remainder of the carbon may be removed by heating in an ignition furnace at a specified temperature if one is available. Use of the furnace allows reproducibility of temperatures and conditions which is not possible for an open flame. If the top of the crucible is to be weighed, the carbon must be burned off before weighing. This can be done by heating to red heat over an open flame.

After removal of all paper and carbon, the crucible should be cooled and stored in a desiccator before weighing.

G. Drying and the Use of Desiccators

Solids such as unknowns and standards must be dried before weighing to rid them of adsorbed moisture. This is usually performed in a drying oven which is used by many students and which is kept between 110 and 115°C. Heating for 1 hr is usually sufficient, but when time is available, solids should be reheated and reweighed until a constant weight is attained. The solid to be dried is placed in a weighing bottle, and the weighing bottle is placed in a beaker covered with a supported watch glass. The contents of the weighing bottle should be marked clearly on the bottle, and the beaker should be marked clearly so that mistakes will not be made in removing the beakers from the oven. The bottle top should be removed from the weighing bottle but should be dried at the same time, as is shown in Fig. 17-10. Crucibles which do not have to be ignited are dried in the same way. The general rule to be followed in any drying or heating operation is that the equipment should be given the same heat treatment before use as it will be given during the procedure being followed.

The desiccator which is shown in Fig. 17-11 is used primarily to store dry materials to keep them dry. It consists of a lower chamber which is filled with a liquid or solid desiccant or drying agent and an upper chamber in which the objects stored are placed on a perforated porcelain plate. It is provided with an airtight ground-glass cover lubricated with stopcock

Figure 17.10. Arrangement for drying solids in drying oven.

Figure 17.11. Storage of dried materials in desiccator.

grease, and the cover is removed by sliding it sidewise. Red-hot objects are always allowed to cool to approximately 100°C or less before being placed in the desiccator. When a very hot object is placed inside, the air becomes heated and expands before the top is replaced; upon cooling, a vacuum is created inside the desiccator. This makes the top difficult to remove, and the rush of air when the top is removed may cause a fine solid to be blown from its container. Since the capacity of the desiccant to maintain a dry atmosphere is limited, the desiccator must be kept closed as much as possible. Leaving the top off for extended periods of time will saturate the desiccant with moisture so that it is no longer effective. Indicating desiccants are available which will change color as the ability of the solid to remove moisture is diminished.

The most popular solid desiccants are anhydrous magnesium perchlorate, activated alumina, and silica gel, all of which can be purchased in the indicating form for admixture with the ordinary product. All of these may also be regenerated at comparatively low temperatures and used over and over again. Silica gel requires 120°C, alumina 175°C, and magnesium perchlorate 240°C for regeneration.

H. Evaporation and Digestion

In many procedures it is necessary to reduce the volume of a solution before the next step in the procedure can be performed. Evaporations are usually carried out on a steam hot plate or over a steam bath in order to maintain as high a temperature as possible without causing the solution to boil. Whenever a solution is allowed to boil, there is loss of the solution itself because some of the microdroplets of spray caused by the ebullition escape from the container. Any solute in these droplets would not be determined in the determination step, and this causes a negative error in the analysis.

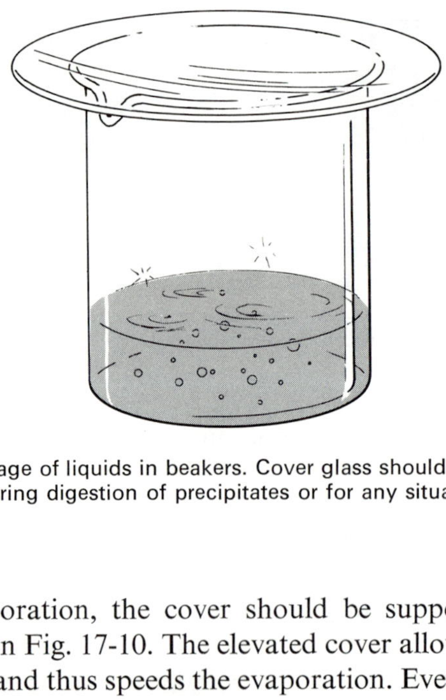

Figure 17.12. Storage of liquids in beakers. Cover glass should rest on beaker and should be in this position during digestion of precipitates or for any situation in which the liquid is boiling.

During evaporation, the cover should be supported by evaporation hooks, as shown in Fig. 17-10. The elevated cover allows the vapors to leave the beaker freely and thus speeds the evaporation. Even though it is undesirable to do so, at times it is necessary to reduce the volume of solution more rapidly than possible by regular evaporation. In this case the solution can be boiled gently with the cover resting on the beaker top, as shown in Fig. 17-12. Vigorous boiling and frothing must be avoided during the evaporation to ensure the smallest possible loss of solute.

It is often necessary to keep a solution at elevated temperature with minimum loss of solvent. Examples are the digestion of a precipitate to increase particle size and the slow dissolution of a solid. In these cases the solution is kept below the boiling point with the beaker covered, as shown in Fig. 17-12.

I. Use of the Volumetric Apparatus

1. General

It is necessary for the student to distinguish carefully between the volume that a vessel holds and the volume that it delivers. If exactly one liter of a liquid is poured into a clean dry flask, the amount of liquid which can be poured from the flask will be less than one liter since an appreciable quantity of the liquid will cling to the sides of the flask. The quantity of liquid delivered by the flask will depend upon the shape of the flask, the cleanliness of the inside surface, the time allowed for drainage, the viscosity of the liquid, the surface tension of the liquid, and the angle at which the flask is held, as well as other factors. If it is desired that the flask deliver exactly one liter, the neck of the flask must be marked at points which will vary depending upon the characteristics of the fluid in the flask. The neck, however, may be marked at a point which shows that the flask contains exactly one liter irrespective of the liquid in the flask. Volumetric flasks are marked in this way—to show the amount which is contained by the flask, not the amount which the flask will deliver. Burettes and transfer (volumetric) pipettes, on the other hand, are marked so that they will deliver a definite volume of liquid.

2. Burettes

Usually two 50-ml burettes are furnished each student in quantitative analysis. One of these will be a Geissler burette with a glass or Teflon stopcock, and the other will be a Mohr burette which is fitted with a glass bead in rubber tubing to control the flow of liquid. The Mohr burette is used for basic titrants such as sodium hydroxide, and the Geissler is used for practically all other titrants.

During titrations, the burette stopcock is normally operated with the left hand, as shown in Fig. 17-13. This leaves the right hand free for stirring the liquid or rotating the titration flask. Left-handed students should reverse this and use the right hand for operating the stopcock. Proper control of the stopcock to ensure delivery of small or large volumes accurately requires relaxed muscular control and extensive practice. Muscular tension often causes the student to push the stopcock out of the burette and thus ruin the titration.

There are many errors which can be involved in the use of burettes, especially by the beginning student in quantitative analysis:

a. Dirty burettes can cause an error because the volume of a liquid which clings to a dirty glass surface is different from the volume which clings to a clean surface. Water and dilute solutions form a smooth unbroken surface on clean glassware but will form droplets on a dirty surface. Burettes

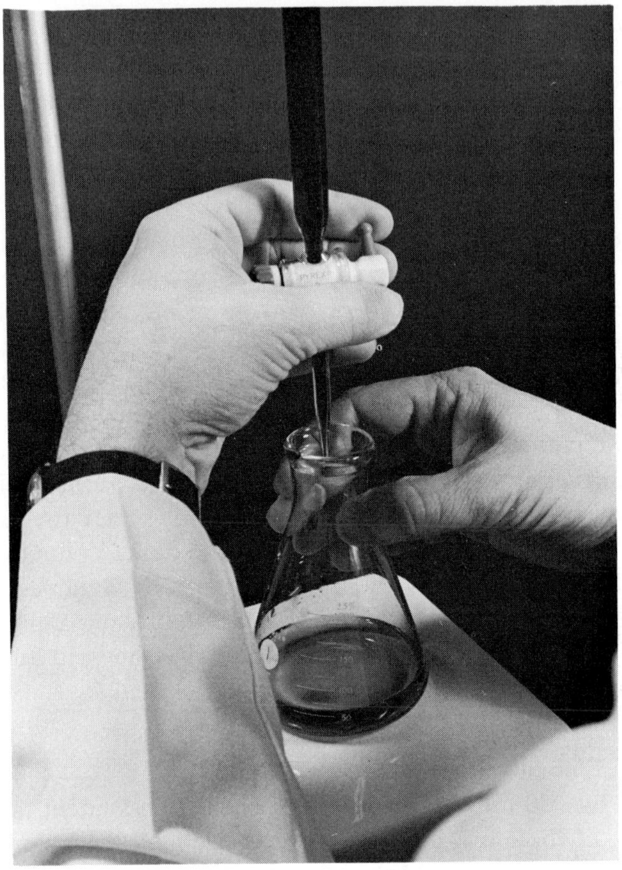

Figure 17.13. Operation of stopcock. Always use left hand for stopcock and right hand to agitate the solution.

should be cleaned periodically whenever it appears that they are not draining evenly. Usually, cleaning with a detergent and a burette brush is satisfactory, but at times it is necessary to use dichromate cleaning solution (concentrated sulfuric acid saturated with potassium dichromate) to get the burette really clean. The student should practice the utmost caution in the use of dichromate cleaning solution since it is very corrosive and will cause serious burns if allowed to remain on the skin for an appreciable length of time. It can also cause almost immediate deterioration of certain types of cloth, which may lead to embarrassing situations if other clothing is not available.

b. If the liquid is allowed to flow out of the burette rapidly, the student should wait at least 1 min before estimating the position of the meniscus to allow for proper drainage. An error of as much as 0.20 ml can result from reading the position of the meniscus too soon.

c. In filling the burette, care must be taken to ensure that no bubble of air remains in the tip. If a bubble is entrapped there and dislodged when the liquid is delivered, the reading will be in error by the volume of the bubble.

d. If a solution of known concentration is poured into a clean burette which is wet with water, the solution is diluted, and the concentration is no longer known exactly. The burette should either be dried before it is filled with the solution or else it should be rinsed at least three times with small volumes of the solution before filling. This rinsing procedure is usually necessary since the burette is seldom dry just before use.

e. An error due to parallax is encountered whenever the eye is not at the exact level of the meniscus, as shown in Fig. 17-14. To offset this error, the student should position his eye so that the back of the milliliter mark above the meniscus appears to be below the front portion and the back of the milliliter mark below the meniscus appears to be above the front portion.

f. Actually any part of the meniscus may be utilized as long as it is reproducible, but the same portion of the meniscus must be read every time the position of the meniscus is estimated or an error will be involved. A piece of white paper or a piece of white paper upon which a large black mark (with smooth upper edge) is made may be placed behind the burette to sharpen the bottom of the meniscus and to aid in the estimation of the correct reading. With intensely colored solutions such as potassium permanganate, the upper part of the meniscus is read.

g. The relative error in measuring the volume of liquid delivered by a burette decreases as the volume increases. Since an error of 0.02 ml is liable

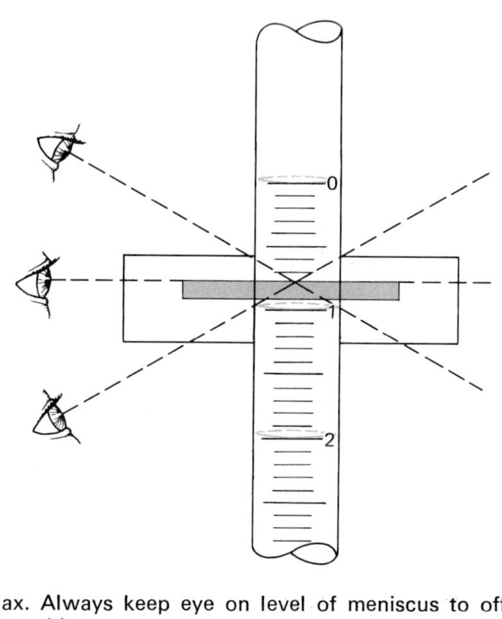

Figure 17.14. Parallax. Always keep eye on level of meniscus to offset parallax during estimation of meniscus position.

in each estimation of the meniscus, an error of 0.04 ml is possible. This represents a relative error of one part per thousand for 40.00 ml, an error of two parts per thousand for 20.00 ml, and an error of 40 parts per thousand for a volume of 1.00 ml. The delivery of more than one burette-full for any one titration is to be avoided since this requires refilling the burette and consequently two additional readings.

h. It is often necessary when approaching the end point of a titration to add a partial drop of titrant. To do this, allow the partial drop to form on the tip of the burette and then touch the tip to the side of the titration flask as shown in Fig. 17-15. The flask can then be titled so that the added amount is mixed with the solution or it can be washed down with water from the wash bottle. The burette tip should never be rinsed since this may dilute the titrant in the tip and cause an error.

i. Burettes will often leak slowly through the stopcock if the stopcock is not greased and inserted properly. The use of a large quantity of grease to offset a leak is not advisable since it can plug the bore of the stopcock or the tip of the burette. The correct way to clean and grease a stopcock is shown in Fig. 17-16. The burette is usually wet when the stopcock is cleaned and replaced, but the stopcock plug and walls must be dry to ensure a satisfactory seal. As a consequence, the burette is always held in a horizontal

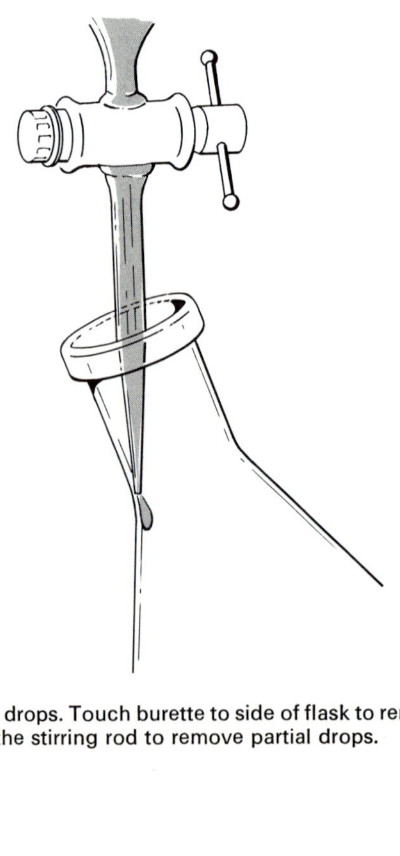

Figure 17.15. Partial drops. Touch burette to side of flask to remove partial drops. If titration vessel is a beaker, use the stirring rod to remove partial drops.

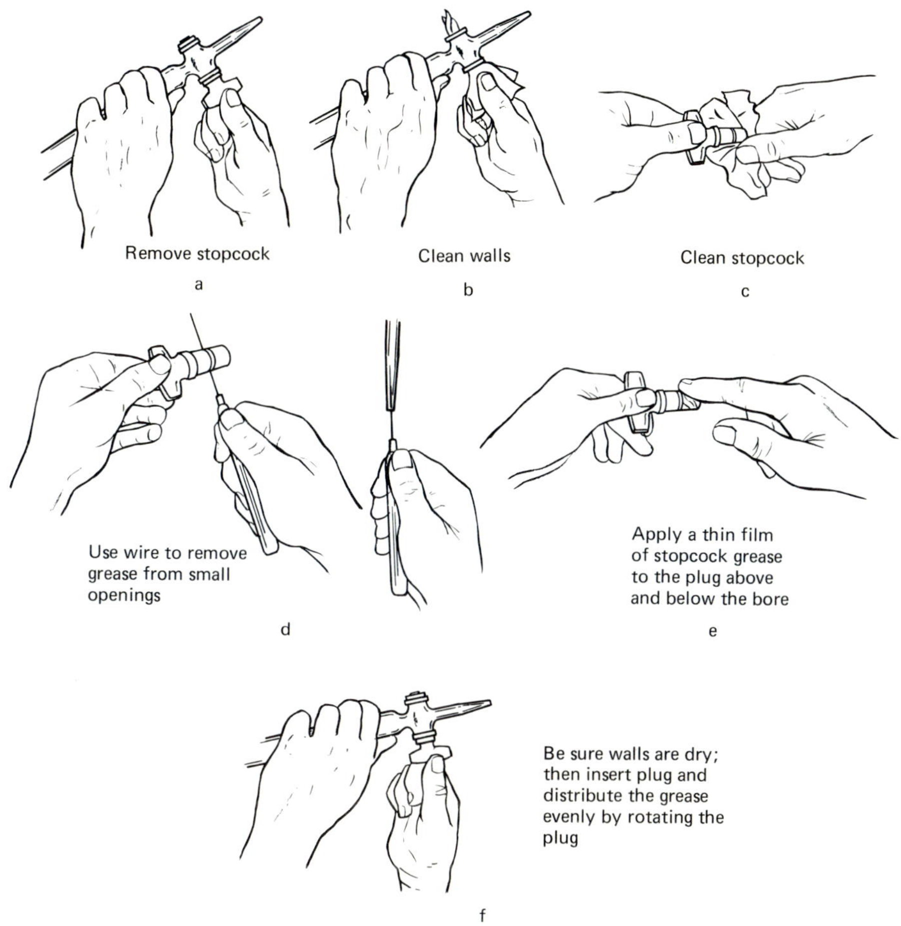

Figure 17.16. Procedure for cleaning and greasing a stopcock.

position during the removal, cleaning, regreasing, and reinsertion of the stopcock plug to ensure that the walls remain dry. The most common error other than getting water between the plug and walls is to use too much stopcock grease, with consequent clogging of the burette during later operation. A satisfactory seal will be transparent, will have no grease in the bore of the stopcock or tip of the burette, and will allow the plug to be rotated easily. A cloudy seal ordinarily indicates that the walls or plug were not dry or else that too much grease was put on the plug.

3. Pipettes

The two main types of pipettes used in the quantitative analysis laboratory are shown in Fig. 17-17. The volumetric or transfer pipette is used to deliver accurately a definite quantity of liquid such as 10.00 or 25.00 ml. The

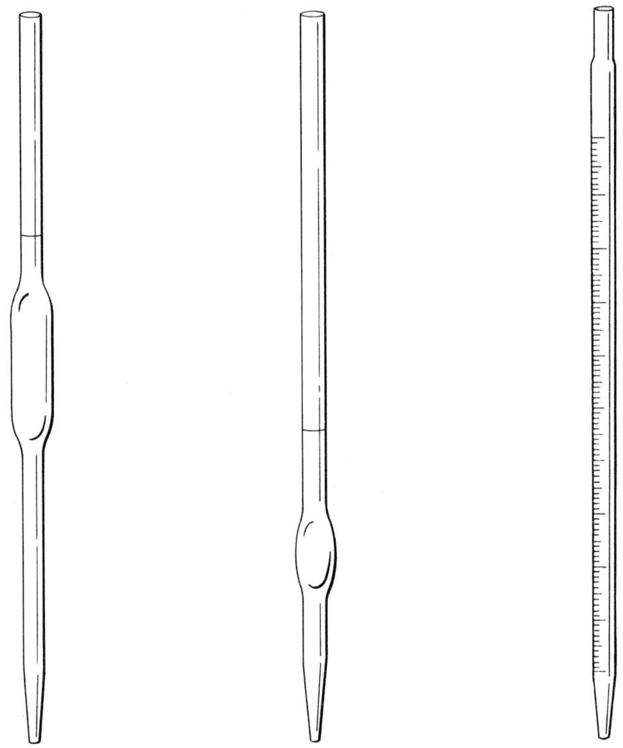

Figure 17.17. Types of pipettes used in quantitative analysis.

Mohr or graduated pipette is graduated so that different volumes can be delivered as desired. The transfer pipette is considered to be the more accurate since only one setting of the meniscus is necessary, while for the Mohr pipette the volume is obtained from the difference between two estimations of the meniscus. Most pipettes are calibrated to deliver a definite volume after a specified drainage time; thus some liquid must be left in the tip of the pipette after delivery. Some pipettes are calibrated, however, so that the last portion must be blown out rather than left in the tip. The blow-out-type pipette is characterized by one or two etched lines around the top of the pipette stem, and the bulb or expanded portion is closer to the tip than in the drainage type. It is seldom that the difference causes an error since most laboratories will have only one of these types on hand, but analysts should be aware that there is a difference and should also be able to identify the different types easily.

To pipette accurately and reproducibly the student must learn how to regulate the flow from the pipette. Since the control of the index finger is usually better than the control of the other fingers, the index finger should be used for this purpose.

The mouth should never be used to produce the necessary suction to pull the liquid into the pipette. A rubber bulb or other suction device should be used for this purpose. Many of the liquids which must be pipetted are corrosive to tissue or poisonous and thus should not be allowed to get into the mouth. No matter how expert an analyst becomes in mouth pipetting, there are times when inadequate control is exerted and a harmful substance is allowed to get in the mouth. The author, even though he knew better, once was unable to taste food for over a week because he used mouth suction for a concentrated sodium hydroxide solution.

The correct sequence to be used in pipetting is shown in Fig. 17-18. The rubber bulb is used to pull the liquid into the stem above the calibration mark, and the right index finger is placed over the top of the stem as the bulb is removed. The pipette is rotated to ensure adequate contact with the finger. The bulb is operated with the left hand. This operation is difficult but can be learned with repeated practice. The thumb should never be used to control the flow. The tip and lower stem of the pipette must be wiped dry with absorbent paper before the liquid level is allowed to fall to the calibration mark since the paper will remove some of the liquid from the tip. Consequently, if the stem and tip are wiped after the liquid level has been adjusted to the calibration mark, too small a volume will be delivered from the pipette. The meniscus is adjusted to the calibration mark by allowing the liquid to flow slowly until the meniscus reaches the calibration mark. Any drop hanging to the tip must be removed by touching the side of the container. The pipette is then moved slowly to a position over or in the receiving container, and the liquid is allowed to drain freely until the last drop falls. For the drainage-type pipettes, the pipette is allowed to drain for a specified length of time, and the tip is then touched to the side of the container to remove any partial drop which has formed. Usually it is sufficient to count to 15 after the last drop falls and then touch the side.

4. Volumetric Flasks

Volumetric flasks are used to prepare solutions of known concentration and are calibrated to contain rather than to deliver. They are designed as shown in Fig. 17-19 with a long neck, which has a much smaller diameter than the body of the flask in order to reduce the error in the estimation of the position of the meniscus. The two methods of preparing a solution of known concentration are the dilution of an accurately known volume of a more concentrated solution and the dissolving of a known weight of a solid in sufficient solvent to make a definite volume of solution.

Solids may be dissolved in another container and then transferred quantitatively to the flask, or else they may be dissolved directly in the

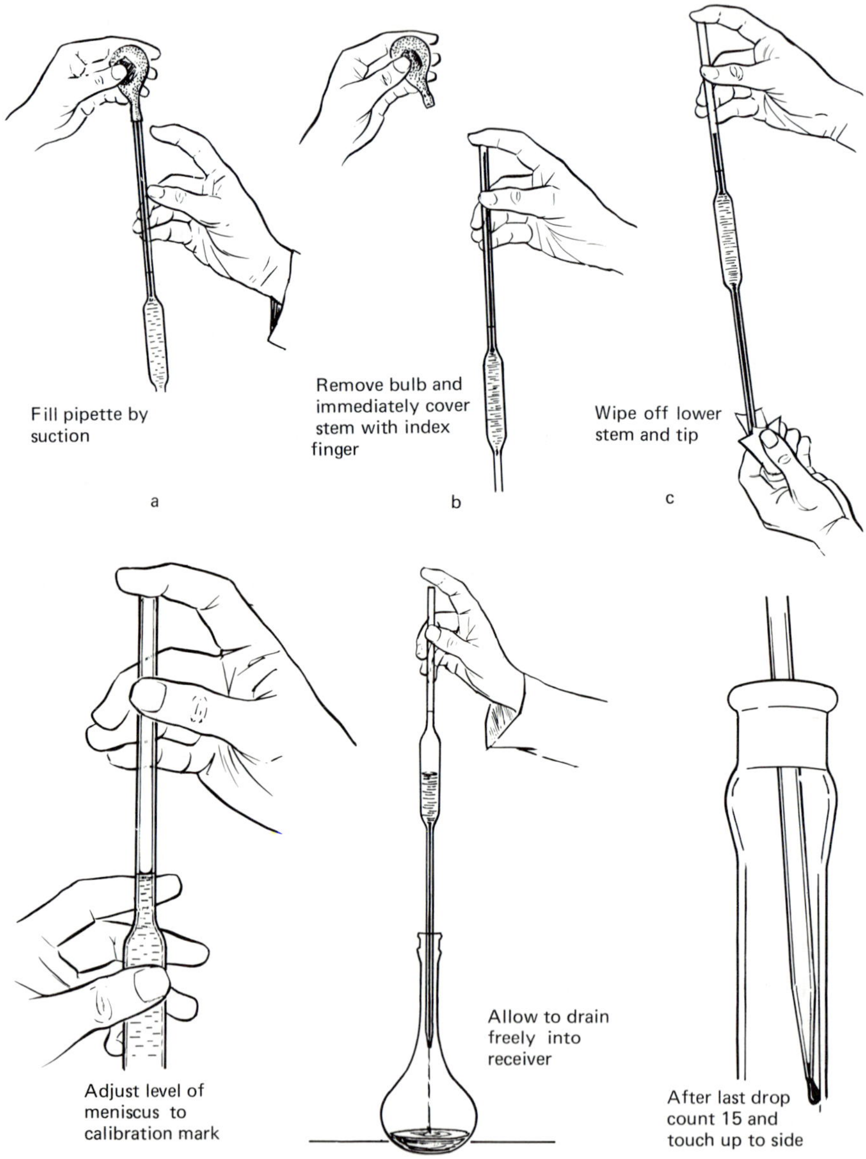

Figure 17.18. Correct procedure for use of pipette.

flask. In the latter case, the flask should be filled approximately three-fourths full and the flask rotated periodically until the solid is dissolved. More solvent can be added if necessary, but the flask should never be filled so full that the liquid cannot be rotated before all the solid is dissolved.

After the solid is dissolved or a liquid pipetted into the flask, it is necessary to fill the flask to the calibration mark and then mix completely before

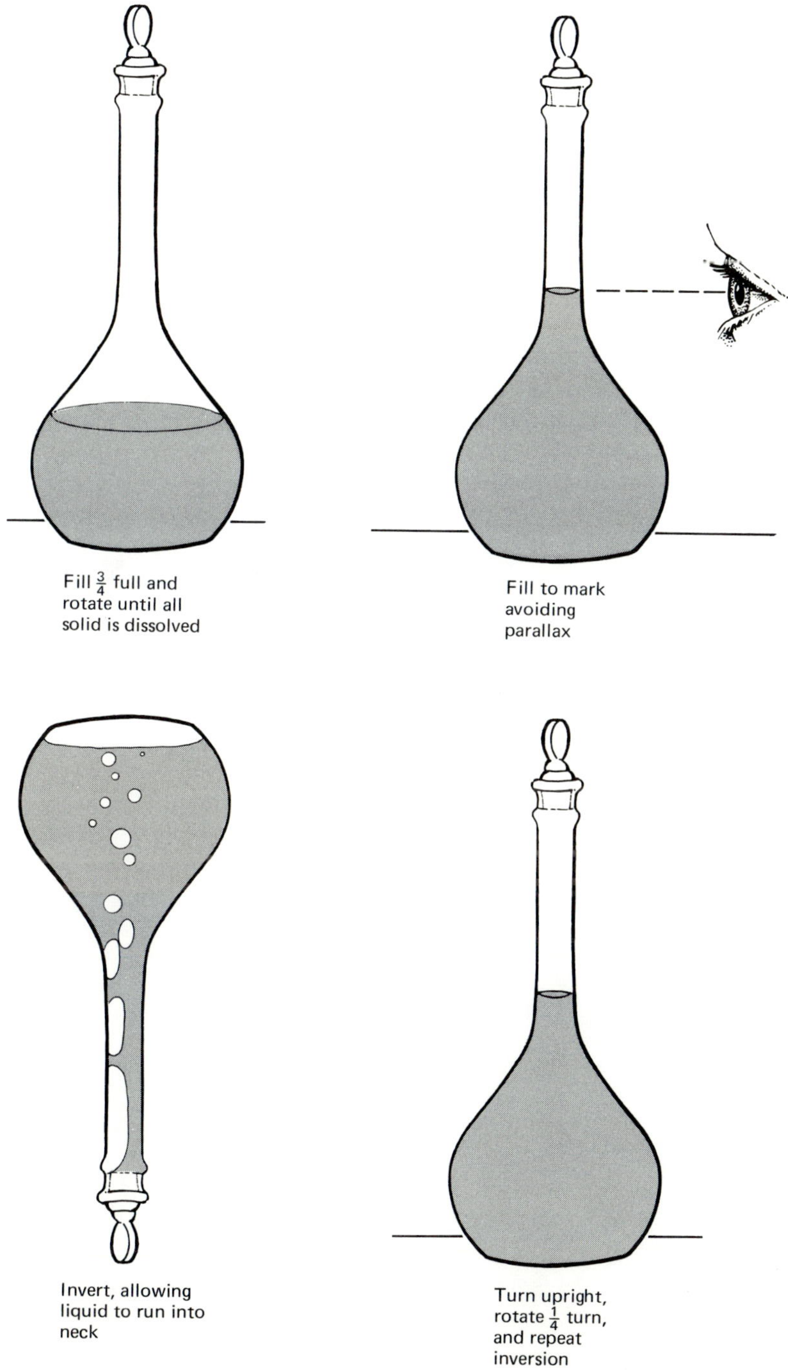

Figure 17.19. Correct procedure for use of volumetric flask.

transferring the solution to a storage vessel. Inadequate mixing after dilution is probably one of the most common errors made by beginning students in quantitative analysis. Adequate mixing is attained if the flask is inverted, turned upright, rotated one-quarter turn, inverted, and again turned upright, this process being repeated through at least four inversions. The proper sequence is shown in Fig. 17-19.

APPENDIX 1

The Equal-Arm Balance

A. Theory

Until recently the most common design of analytical balance in use in universities and in industrial laboratories was the equal-arm or two-pan balance (see Fig. 2-2). This type of balance is designed to weigh the object of unknown mass by direct comparison with an external set of standards or "weights" of known mass.

The sensitivity of the equal-arm balance varies with the amount of load on the pans. The variation is probably due to frictional effects which increase as the weight on the pans increases. The substitution or single-pan balance operates at constant load and thus at constant sensitivity.

The equal-arm balance is a lever of the first class, i.e., the fulcrum is between the two acting forces, the unknown object, and the weights (see Fig. 2-1). It is extremely important for a balance of this type to have the two lever arms equal.

B. Construction

The main components of the equal-arm balance, which are also its most obvious features (see Fig. A1-1), are (1) beam and knife edges, (2) stirrups and pans, (3) pointer and scale, and (4) release and support mechanisms.

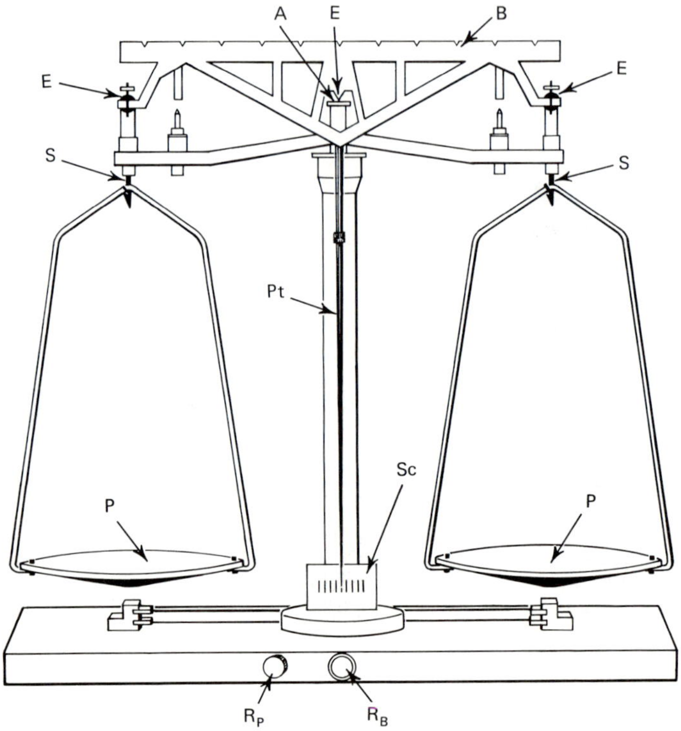

B	Beam
E	Agate knife edge
A	Agate plate
S	Stirrup
P	Pan
Pt	Pointer
Sc	Pointer scale
R_B	Beam release knob
R_P	Pan release knob

Figure A1.1. Construction details of equal-arm balance.

1. Beam and Knife Edges

When the balance is in operation, the beam is supported by a central knife edge on a plane on the beam post. Both the central knife edge and the plane are made of either agate or sapphire to reduce pivotal friction and to provide hard, durable surfaces. This central knife edge is the fulcrum of the lever. At each end of the lever (beam) there is another support mechanism similar to the central knife edge in construction. These two terminal knife edges are designed to be equidistant from the central knife edge. The distance from the central to the terminal knife edge is called the arm of the balance. All knife edges must be parallel and in the same horizontal plane. The beam is constructed of a rigid lightweight metal. The rigidity of the beam is maintained by a series of trusses that support the beam but keep the mass of the beam low. It is desirable to keep the mass of the beam as small as is feasible.

2. Strirrups and Pans

Suspended from the terminal knife edges are the stirrups. These stirrups hold the pans upon which the objects and weights are placed.

3. Pointer and Scale

The pointer is a long rod attached to the beam so that it extends downward to a scale placed on the bottom of the beam post. The scale consists of lines drawn vertically and equally spaced, with the center line labeled zero and the lines numbered outward from the center on either side. As the beam deviates from the horizontal its degree of deflection is indicated by the pointer position on the scale.

4. Release and Support Mechanisms

The mechanical release and support mechanisms for the beam and the pans are operated from outside the balance case. When the beam is supported, the central knife edge is raised off the agate plane a slight distance. In this position the knife edge is protected from harm if the beam receives a serious jolt. Gentle release of the beam mechanism is essential to the continued good working of a balance. Release of the beam first lowers the knife edge to the agate plane and then releases the beam arms to allow free movement. The pan supports are located underneath the pan and when released are moved downward to allow free movement of the pans.

5. Damping

The release and support mechanisms just described are designed to relieve the knife edges from undue strain that could result in their damage. A feature that is not necessary to the working of the analytical balance but which few chemists would wish to relinquish is the *damping* mechanism found on double-pan balances today. Prior to the introduction of damping methods, it was necessary to weigh by the *swing method.* This method, tedious and time-consuming, is described in Section D. The low pivotal friction of the central knife edge results in extended oscillations of the beam assembly back and forth across the equilibrium position. Damping of the balance arm reduces the number of oscillations to just one or two. The equilibrium position of undamped balances may be calculated without the beam coming to a complete rest. Figure A1-2 shows a typical damping mechanism which makes use of magnets to slow the oscillations of the beam without interfering with the normal attainment of the equilibrium position. Air damping is used on some balances.

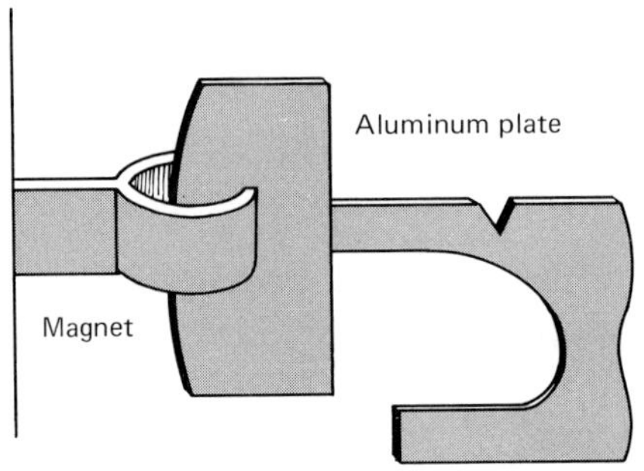

Figure A1.2. Magnetic damper.

6. Chainweight Balances

Earlier double-pan balances used weights ranging from 0.010 to 0.500 g. These small weights were placed on the right-hand pan along with the larger weights. A small rider which was moved along the beam was used for milligram values, and the mass value to the nearest $\frac{1}{10}$th of a milligram was calculated using the sensitivity value of the balance.

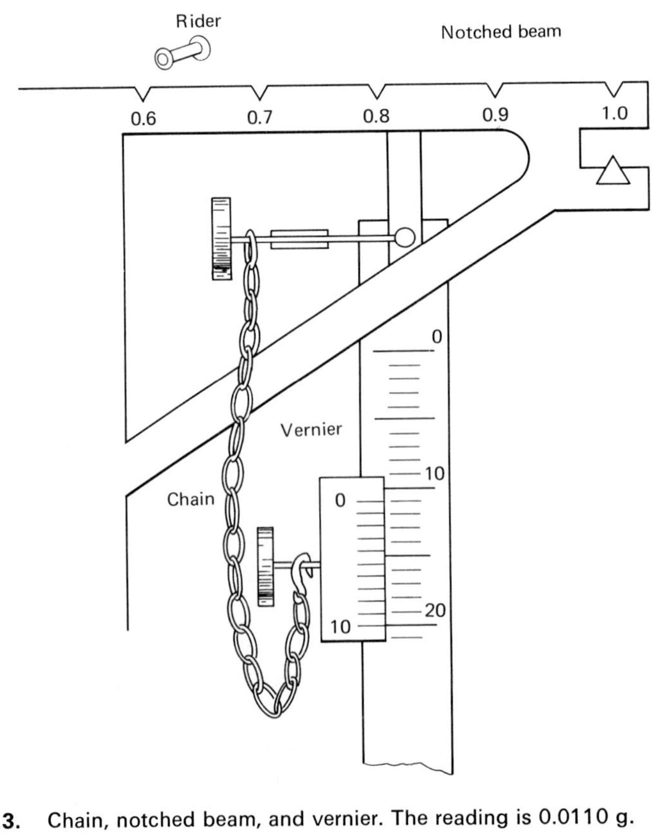

Figure A1.3. Chain, notched beam, and vernier. The reading is 0.0110 g.

A major improvement of the two-pan balance occurred when a fine gold chain was suspended from the right-hand side of the beam to take care of the weight reading from 0.0001 to 0.0999 g. Determination of tenth gram values is carried out on these balances by means of a movable rider which may be moved from 0.0000 to 1.0000 g in 0.1000-g intervals by placing it in the respective notch in the beam, as shown in Fig. A1-3.

Lengthening of the chain increases the weight on the right-hand pan by the value given on the fixed scale and vernier. The rider is calibrated in tenths of a gram, the fixed scale provides the hundredths and thousandths of a gram figures, and the vernier reading gives the tenths of a milligram value. The chain is raised or lowered in conjunction with the vernier by means of a wheel outside the case. The mechanism described eliminates the troublesome small weights. Refer to Fig. A1-3 for illustration.

The way that the chain is attached may vary from manufacturer to manufacturer, but the general form is as described above. The chief disadvantage of the chainomatic balance is the difficulty in making a chain that is completely uniform. However, for most work the reduced precision of the chain-type balance is more than compensated for by the convenience and simplicity of operation.

C. Analytical Weights

The double-pan balance requires a set of comparison masses (weights) or analytical weights. A set for the chainomatic balance usually consists of the following denominations: 100, 50, 20, 10, 10, 5, 2, 2, and 1 g. A non-chainomatic balance would require additional weights of 500, 200, 200, 100, 50, 20, 20, and 10 mg. Other combinations are of course possible which would allow a complete sequence of values.

Manufacturers of weights have two very important considerations in selecting materials. The weight must be made of relatively dense material so that it is of convenient size and the weight must be made of very non-corrosive material so that its mass will remain constant. Several metals satisfy these requirements to a degree which is directly related to their values; i.e., gold or platinum would be the best metals to use except for their very high cost. Compromises are therefore necessary. The National Bureau of Standards provides standards of construction and tolerances for weights. Specifications for various classes of weights that are in use in quantitative analysis are included in the National Bureau of Standards Circular 547, *Precision Laboratory Standards of Mass and Laboratory Weights.* The NBS tolerances are shown in Table A1-1.

TABLE A1.1 Tolerances in Milligrams Applying to National Bureau of Standards Weight Classes

Denomination	Class M	Class S	Class S-1	Class P
1 kg	5.0	2.5	10	20
500 g	2.5	1.2	5.0	10
300 g	1.5	0.75	3.0	6.0
200 g	1.0	0.50	2.0	4.0
100 g	0.50	0.25	1.0	2.0
50 g	0.25	0.12	0.60	1.2
30 g	0.15	0.074	0.45	0.90
20 g	0.10	0.074	0.35	0.70
10 g	0.050	0.074	0.25	0.50
5 g	0.034	0.054	0.18	0.36
3 g	0.034	0.054	0.15	0.30
2 g	0.034	0.054	0.13	0.26
1 g	0.034	0.054	0.10	0.20

Classes of analytical weights are identified by descriptive title and letter designation and comprise major classes M, S, S-1, and P. Principle characteristics of these classes are as follows.

Class M, high-precision scientific standards: designed for use as reference standards, for analytical work of the highest precision and for specialized work requiring the most accurate determinations of mass. Weights of this class are made in one piece so that their effective density may be very exactly determined. The gram weights are normally made of brass and are plated with gold, platinum, or rhodium.

Class S, scientific working standards and precise analytical weights: for use as working standards in the calibration of other weights, for exact analytical weighing without necessity for the application of correction factors, and for keyboard or dial-controlled weights in rapid weighing mechanized balances. Class S weights of 1 g and larger are normally made in two pieces, the knob being separable and serving to close an adjusting cavity beneath it. Lead may not be used as the adjusting material. The gram weights are made of brass, bronze, or stainless seel, brass or bronze being plated with platinum or rhodium or lacquered.

Class S-1, routine analytical weights: for general application as precise balance weights in quantitative analysis. With respect to design and materials, this class is identical in specifications with class S but is held to somewhat less exact adjustment tolerances in order to facilitate economy of manufacture.

Class P (formerly class S-2), rough analytical weights: for use as student weights with analytical balances and for routine analytical weighings in connection with methods not requiring or not capable of high precision. General requirements are similar to those of classes S and S-1 but less discriminatory in respect to materials of construction, finish, and packaging.

D. Weighing by the Swing Method

Most modern equal-arm balances have a notched beam for the determination of the weight to the closest 0.1 g and a movable chain which is used to obtain the weight to the closest 0.0001 g or 0.1 mg. Many of these balances are also damped so that weights can be obtained by reproduction of the equilibrium point of the balance by adjustment of the chain position. For those balances which are not damped and those which do not have a notched beam, it is usually preferable to weigh an object to the closest milligram and to calculate the weight to the closest 0.1 mg by use of the sensitivity, the equilibrium point of the balance, and the rest points of two weights which differ by 1 mg. This method of calculating the weight to the closest 0.1 mg is known as weighing by the swing method and is described in this appendix. The student should be familiar with the various aspects of construction and operation of equal-arm balances discussed in Chapter 2.

1. Rest Points and Turning Points

When the beam of an equal-arm balance is set in motion, the pointer moves to one side of the scale, reverses its direction, and moves to the other side. It continues moving in this manner until the beam motion finally ceases, and the balance comes to rest. The position on the scale at which the pointer reverses its direction is known as the turning point (TP) of the swing, while the position at which the pointer comes to rest is called the rest point (RP) of the balance. The position of the pointer is estimated to the closest tenth of a scale division, and the side of zero on which it falls is specified by using the letter R for right and the letter L for left. Examples are shown in Fig. A1-4. The rest point determined with the balance empty is known as the rest point zero (RP0) and is also often called the equilibrium point of the balance. Whenever the RP0 is reproduced during a weighing, the weights on both arms of the balance are equal. In determining the weight of an object, the object is placed on one pan and weights are added to the other pan until RP0 is reproduced. The sum of such weights is the weight of the object being weighed. If the balance being used in a damped balance, the RP0 is read directly as soon as the beam motion has ceased and the pointer has come to rest. In any undamped balance, the time necessary for the beam motion to cease may be many minutes, and consequently the RP0 must be calculated rather than read directly. This calculation is made from several successive turning points.

a. CALCULATION OF REST POINT. To calculate the rest point from a series of successive swings, at least five successive turning points must be used which have equal differences on each and both sides. The turning points on each side are averaged, and the difference between these averages is

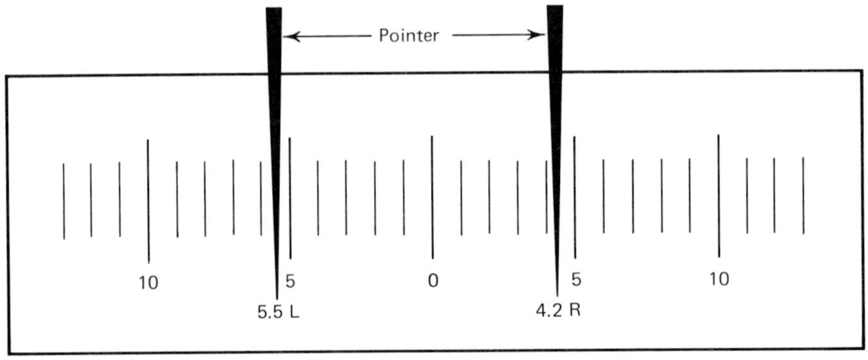

Figure A1.4. Turning points for equal-arm balance. The scale is located at the bottom of the beam post and has equally spaced lines with the center line labeled zero and the lines numbered outward from the center on both sides. Turning points of 5.5 L and 4.2 R are illustrated.

obtained by subtraction. This difference between the averages is called the deflection of the balance and may be defined as the difference between turning points when one turning point is zero. The rest point will be halfway between these two turning points and thus will be one-half the deflection. The rest point is always expressed to the closest tenth of a scale division. As an example, let us assume that the following turning points were obtained during the weighing of an object:

	L	R
	4.6	
		4.0
	4.4	
		3.8
	4.2	
Total	13.2	7.8
Average	4.4	3.9

Deflection = 4.4 − 3.9 = 0.5 L

Rest Point = 0.5 L ÷ 2 = 0.25 L. Round off to closest tenth of a unit = 0.3 L

Best results are obtained when the turning points on at least one side are between the third and fifth divisions or when the total amplitude is between six and ten divisions. The first two turning points of any set of swings should not be used.

2. Sensitivity

The sensitivity of a balance is the weight necessary to cause the rest point to move one division on the scale and is expressed as milligrams per division or as grams per division. The reciprocal sensitivity is sometimes used and may be defined as the displacement of the rest point per unit of weight. It is usually expressed in terms of divisions per milligram. The sensitivity varies with the amount of load on the pans. If many weighings are to be made by the swing method, it is best to determine the sensitivity at various weights and draw a graph of sensitivity versus weight from which the actual sensitivity for any given weighing may be read. If, however, only a few weighings are to be made by the swing method, it is preferable to calculate the sensitivity from data taken during the weighing operation. To calculate the sensitivity, two rest points differing by one milligram of weight are needed.

When weighing by the swing method, it is difficult and time-consuming to try to reproduce the exact RP0. If a rest point close to the RP0 is obtained and the sensitivity is known, the amount of weight necessary to be added to the known weight to reproduce RP0 can be calculated. An example showing the method of calculating the sensitivity and the weight of an object to the closest tenth of a milligram by the swing method is given below:

RP0 — 0.1 L

RP at load 16.475 (low load) — 1.3 R

RP at load 16.476 (low load + 1 mg) — 1.2 L

a. CALCULATION OF SENSITIVITY. Number of divisions RP moved by 1 mg of weight:

$$1.3 \text{ R to } 1.2 \text{ L} = 2.5 \text{ div.}$$

Weight necessary to move RP one division:

$$\frac{0.001 \text{ g}}{2.5 \text{ div.}} = 0.00040 \text{ g/div.}$$

Sensitivity is 0.00040 g/div. Sensitivities calculated in this way should be expressed to the closest hundredth of a milligram or hundred thousandth of a gram.

b. CALCULATION OF WEIGHT OF OBJECT TO THE CLOSEST TENTH OF A MILLIGRAM. It is necessary to find what fraction of a gram must be added to the weight at low load to move the RP to the RP0 position. To do this, determine the number of divisions the RP must be moved and multiply by the sensitivity. Add this amount to the weight at low load. In the above example, the RP must be moved from 1.3 R to 0.1 L, which is 1.4 divisions.

$$1.4 \text{ div.} \times 0.00040 \text{ g/div.} = 0.00056 \text{ g}$$

Before adding this to the weight at low load, round off to the closest tenth of a milligram or 0.00056 rounds off to 0.0006.

$$16.475 + 0.0006 = 16.4756$$

which is the weight of the object to the closest tenth of a milligram.

E. Problems

1. Calculate the rest point of a balance from the following swings:
   (a) 3.0 L – 3.7 R – 2.8 L – 3.5 R – 2.6 L
   (b) 3.0 R – 3.9 L – 2.8 R – 3.7 L – 2.6 R
   (c) 4.0 L – 5.5 R – 3.8 L – 5.3 R – 3.6 L
   (d) 2.5 R – 6.4 L – 2.3 R – 6.2 L – 2.1 R
   (e) 4.8 L – 4.7 R – 4.6 L – 4,5 R – 4.4 L
   (f) 9.8 R – 5.7 L – 9.4 R – 5.3 L – 9.0 R

   Ans. (b) 0.5 L

2. Calculate (a) the sensitivity and (b) the correct weight to the closest 0.0001 g for the following weighings; RPLL is the RP at weight LL, and RPLL + 1 is the rest point at weight LL plus 1.0 mg:

	RPO	RPLL	LL	RPLL + 1
(a)	0.2 R	1.0 R	5.526	1.0 L
(b)	0.2 R	1.0 R	7.653	1.5 L
(c)	0.2 L	0.5 R	6.373	1.5 L
(d)	0.5 L	1.5 R	8.379	1.5 L
(e)	0.8 L	0.0	11.876	2.8 L
(f)	0.0	0.1 L	9.999	1.9 L

Ans. (a) 0.00040 g/div (b) 7.6533 g

APPENDIX 2

Inorganic Nomenclature

Several methods of naming inorganic compounds have been utilized in the past, but the method used to the greatest extent today is the system proposed by a committee of the International Union of Pure and Applied Chemistry. This system is described below and is correlated with some of the older systems still in use. The naming of binary compounds is discussed before ternary compounds.

A. Binary Compounds

Binary compounds contain two different elements, and the most electropositive element is named first since it is written first in the formula. To name a binary compound, give the name of the most electropositive element followed by the name of the most electronegative element ending in "ide." The majority of such compounds are those formed by reaction of a metal with a nonmetal and thus are classified as being salts. For metals with variable valence, the oxidation number is shown by use of a roman numeral in parenthesis following the name of the metal. Examples are

NaCl	sodium chloride	$FeCl_2$	iron(II) chloride
Al_2O_3	aluminum oxide	Fe_2O_3	iron(III) oxide
Ca_3P_2	calcium phosphide	$SnBr_4$	tin(IV) bromide
CdS	cadmium sulfide	Cu_2O	copper(I) oxide

In the older system, variable valence of metals was indicated by use of the suffixes " ous " and " ic " on the name of the metal, with " ous " indicating the lower and " ic " the higher valence. Thus $FeCl_2$ is ferrous chloride and $FeCl_3$ is ferric chloride.

Binary compounds may also be formed between two nonmetals with the most electropositive being written first and named first. For elements of variable valence which can form more than one compound, prefixes are used with each nonmetal to indicate the number of atoms of each element present. Prefixes in common use are " mono " for one, " di " for two, " tri " for three, " tetra " for four, " penta " for five, " hexa " for six, " hepta " for seven, and " octa " for eight. The prefix " mono " is understood if no prefix is used. Examples are (older methods of naming are given in parentheses when there is a difference)

N_2O	dinitrogen monoxide (nitrous oxide)
NO	nitrogen oxide or mononitrogen monoxide (nitric oxide)
NO_2	nitrogen dioxide
N_2O_3	dinitrogen trioxide (nitrogen trioxide)
SO_2	sulfur dioxide
SO_3	sulfur trioxide
IF_7	iodine heptafluoride
BrF_5	bromine pentafluoride

A few polyatomic ions have special names and are treated as if they were single atoms or elements in naming their compounds. Examples are " OH^- " (hydroxide), " CN^- " (cyanide), and " NH_4^+ " (ammonium). Thus NaOH is sodium hydroxide, KCN is potassium cyanide, and NH_4Cl is ammonium chloride.

If a binary hydrogen compound acts as an acid when it is dissolved in water it is usually named as an acid with the prefix " hydro " and the suffix " ic." Thus HCl is hydrochloric acid, H_2S is hydrosulfuric acid, HCN is hydrocyanic acid, HI is hydroiodic acid, and HBr is hydrobromic acid. These compounds also may be named as hydrogen chloride, hydrogen sulfide, hydrogen cyanide, etc. This type of name is usually used when referring to the actual compound, while the acid name is used to refer to the water solution of the compound. Thus we speak of hydrogen chloride gas, the water solution of which is hydrochloric acid.

B. Ternary Compounds

Ternary compounds are those containing three different elements. It has already been noted that some ternary compounds, such as NH_4Cl, KOH, and HCN, are named as if they were binary compounds. Chlorine, nitrogen, sulfur, phosphorus, and several other elements each form oxyacids (ternary compounds with hydrogen and oxygen) which differ from each other in their oxygen content. Usually the most common acid of a series bears the name of the acid-forming element ending with the suffix "ic." This may be noted in the names chloric acid ($HClO_3$), sulfuric acid (H_2SO_4), nitric acid (HNO_3), and phosphoric acid (H_3PO_4). The names of acids containing one oxygen atom more than the "ic" acid retain the suffix "ic" and have the prefix "per" added. The name perchloric acid for $HClO_4$ illustrates this rule. An acid which contains one less oxygen atom than the "ic" acid is named with the suffix "ous." Examples are chlorous acid ($HClO_2$), sulfurous acid (H_2SO_3), nitrous acid (HNO_2), and phosphorous acid (H_3PO_3). Acids with one less oxygen atom than the "ous" acid are named by adding the prefix "hypo" and retaining the ending "ous." Thus, $HClO$ is hypochlorous acid, and H_2SO_2 is hyposulfurous acid.

Metal salts of the oxyacids (compounds in which a metal replaces the hydrogen of the acid) are named by naming the metal and then the acid radical. The ending "ic" of the oxyacid name is changed to "ate" and the ending "ous" of the acid is changed to "ite" for the salt. The salts of perchloric acid are perchlorates, those of sulfuric acid are sulfates, those of nitrous acid are nitrites, and those of hyposulfurous acid are hyposulfites. This system of naming applies to all inorganic oxyacids and their salts. Examples of the names of the oxyacids of chlorine and the corresponding sodium salts are given below:

Acids		Salts	
$HClO$	hypochlorous acid	$NaClO$	sodium hypochlorite
$HClO_2$	chlorous acid	$NaClO_2$	sodium chlorite
$HClO_3$	chloric acid	$NaClO_3$	sodium chlorate
$HClO_4$	perchloric acid	$NaClO_4$	sodium perchlorate

If an acid contains more than one hydrogen, it may react partially with a base to form acid salts which still contain some hydrogen. Examples are $NaHS$, $NaHCO_3$, KH_2PO_4, and $CaHPO_4$. Such salts can be named by stating the electropositive elements present followed by the usual name of the acid radical; thus $NaHS$ is sodium hydrogen sulfide, $NaHCO_3$ sodium hydrogen carbonate, KH_2PO_4 is potassium dihydrogen phosphate, and $CaHPO_4$ is calcium hydrogen phosphate. If the acid which produces the salt contains two hydrogens (is diprotic), the salt may be named by using the prefix "bi" in front of the name of the acid radical; thus $NaHS$ is sodium bisulfide, $NaHCO_3$ is sodium bicarbonate, $NaHSO_4$ is sodium bisulfate, and $Mg(HSO_3)_2$ is magnesium bisulfite. If the oxyacid contains three hydrogens (is triprotic), the acid salts are often called the primary and secondary salts. Thus NaH_2PO_4 is primary sodium phosphate, and Li_2HPO_4 is secondary lithium phosphate.

APPENDIX 3

General Solubility Rules

It should be realized that these generalizations apply to the simple compounds of the more common metals and acids. There are many exceptions among the less common metals and complex compounds.

1. Practically all the ammonium and alkali metal salts are soluble. The alkali metals are the Group I metals: lithium, sodium, potassium, rubidium, and cesium.

2. Most nitrates, chlorates, and acetates are soluble.

3. All chlorides are soluble except silver, mercurous, and lead chlorides. Lead chloride is soluble in hot water.

4. All sulfates except barium, strontium, and lead sulfates are soluble Calcium sulfate is only moderately soluble.

5. All carbonates, phosphates, arsenates, and sulfides are insoluble with the exception of the ammonium and alkali salts. The alkaline earth sulfides are hydrolyzed in water.

6. All hydroxides except sodium, potassium, ammonium, barium, and strontium are insoluble. Calcium hydroxide is only moderately soluble.

APPENDIX 4

Density of Some Common Substances

Substance	Density (g/cm^3)
Air	0.0012
Aluminum	2.7
Brass	8.4
Copper	8.9
Glass	2.6
Gold	19.3
Iron	7.9
Lead	11.3
Mercury	13.6
Nickel	8.9
Platinum	21.4
Porcelain	2.4
Silver	10.5
Stainless steel	8.0
Water	1.0

APPENDIX **5**

Literature of Analytical Chemistry

The literature of analytical chemistry is constantly growing as new methods are developed and new principles are discovered which can form the basis of an analytical procedure. Most of the new information is contained in research articles published in various journals or periodicals. Since it is impossible to read all the analytical research papers, even those dealing with relatively narrow fields, many analytical chemists read only a few articles and keep abreast of the literature by reading abstracts and reviews. The best sources of abstracts are *Chemical Abstracts*, *Analytical Abstracts*, *Chemisches Zentralblatt*, and *Gas Chromatography Abstracts*. Each issue of some of the analytical periodicals have rather complete abstracts of the articles in that issue as well as abstracts of papers in other journals.

Review articles are probably the easiest and quickest way to keep up with the progress in any particular field. *Analytical Chemistry* devotes one issue each year to very complete biennial reviews in various fields of pure and applied analytical chemistry. There are also annual "reviews" and "advances" in the various aspects of analysis and instrumentation which can be beneficial in obtaining information on recent developments.

Only a few of the periodicals and books devoted to analytical chemistry are listed in order to furnish the student with an introduction to the literature available and to suggest sources for further reading and study. Many excellent sources of information must be omitted in any such list.

A. Periodicals

Analytical Chemistry (U.S.A.)
Analytica Chimica Acta (International)
Analyst, The (British)
Chemist-Analyst (U.S.A.)
Journal of Chemical Education
Separation Science
Talanta (International)
Zeitschrift für Analytische Chemie (German)
Zhurnal Analiticheskoi Khimii (Russian; English translation available)

B. Multivolume Collections of Principles and Methods

N. H. Furman, ed., Vol I, F. J. Welcher, ed., Vol II, *Standard Methods of Chemical Analysis*, 6th ed., Van Nostrand, Princeton, N.J., 1963.
I. M. Kolthoff and P. J. Elving, ed., *Treatise on Analytical Chemistry*, Wiley-Interscience, New York, 1959.
F. J. Welcher, *Organic Analytical Reagents*, 4 vols, Van Nostrand, Princeton, N.J., 1947.
C. L. Wilson and D. W. Wilson eds., *Comprehensive Analytical Chemistry*, Van Nostrand, Princeton, N.J., 1959.

C. Official Methods

A.S.T.M. Methods for Chemical Analysis of Metals, American Society for Testing Materials, Philadelphia. Latest edition.
W. Horwitz, *Official Methods of Analysis*, Association of Official Analytical Chemists, Washington, D.C., Latest edition.
E. Wichers et al., *Reagent Chemicals*; *American Chemical Society*, American Chemical Society, Washington, D.C. Latest edition.
E. K. Kline et al., *Standard Methods for the Examination of Water, Sewage and Industrial Wastes*, American Public Health Association, New York. Latest edition.

D. Texts and Reference Books

1. Texts

G. H. Ayres, *Quantitative Chemical Analysis*, 2nd ed., Harper & Row, New York, 1968.
R. B. Fischer and D. G. Peters, *Quantitative Chemical Analysis*, 3rd ed., Saunders, Philadelphia, 1968.
J. S. Fritz and G. H. Schenk, *Quantitative Analytical Chemistry*, 2nd ed., Allyn & Bacon, Boston, 1969.

L. E. Hamilton and S. G. Simpson, *Quantitative Chemical Analysis*, 12th ed., Macmillan, New York, 1964.
I. M. Kolthoff, E. B. Sandell, E. J. Meehan, and S. Bruckenstein, *Quantitative Chemical Analysis*, 4th ed., Macmillan, New York, 1969.
H. A. Laitinen, *Chemical Analysis*, McGraw-Hill, New York, 1960.
L. Meites and H. C. Thomas, *Advanced Analytical Chemistry*, McGraw-Hill, New York, 1958.
D. A. Skoog and D. M. West, *Fundamentals of Analytical Chemistry*, 2nd ed., Holt, Rinehart and Winston, New York, 1969.
A. I. Vogel, *Quantitative Inorganic Analysis*, 3rd ed., Wiley, New York, 1961.

2. Instrumental Analysis

P. Delahay, *Instrumental Analysis*, Macmillan, New York, 1957.
G. Ewing, *Instrumental Methods of Chemical Analysis*, 3rd ed., McGraw-Hill, New York, 1969.
C. E. Meloan and R. W. Kiser, *Problems and Experiments in Instrumental Analysis*, Merrill, Columbus, Ohio, 1963.
H. H. Willard, L. L. Merritt, Jr., and J. A. Dean, *Instrumental Methods of Analysis*, 4th ed., Van Nostrand, Princeton, N.J., 1965.

3. Theory and Special Topics

R. G. Bates, *Determination of pH*, Wiley, New York, 1964.
E. W. Berg, *Physical and Chemical Methods of Separation*, McGraw-Hill, New York, 1963.
D. F. Boltz, *Colorimetric Determination of Nonmetals*, Wiley-Interscience, New York, 1958.
S. Dal Nogare and R. S. Juvet, *Gas-Liquid Chromatography*, Wiley-Interscience, New York, 1962.
C. Duval, *Inorganic Thermogravimetric Analysis*, 2nd ed., Elsevier, Amsterdam, 1963.
G. Eisenman, R. G. Bates, and G. Mattock, *The Glass Electrode*, Wiley-Interscience, New York, 1966.
F. Feigl, *Spot Tests*, 5th ed., Elsevier, Amsterdam, 1962.
J. S. Fritz and G. S. Hammond, *Quantitative Organic Analysis*, Wiley, New York, 1957.
L. Gordon, M. L. Salutsky, and H. H. Willard, *Precipitation from Homogeneous Solution*, Wiley, New York, 1959.
L. F. Hamilton, S. G. Simpson, D. W. Ellis, *Calculations of Analytical Chemistry*, 7th ed., McGraw-Hill, New York, 1969.
E. Heftman, *Chromatography*, 2nd ed., Reinhold, New York, 1967.
F. Helfferich, *Ion Exchange*, McGraw-Hill, New York, 1962.
J. J. Lingane, *Electroanalytical Chemistry*, Wiley-Interscience, New York, 1953.
G. E. F. Lundell, J. I. Hoffman, and H. A. Bright, *Chemical Analysis of Iron and Steel*, Wiley, New York, 1931.
M. G. Mellon, *Analytical Absorption Spectroscopy*, Wiley, New York, 1950.
G. H. Morrison and H. Freiser, *Solvent Extraction in Analytical Chemistry*, Wiley, New York, 1957.
A. Ringbom, *Complexation in Analytical Chemistry*, Wiley-Interscience, New York, 1963.

O. Samuelson, *Ion Exchange Separations in Analytical Chemistry*, Wiley, New York, 1963.

G. Schwarzenbach, *Complexometric Titrations*, transl. by H. Irving, Wiley-Interscience, New York, 1957.

S. Siggia, *Quantitative Organic Analysis via Functional Groups*, 3rd ed., Wiley, New York, 1963.

L. G. Sillen and A. E. Martell, *Stability Constants of Metal-ion Complexes. I, Inorganic Ligands*; *II, Organic Ligands*, The Chemical Society, London, 1964.

H. F. Walton, *Principles and Methods of Chemical Analysis*, Van Nostrand, Princeton, N.J., 1964.

F. J. Welcher, *The Analytical Uses of Ethylenediaminetetraacetic Acid*, Van Nostrand, Princeton, N.J., 1958.

A. L. Winton and K. B. Winton, *The Analysis of Foods*, Wiley, New York, 1945.

APPENDIX 6

Solubility Product Constants of Some Salts Arranged by Alphabetical Order of the Name of the Metal

Salt	K_{sp}	pK_{sp}
Aluminum		
Hydroxide, $Al(OH)_3$	4.6×10^{-33}	32.34
Phosphate, $AlPO_4$	6.2×10^{-19}	18.20
Barium		
Carbonate, $BaCO_3$	5.1×10^{-9}	8.29
Chromate, $BaCrO_4$	1.2×10^{-10}	9.93
Fluoride, BaF_2	1.0×10^{-6}	6.00
Iodate, $Ba(IO_3)_2$	6.0×10^{-10}	9.22
Oxalate, BaC_2O_4	1.5×10^{-8}	7.82
Sulfate, $BaSO_4$	1.0×10^{-10}	10.00
Bismuth		
Phosphate, $BiPO_4$	1.3×10^{-23}	22.89
Sulfide, Bi_2S_3	1.0×10^{-96}	96.00
Cadmium		
Hydroxide, $Cd(OH)_2$	1.2×10^{-14}	13.93
Sulfide, CdS	7.1×10^{-17}	16.15

Salt	K_{sp}	pK_{sp}
Calcium		
Carbonate, $CaCO_3$	1.2×10^{-8}	7.92
Fluoride, CaF_2	3.5×10^{-11}	10.46
Iodate, $Ca(IO_3)_2$	6.5×10^{-7}	6.19
Oxalate, CaC_2O_4	2.3×10^{-9}	8.64
Phosphate, $Ca_3(PO_4)_2$	2.0×10^{-29}	28.70
Cerium(III) (cerous)		
Hydroxide, $Ce(OH)_3$	6.3×10^{-21}	20.2
Chromium(III) (chromic)		
Hydroxide, $Cr(OH)_3$	6.0×10^{-31}	30.22
Phosphate, $CrPO_4$	2.4×10^{-23}	22.62
Cobalt(II) (cobaltous)		
Hydroxide, $Co(OH)_2$	2.5×10^{-16}	15.60
Sulfide, CoS	3×10^{-26}	25.52
Copper(I) (cuprous)		
Iodide, Cu_2I_2	1.1×10^{-12}	11.93
Sulfide, Cu_2S	1×10^{-49}	49.00
Thiocyanate, $Cu_2(SCN)_2$	1.9×10^{-13}	12.73
Copper(II) (cupric)		
Carbonate, $CuCO_3$	2.5×10^{-10}	9.60
Hydroxide, $Cu(OH)_2$	2.2×10^{-20}	19.66
Oxalate, CuC_2O_4	2.9×10^{-8}	7.54
Sulfide, CuS	8×10^{-36}	35.1
Iron(II) (ferrous)		
Hydroxide, $Fe(OH)_2$	3.2×10^{-14}	13.50
Sulfide, FeS	5.0×10^{-18}	17.30
Iron(III) (ferric)		
Hydroxide, $Fe(OH)_3$	2.5×10^{-39}	38.60
Phosphate, $FePO_4$	1.4×10^{-22}	21.87
Lead		
Carbonate, $PbCO_3$	3.4×10^{-14}	13.48
Chromate, $PbCrO_4$	1.8×10^{-14}	13.75
Fluoride, PbF_2	2.7×10^{-8}	7.57
Hydroxide, $Pb(OH)_2$	1.2×10^{-15}	14.93
Iodide, PbI_2	6.5×10^{-9}	8.19
Oxalate, PbC_2O_4	8.3×10^{-12}	11.08
Phosphate, $Pb_3(PO_4)_2$	8.0×10^{-43}	42.10
Sulfate, $PbSO_4$	1.0×10^{-8}	8.00
Sulfide, PbS	7.1×10^{-29}	28.15
Magnesium		
Ammonium phosphate, $MgNH_4PO_4$	2.5×10^{-13}	12.60
Carbonate, $MgCO_3$	1×10^{-5}	5.00
Fluoride, MgF_2	6.4×10^{-9}	8.19
Hydroxide, $Mg(OH)_2$	1.2×10^{-11}	10.92
Oxalate, MgC_2O_4	8.6×10^{-5}	4.07
Manganese(II) (manganous)		
Hydroxide, $Mn(OH)_2$	1.9×10^{-13}	12.72
Oxalate, MnC_2O_4	1.1×10^{-15}	14.96
Sulfide, MnS	1.1×10^{-15}	14.96
Mercury(I) (mercurous)		
Bromide, Hg_2Br_2	1.3×10^{-21}	20.89
Chloride, Hg_2Cl_2	1.3×10^{-18}	17.88

Salt	K_{sp}	pK_{sp}
Iodide, Hg_2I_2	1.2×10^{-28}	27.92
Sulfide, Hg_2S	1.0×10^{-48}	48.00
Mercury(II) (mercuric)		
Hydroxide, $Hg(OH)_2$	3.0×10^{-26}	25.52
Iodide, HgI_2	8.8×10^{-12}	11.06
Sulfide, HgS	3.0×10^{-52}	51.52
Nickel(II) (nickelous)		
Hydroxide, $Ni(OH)_2$	6.2×10^{-16}	15.21
Sulfide, NiS	3.0×10^{-21}	20.52
Silver		
Bromide, AgBr	4.9×10^{-13}	12.31
Carbonate, Ag_2CO_3	8.1×10^{-12}	11.09
Chloride, AgCl	1.56×10^{-10}	9.81
Chromate, Ag_2CrO_4	1.3×10^{-12}	11.89
Cyanide, AgCN	1.6×10^{-14}	13.80
Hydroxide, AgOH	2.6×10^{-8}	7.59
Iodate, $AgIO_3$	3.1×10^{-8}	7.51
Iodide, AgI	8.3×10^{-17}	16.08
Nitrite, $AgNO_2$	1.6×10^{-4}	3.80
Oxalate, $Ag_2C_2O_4$	1.1×10^{-11}	10.96
Phosphate, Ag_3PO_4	1.3×10^{-20}	19.89
Sulfide, Ag_2S	3.3×10^{-52}	51.48
Thiocyanate, AgSCN	1.4×10^{-12}	11.85
Strontium		
Carbonate, $SrCO_3$	1.1×10^{-10}	9.96
Chromate, $SrCrO_4$	3.6×10^{-5}	4.44
Fluoride, SrF_2	2.5×10^{-9}	8.60
Iodate, $Sr(IO_3)_2$	3.3×10^{-7}	6.48
Oxalate, SrC_2O_4	5.0×10^{-8}	7.30
Phosphate, $Sr_3(PO_4)_2$	1.0×10^{-31}	31.00
Sulfate, $SrSO_4$	2.8×10^{-7}	6.56
Thallium(I) (thallous)		
Bromide, TlBr	2.0×10^{-6}	5.70
Chloride, TlCl	1.5×10^{-4}	3.82
Iodide, TlI	2.8×10^{-8}	7.55
Sulfide, Tl_2S	1.0×10^{-21}	21.00
Tin(II) (stannous)		
Hydroxide, $Sn(OH)_2$	3.2×10^{-26}	25.50
Sulfide, SnS	1.0×10^{-26}	26.00
Zinc		
Carbonate, $ZnCO_3$	2.1×10^{-11}	10.68
Hydroxide, $Zn(OH)_2$	3.3×10^{-17}	16.48
Phosphate, $Zn_3(PO_4)_2$	1.0×10^{-32}	32.00
Sulfide, ZnS	1.6×10^{-24}	23.80

APPENDIX 7

Ionization Constants of Weak Acids and Bases

In general, the value of the first ionization constant for inorganic acids which contain oxygen depends upon the ratio of oxygen to hydrogen. For acids with the general formula H_nXO_n, K_1 is approximately 10^{-7} or less, and the acid is very weak; for H_nXO_{n+1} acids, K_1 is in the range of 10^{-2}–10^{-5} and K_2 is often approximately 10^{-7} so that the acid is considered to be weak or moderately strong; for acids with the formula H_nXO_{n+2}, K_1 is usually about 10^3 and K_2 is approximately 10^{-2} so that the acid is classified as a strong acid; K_1 for acids with the formula H_nXO_{n+3} is usually infinitely large and the acids are known as very strong acids. Also, the difference in successive ionization constants for the inorganic oxygen acids generally is approximately 10^{-4}. It should be realized that there are many exceptions to these general statements and that the statements do not apply to organic acids. Actually, the only acids in common laboratory use which are considered to be strong acids are hydrochloric, nitric, perchloric, and sulfuric.

ACIDS

Acid	Equilibrium	K_a	pK_a
Acetic	$HC_2H_3O_2 \rightleftharpoons H^+ + C_2H_3O_2^-$	1.8×10^{-5}	4.76
Arsenic	(1) $H_3AsO_4 \rightleftharpoons H^+ + H_2AsO_4^-$	6.0×10^{-3}	2.22
	(2) $H_2AsO_4^- \rightleftharpoons H^+ + HAsO_4^{2-}$	1.1×10^{-7}	6.98
	(3) $HAsO_4^{2-} \rightleftharpoons H^+ + AsO_4^{3-}$	3.0×10^{-12}	11.53
Arsenious	$HAsO_2 \rightleftharpoons H^+ + AsO_2^-$	6×10^{-10}	9.2
Benzoic	$HC_7H_5O_2 \rightleftharpoons H^+ + C_7H_5O_2^-$	6.31×10^{-5}	4.20

Acid	Equilibrium	K_a	pK_a
Boric	$HBO_2 \rightleftharpoons H^+ + BO_2^-$	6.0×10^{-10}	9.22
Butyric	$HC_4H_7O_2 \rightleftharpoons H^+ + C_4H_7O_2^-$	1.51×10^{-5}	4.82
Caproic	$HC_6H_{11}O_2 \rightleftharpoons H^+ + C_6H_{11}O_2^-$	1.32×10^{-5}	4.88
Carbonic	(1) $H_2CO_3 \rightleftharpoons H^+ + HCO_3^-$	4.3×10^{-7}	6.37
	(2) $HCO_3^- \rightleftharpoons H^+ + CO_3^{2-}$	5.6×10^{-11}	10.25
Chromic	(2) $HCrO_4^- \rightleftharpoons H^+ + CrO_4^{2-}$	3.2×10^{-7}	6.50
Citric	(1) $H_3C_6H_5O_7 \rightleftharpoons H^+ + H_2C_6H_5O_7^-$	8.3×10^{-4}	3.08
	(2) $H_2C_6H_5O_7^- \rightleftharpoons H^+ + HC_6H_5O_7^{2-}$	2.2×10^{-5}	4.66
	(3) $HC_6H_5O_7^{2-} \rightleftharpoons H^+ + C_6H_5O_7^{3-}$	4.0×10^{-7}	6.40
Cyanic	$HCNO \rightleftharpoons H^+ + CNO^-$	2.2×10^{-4}	3.66
Cyanuric	$HC_3H_2N_3O_3 \rightleftharpoons H^+ + C_3H_2N_3O_3^-$	1.8×10^{-7}	6.74
EDTA	(1) $H_4Y \rightleftharpoons H^+ + H_3Y^-$	1×10^{-2}	2.0
	(2) $H_3Y^- \rightleftharpoons H^+ + H_2Y^{2-}$	2.1×10^{-3}	2.67
	(3) $H_2Y^{2-} \rightleftharpoons H^+ + HY^{3-}$	6.9×10^{-7}	6.16
	(4) $HY^{3-} \rightleftharpoons H^+ + Y^{4-}$	5.5×10^{-11}	10.26
Formic	$HCOOH \rightleftharpoons H^+ + HCOO^-$	1.7×10^{-4}	3.77
Fumaric	(1) $H_2C_4H_2O_4 \rightleftharpoons H^+ + HC_4H_2O_4^-$	9.6×10^{-4}	3.02
	(2) $HC_4H_2O_4^- \rightleftharpoons H^+ + C_4H_2O_4^{2-}$	4.1×10^{-5}	4.39
Hydrazoic	$HN_3 \rightleftharpoons H^+ + N_3^-$	1.9×10^{-5}	4.72
Hydrocyanic	$HCN \rightleftharpoons H^+ + CN^-$	4.9×10^{-10}	9.31
Hydrofluoric	$HF \rightleftharpoons H^+ + F^-$	2.4×10^{-4}	3.62
Hydrosulfuric	(1) $H_2S \rightleftharpoons H^+ + HS^-$	1.0×10^{-7}	7.00
	(2) $HS^- \rightleftharpoons H^+ + S^{2-}$	1.2×10^{-13}	12.92
Hypobromous	$HBrO \rightleftharpoons H^+ + BrO^-$	2.5×10^{-9}	8.60
Hypochlorous	$HClO \rightleftharpoons H^+ + ClO^-$	3.0×10^{-8}	7.53
Hypoiodous	$HIO \rightleftharpoons H^+ + IO^-$	5×10^{-13}	12.3
Lactic	$HC_3H_5O_3 \rightleftharpoons H^+ + C_3H_5O_3^-$	1.39×10^{-4}	3.86
Maleic	(1) $H_2C_4H_2O_4 \rightleftharpoons H^+ + HC_4H_2O_4^-$	1.0×10^{-2}	2.00
	(2) $HC_4H_2O_4^- \rightleftharpoons H^+ + C_4H_2O_4^{2-}$	5.5×10^{-7}	6.26
Malic	(1) $H_2C_4H_4O_5 \rightleftharpoons H^+ + HC_4H_4O_5^-$	4.0×10^{-4}	3.40
	(2) $HC_4H_4O_5^- \rightleftharpoons H^+ + C_4H_4O_5^{2-}$	8.9×10^{-6}	5.05
Malonic	(1) $H_2C_3H_2O_4 \rightleftharpoons H^+ + HC_3H_2O_4^-$	1.40×10^{-3}	2.85
	(2) $HC_3H_2O_4^- \rightleftharpoons H^+ + C_3H_2O_4^{2-}$	8.0×10^{-7}	6.10
Nitrous	$HNO_2 \rightleftharpoons H^+ + NO_2^-$	5.1×10^{-4}	3.29
Oxalic	(1) $H_2C_2O_4 \rightleftharpoons H^+ + HC_2O_4^-$	5.6×10^{-2}	1.25
	(2) $HC_2O_4^- \rightleftharpoons H^+ + C_2O_4^{2-}$	5.2×10^{-5}	4.28
Phenol	$C_6H_5OH \rightleftharpoons H^+ + C_6H_5O^-$	1.3×10^{-10}	9.89
Phosphoric	(1) $H_3PO_4 \rightleftharpoons H^+ + H_2PO_4^-$	7.5×10^{-3}	2.12
	(2) $H_2PO_4^- \rightleftharpoons H^+ + HPO_4^{2-}$	6.2×10^{-8}	7.21
	(3) $HPO_4^{2-} \rightleftharpoons H^+ + PO_4^{3-}$	4.7×10^{-13}	12.33
Phthalic	(1) $H_2C_8H_4O_4 \rightleftharpoons H^+ + HC_8H_4O_4^-$	8.0×10^{-4}	3.10
	(2) $HC_8H_4O_4^- \rightleftharpoons H^+ + C_8H_4O_4^{2-}$	4.0×10^{-6}	5.40
Propionic	$HC_3H_5O_2 \rightleftharpoons H^+ + C_3H_5O_2^-$	1.34×10^{-5}	4.87
Succinic	(1) $H_2C_4H_4O_4 \rightleftharpoons H^+ + HC_4H_4O_4^-$	6.5×10^{-5}	4.19
	(2) $HC_4H_4O_4^- \rightleftharpoons H^+ + C_4H_4O_4^{2-}$	3.3×10^{-6}	5.48
Sulfuric	(2) $HSO_4^- \rightleftharpoons H^+ + SO_4^{2-}$	1.0×10^{-2}	2.00
Sulfurous	(1) $H_2SO_3 \rightleftharpoons H^+ + HSO_3^-$	1.3×10^{-2}	1.89
	(2) $HSO_3^- \rightleftharpoons H^+ + SO_3^{2-}$	6.3×10^{-8}	7.20
Tartaric	(1) $H_2C_4H_4O_6 \rightleftharpoons H^+ + HC_4H_4O_6^-$	3.0×10^{-3}	2.52
	(2) $HC_4H_4O_6^- \rightleftharpoons H^+ + C_4H_4O_6^{2-}$	6.9×10^{-5}	4.16

BASES

Base	Equilibrium	K	pK_b
Ammonia	$NH_3 + H_2O \rightleftharpoons NH_4OH \rightleftharpoons NH_4^+ + OH^-$	1.8×10^{-5}	4.74
Aniline	$C_6H_5NH_2 + H_2O \rightleftharpoons C_6H_5NH_3^+ + OH^-$	4.6×10^{-10}	9.34
Benzylamine	$C_7H_7NH_2 + H_2O \rightleftharpoons C_7H_7NH_3^+ + OH^-$	2.0×10^{-5}	4.70
Ethylamine	$C_2H_5NH_2 + H_2O \rightleftharpoons C_2H_5NH_3^+ + OH^-$	5.6×10^{-4}	3.25
Hydrazine	$N_2H_4 + H_2O \rightleftharpoons N_2H_5^+ + OH^-$	3×10^{-6}	5.5
Hydroxylamine	$NH_2OH \rightleftharpoons NH_2^+ + OH^-$	6.6×10^{-9}	8.18
Methylamine	$CH_3NH_2 + H_2O \rightleftharpoons CH_3NH_3^+ + OH^-$	5×10^{-4}	3.3
α-Naphthylamine	$C_{10}H_7NH_2 + H_2O \rightleftharpoons C_{10}H_7NH_3^+ + OH^-$	9.9×10^{-11}	10.01
β-Naphthylamine	$C_{10}H_7NH_2 + H_2O \rightleftharpoons C_{10}H_7NH_3^+ + OH^-$	2×10^{-10}	9.70
Phenylhydrazine	$C_6H_5N_2H_3 + H_2O \rightleftharpoons C_6H_5N_2H_4^+ + OH^-$	1.6×10^{-9}	8.80
Pyridine	$C_5H_5N + H_2O \rightleftharpoons C_5H_5NH^+ + OH^-$	2.3×10^{-9}	8.64
Quinoline	$C_9H_7N + H_2O \rightleftharpoons C_9H_7NH^+ + OH^-$	1×10^{-9}	9.0
Silver hydroxide	$AgOH \rightleftharpoons Ag^+ + OH^-$	1.1×10^{-4}	3.96
Triethylamine	$(C_2H_5)_3N + H_2O \rightleftharpoons (C_2H_5)_3NH^+ + OH^-$	2.6×10^{-4}	3.58

APPENDIX **8**

Stability (Formation) Constants of Complexes

The table value is the logarithm of the stability (formation) constant. Stepwise constants are designated by $k_1, k_2, \ldots, k_n$. The cumulative constant is the product of the stepwise constants, i.e., $K_{stab} = k_1 k_2 \cdots k_n$; hence $\log K_{stab} = \log k_1 + \log k_2 + \cdots + \log k_n$. Where values of the stepwise constants are not given, the parenthetical number before $\log K_{stab}$ indicates the value of n for the overall reaction $M + nL \rightleftharpoons ML_n$.

Ligand	Cation	log k_1	log k_2	log k_3	log k_4	log K_{stab}
NH_3	Ag^+	3.34	3.83			7.17
	Cd^{2+}	2.51	1.96	1.30	0.79	6.56
	Co^{2+}	2.07	1.60	0.93	0.64	5.24
	Cu^{2+}	3.99	3.34	2.73	1.97	12.03
	Ni^{2+}	2.67	2.12	1.61	1.07	7.47
	Zn^{2+}	2.18	2.25	2.31	1.96	8.70
CN^-	Ag^+					(2) 21.1
	Cd^{2+}	5.18	4.42	4.32	3.19	17.11
	Co^{2+}					(6) 19.09
	Cu^+	$\log(k_1k_2) = 24$		4.59	1.70	30.3
	Fe^{2+}					(6) 24
	Fe^{3+}					(6) 31
	Ni^{2+}					(4) 31.3
	Zn^{2+}					(4) 16.7
$C_2O_4{}^{2-}$	Fe^{3+}	9.40	6.78	4.00		20.18
	Mg^{2+}	3.82	1.43			5.25
	Pb^{2+}					(2) 6.54
F^-	Al^{3+}	6.13	5.02	3.85	2.74	(6) 19.84
	Fe^{3+}	5.17	3.92	2.91		(6) 15.3
	Th^{4+}	7.70	5.78	4.51		(3) 17.99
	Zr^{4+}	8.73	7.34	5.83		(3) 21.90
Cl^-	Ag^+					(2) 5.28
	Cd^{2+}	2.00	0.60	0.10	0.30	3.00
	Cu^+					(2) 5.54
	Hg^{2+}	6.74	6.48	0.85	1.00	15.07
	Pd^{2+}	6.1	4.6	2.4	2.6	15.7
	Zn^{2+}					(4) 0.20
I^-	Ag^+					(2) 13.7
	Cd^{2+}	2.28	1.64	1.08	1.10	6.10
	Hg^{2+}	12.87	10.95	3.78	2.23	29.83
SCN^-	Ag^+					(2) 8.78
	Fe^{3+}	3.03				
	Hg^{2+}	9.48	10.30	1.70		(4) 21.9
$S_2O_3^{2-}$	Ag^+					(2) 13.22
EDTA[a]	Ag^+	7.30				
	Ba^{2+}	7.76				
	Ca^{2+}	11.00				
	Cd^{2+}	16.46				
	Co^{2+}	16.31				
	Cu^{2+}	18.79				
	Fe^{2+}	14.33				
	Fe^{3+}	24.23				
	Mg^{2+}	8.69				
	Mn^{2+}	13.79				
	Ni^{2+}	18.56				
	Pb^{2+}	18.04				
	Sr^{2+}	8.63				
	Zn^{2+}	16.26				

Ligand	Cation	Equilibrium[b]	log K_{stab}
OH^-	Al^{3+}	$Al(OH)_3 + OH^- \rightleftharpoons Al(OH)_4^-$	2.60
	Cr^{3+}	$Cr(OH)_3 + OH^- \rightleftharpoons Cr(OH)_4^-$	−2.0
	Pb^{2+}	$Pb(OH)_2 + OH^- \rightleftharpoons Pb(OH)_3^-$	−1.70
	Sn^{2+}	$Sn(OH)_2 + OH^- \rightleftharpoons Sn(OH)_3^-$	−3.30
	Zn^{2+}	$Zn(OH)_2 + 2OH^- \rightleftharpoons Zn(OH)_4^{2-}$	−1.0

[a] EDTA forms only 1 : 1 complexes with the cations.

[b] The values given for the various cations with OH^- as the ligand are for the equilibrium between the solid amphoteric hydroxide and the anionic complex.

APPENDIX 9

Selected Standard Electrode (Half-Cell) Reduction Potentials

The direction of the half-reaction equation and the sign of the potential conform to the IUPAC (Stockholm) conventions.

Half reaction	E(V)
$F_2 + 2e \rightleftharpoons 2F^-$	+2.65
$O_3 + 2H^+ + 2e^- \rightleftharpoons O_2 + H_2O$	+2.07
$S_2O_8^{2-} + 2e^- \rightleftharpoons 2SO_4^{2-}$	+2.01
$H_2O_2 + 2H^+ + 2e^- \rightleftharpoons 2H_2O$	+1.77
$Ce^{4+} + e^- \rightleftharpoons Ce^{3+}$ (perchlorate solution)	+1.70
$MnO_4^- + 4H^+ + 3e^- \rightleftharpoons MnO_2 + 2H_2O$	+1.695
$Ce^{4+} + e^- \rightleftharpoons Ce^{3+}$ (nitrate solution)	+1.61
$BrO_3^- + 6H^+ + 5e^- \rightleftharpoons \frac{1}{2}Br_2 + 3H_2O$	+1.52
$MnO_4^- + 8H^+ + 5e^- \rightleftharpoons Mn^{2+} + 4H_2O$	+1.51
$Au^{3+} + 3e^- \rightleftharpoons Au$	+1.50
$ClO_3^- + 6H^+ + 5e^- \rightleftharpoons \frac{1}{2}Cl_2 + 3H_2O$	+1.47
$BrO_3^- + 6H^+ + 6e^- \rightleftharpoons Br^- + 3H_2O$	+1.45
$Ce^{4+} + e^- \rightleftharpoons Ce^{3+}$ (sulfate solution)	+1.44
$Cl_2 + 2e^- \rightleftharpoons 2Cl^-$	+1.359
$Cr_2O_7^{2-} + 14H^+ + 6e^- \rightleftharpoons 2Cr^{3+} + 7H_2O$	+1.33
$Ce^{4+} + e^- \rightleftharpoons Ce^{3+}$ (chloride solution)	+1.28
$MnO_2 + 4H^+ + 2e^- \rightleftharpoons Mn^{2+} + 2H_2O$	+1.23
$O_2 + 4H^+ + 4e^- \rightleftharpoons 2H_2O$	+1.229
$IO_3^- + 6H^+ + 5e^- \rightleftharpoons \frac{1}{2}I_2 + 3H_2O$	+1.20
$IO_3^- + 6H^+ + 6e^- \rightleftharpoons I^- + 3H_2O$	+1.087

Half reaction	E(V)
Br_2 (liquid) $+ 2e^- \rightleftharpoons 2Br^-$	+1.065
$OP + e^- \rightleftharpoons OP'$ (orthophenanthroline)	+1.06
$AuCl_4^- + 3e^- \rightleftharpoons Au + 4Cl^-$	+1.00
$NO_3^- + 4H^+ + 3e^- \rightleftharpoons NO + 2H_2O$	+0.96
$2Hg^{2+} + 2e^- \rightleftharpoons Hg_2^{2+}$	+0.920
$Cu^{2+} + I^- + e^- \rightleftharpoons CuI$	+0.85
$DPAS + e^- \rightleftharpoons DPAS$ (red)(diphenylamine sulfonate)	+0.84
$\frac{1}{2}O_2 + 2H^+(10^{-7}\ M) + 2e^- \rightleftharpoons H_2O$	+0.815
$Ag^+ + e^- \rightleftharpoons Ag$	+0.799
$Hg_2^{2+} + 2e^- \rightleftharpoons 2Hg$	+0.789
$Fe^{3+} + e^- \rightleftharpoons Fe^{2+}$	+0.771
$Sb^{5+} + 2e^- \rightleftharpoons Sb^{3+}$	+0.75
Q (saturated solution) $+ 2H^+ + 2e^- \rightleftharpoons H_2Q$ (saturated solution) (quinhydrone electrode)	+0.700
$O_2 + 2H^+ + 2e^- \rightleftharpoons H_2O_2$	+0.682
$2HgCl_2 + 2e^- \rightleftharpoons Hg_2Cl_2 + 2Cl^-$	+0.63
$H_3AsO_4 + 2H^+ + 2e^- \rightleftharpoons H_3AsO_3 + H_2O$	+0.559
$I_2(I_3^-) + 2e^- \rightleftharpoons 2I^-(3I^-)$	+0.535
$Fe(CN)_6^{3-} + e^- \rightleftharpoons Fe(CN)_6^{4-}$	+0.36
$Cu^{2+} + 2e^- \rightleftharpoons Cu$	+0.337
$UO_2^{2+} + 4H^+ + 2e^- \rightleftharpoons U^{4+} + 2H_2O$	+0.334
$BiO^+ + 2H^+ + 3e^- \rightleftharpoons Bi + H_2O$	+0.32
$Hg_2Cl_2 + 2e^- \rightleftharpoons 2Hg + 2Cl^-$ (1 M KCl)(calomel cell)	+0.285
$Hg_2Cl_2 + 2e^- \rightleftharpoons 2Hg + 2Cl^-$ (saturated KCl)(calomel cell)	+0.246
$AgCl + e^- \rightleftharpoons Ag + Cl^-$	+0.222
$SbO^+ + 2H^+ + 3e^- \rightleftharpoons Sb + H_2O$	+0.212
$S_4O_6^{2-} + 2e^- \rightleftharpoons 2S_2O_3^{2-}$	+0.17
$SO_4^{2-} + 4H^+ + 2e^- \rightleftharpoons H_2SO_3 + H_2O$	+0.17
$Sn^{4+} + 2e^- \rightleftharpoons Sn^{2+}$	+0.15
$TiO^{2+} + 2H^+ + e^- \rightleftharpoons Ti^{3+} + H_2O$	+0.1
$AgBr + e^- \rightleftharpoons Ag + Br^-$	+0.095
$2H^+ + 2e^- \rightleftharpoons H_2$	+0.000
$Pb^{2+} + 2e^- \rightleftharpoons Pb$	−0.126
$Sn^{2+} + 2e^- \rightleftharpoons Sn$	−0.136
$AgI + e^- \rightleftharpoons Ag + I^-$	−0.151
$Ni^{2+} + 2e^- \rightleftharpoons Ni$	−0.24
$V^{3+} + e^- \rightleftharpoons V^{2+}$	−0.26
$Co^{2+} + 2e^- \rightleftharpoons Co$	−0.28
$Tl^+ + e^- \rightleftharpoons Tl$	−0.34
$PbSO_4 + 2e^- \rightleftharpoons Pb + SO_4^{2-}$	−0.36
$Cd^{2+} + 2e^- \rightleftharpoons Cd$	−0.403
$Cr^{3+} + e^- \rightleftharpoons Cr^{2+}$	−0.41
$2H^+(10^{-7}\ M) + 2e^- \rightleftharpoons H_2$	−0.414
$Fe^{2+} + 2e^- \rightleftharpoons Fe$	−0.440
$2CO_2 + 2H^+ + 2e^- \rightleftharpoons H_2C_2O_4$	−0.49
$S + 2e^- \rightleftharpoons S^{2-}$	−0.51
$AsO_4^{3-} + 3H_2O + 2e^- \rightleftharpoons H_2AsO_3^- + 4OH^-$	−0.67
$Cr^{3+} + 3e^- \rightleftharpoons Cr$	−0.74
$Zn^{2+} + 2e^- \rightleftharpoons Zn$	−0.763
$Mn^{2+} + 2e^- \rightleftharpoons Mn$	−1.18
$Al^{3+} + 3e^- \rightleftharpoons Al$	−1.67

Half reaction	E(V)
$AlO_2^- + 2H_2O + 3e^- \rightleftharpoons Al + 4OH^-$	−2.35
$Mg^{2+} + 2e^- \rightleftharpoons Mg$	−2.37
$Na^+ + e^- \rightleftharpoons Na$	−2.714
$Ca^{2+} + 2e^- \rightleftharpoons Ca$	−2.87
$Sr^{2+} + 2e^- \rightleftharpoons Sr$	−2.89
$Ba^{2+} + 2e^- \rightleftharpoons Ba$	−2.90
$K^+ + e^- \rightleftharpoons K$	−2.925

APPENDIX 10

Use of Logarithms

A. General

The use of logarithms simplifies many calculations which must be performed in analytical chemistry, and as a consequence students who learn to use logarithms correctly will benefit greatly from the knowledge. A logarithm (often abbreviated log) of a number is the power (or exponent) of ten which will give the number or

$$\text{Number} = 10^{\text{logarithm of number}}$$

and

$$253 = 10^{\text{log of } 253}$$

In many cases it is advantageous to express numbers, especially very large or very small numbers, in the exponential form by showing them as numbers between 1 and 10 multiplied by 10 raised to the appropriate power, as shown in

$$\begin{aligned} 25{,}300{,}000 &= 2.53 \times 10^{7} \\ 25{,}300 &= 2.53 \times 10^{4} \\ 253 &= 2.53 \times 10^{2} \\ 0.253 &= 2.53 \times 10^{-1} \\ 0.0000253 &= 2.53 \times 10^{-5} \end{aligned}$$

By expressing the number in this way, the magnitude of the number is shown by the exponent and thus is easy to determine. The exponent simply represents the number of times the number 1 must be multiplied or divided by 10, depending upon whether the exponent is positive or negative. Thus

$$10^2 = 1 \times 10 \times 10 = 100$$

and

$$10^{-3} = \frac{1}{10 \times 10 \times 10} = \frac{1}{1000} = 0.001$$

Since the log of a number is the power to which 10 must be raised to equal the number, the

$$\text{log of } 100 = \text{log of } 10^2 = 2$$

and

$$\text{log of } 0.001 = \text{log of } 10^{-3} = -3$$

Similarly, the log of 1 is 0, the log of 10 is 1, and any number between 1 and 10 will have a logarithm between 0 and 1. Since most numbers are not exact powers of 10, most logs are not whole numbers but are fractional numbers composed of an integer (known as the *Characteristic*) and a decimal fraction (known as the *mantissa*). The mantissa is found in a logarithm table and the characteristic is obtained by inspection since the characteristic serves only to locate the decimal point. The logarithms of all numbers which have the same integers in the same sequence all have the same mantissa but will have different characteristics depending upon the magnitude of the number. For example, using the integer sequence 34567 for which the mantissa is 0.53866,

$$\text{log of } 345{,}670 = 5.53866$$
$$\text{log of } 34.567 = 1.53866$$
$$\text{log of } 0.34567 = \bar{1}.53866 \quad \text{or} \quad 9.53866 - 10 \quad \text{or} \quad -0.46134$$
$$\text{log of } 0.00034567 = \bar{4}.53866 \quad \text{or} \quad 6.53866 - 10 \quad \text{or} \quad -3.46134$$

For numbers less than 1, the negative sign is usually written over the integer to show that the characteristic is negative, while the mantissa is positive. The logs of small numbers of this type may also be written to show the addition of 10 followed by the subtraction of 10. In the last example above, the characteristic is calculated by the addition of -4 to $+10$ to obtain $+6$. In some calculations it is desirable to show the logarithms of fractional numbers as completely negative numbers rather than as a negative integer followed by a positive decimal fraction. Expressed this way, the logarithm of 0.00034567 is —3.46139, which is obtained by subtracting the mantissa from 1.00000 and reducing the characteristic by 1. It is probably easier to express the logs in this way if the number is first written as a number between 1 and 10 multiplied by 10 with a whole number negative exponent. Using the example above, first convert 0.00034567 to 3.4567×10^{-4} and then convert 3.4567 to 10 with an exponent, or $10^{0.53866}$. Addition of exponents, equivalent to multiplication, gives $10^{-3.46134}$ and the log of $10^{-3.46134}$ is —3.4614.

B. Use of Five-Place Logarithm Table

To find the logarithm of a number, first determine the integer sequence irrespective of the position of the decimal point. To do this, do not count zeros in front of the first numeral but do count zeros occurring in the integer sequence or at the end of the integer sequence. Find the first three numerals in the integer sequence under column N in the log table. The fourth integer in the sequence should then be located under the proper column headed 1 through 9. If the number is expressed as five significant figures (is a number with a five-integer sequence), the logarithm to the fifth place can be estimated from the proportional parts column. As an example, determine the logarithm of the numbers 12300, 12340, and 12345.

$$\log \text{ of } 12300 = 4.08991$$
$$\log \text{ of } 12340 = 4.09132$$
$$\log \text{ of } 12345 = 4.09150$$

In looking up the log of the second number (12340), it is noted that the number under column 4 has an asterisk. This asterisk indicates that the first two integers of the log are on the line below rather than above. To obtain the log of the third number (12345), the mantissa of 1234 is 0.09132 and of 1235 is 0.09167. The difference between these is 0.00035. The value to be added to the log of 1234 to obtain the log of 12345 is obtained from the proportional parts column under 35. Since the fifth integer is 5, the value to be added is 0.000175 which rounds off to 0.00018. Addition of this amount to 0.09132 gives 0.09150 as the correct mantissa. The characteristic is 4 for all the numbers since there are five integers in each of the numbers.

To find the number corresponding to a logarithm (often called the antilogarithm), first be sure that the mantissa as written is positive rather than negative. If the mantissa is negative, subtract it from 1.00000 to find the mantissa to look up in the log table. Find this mantissa in the table and the first three integers of the number will be given in the column headed *N*. If the last three numerals of the mantissa agree exactly with the number in the one of the vertical columns, the heading of the column will be the fourth integer and the fifth integer will be zero. If the last three numerals in the mantissa do not correspond exactly with those shown in one of the vertical columns, the fourth and fifth integers must be estimated using the proportional parts column. As an example, find the number which corresponds to the logarithm 3.23060. To look up mantissa 0.23060, first find 23 in the column just to the right of the *N* column and then, going across the columns, find a three-digit number which is identical with 060. If no such number occurs, find the two numbers which bracket 060. The first three integers are obtained from the column headed *N* and found to be 170. The number under column 0 is 045 and that under column 1 is 0.70. The difference between these two is 25, so go to the proportional parts column headed 25. The difference between the column value of 045 and the actual logarithm is 15. Under the proportional parts column headed 25, the number 15 appears opposite 6. The correct sequence of integers then is 17006, and the number is 1700.6. To become facile in the use of logarithm tables, the student should practice looking up logs of numbers and converting logs to numbers until he is sure of the procedures involved and can perform the manipulations rapidly and accurately.

C. Numerical Operations Which Can Be Performed Simply by Logarithms

1. Numerical Operations

The numerical operations which are often performed by logarithms to simplify calculations may be summarized as

Numerical operation	Corresponding manipulation with logrithms
Multiplication	Add logarithms
Division	Subtract logarithms
Raise to a power	Multiply logarithm by desired power
Extract a root	Divide logarithm by desired root
Addition	*Cannot be Performed by Logarithms*
Subtraction	*Cannot be Performed by Logarithms*

2. Examples

a. MULTIPLICATION AND DIVISION. It is often necessary in analytical calculations to multiply several numbers together and then divide by the product of several other numbers. As an example, the calculation of the per cent of chloride in a sample by a direct titrimetric method upon an aliquot of the sample solution is shown below. In this example, 35.23 ml of 0.1037 N silver nitrate was required for a 25.00-ml aliquot of a sample weighing 2.3075 g which was dissolved and diluted to 250.0 ml. The numerical set up for such a problem is

$$\% \text{Cl} = \frac{35.23 \times 0.1037 \times 0.035453 \times 100}{2.3075 \times 25/250}$$

To perform this by logarithms, first look up and add the logarithms of the numbers in the numerator.

log of 35.23 = 1.54691	or	11.54691 − 10		
log of 0.1037 = $\bar{1}$.01578	or	9.01578 − 10		
log of 0.035453 = $\bar{2}$.54966	or	8.54966 − 10		
log of 100 = 2.00000	or	12.00000 − 10		
log of product = 1.11235	or	41.11235 − 40	or	1.11235

Next, simplify the denominator if possible, look up the logs, and add to obtain the logarithm of the denominator. Since 25/250 is 0.1, the denominator is 2.3075 × 0.1 and

log of 2.3075 = 0.36315	or	10.36315 − 10		
log of 0.1 = $\bar{1}$.00000	or	9.00000 − 10		
log of product = $\bar{1}$.36315	or	19.36315 − 20	or	9.36315 − 10

Now subtract the logarithm of the denominator from the logarithm of the numerator to carry out the division.

log of numerator = 1.11235	or	11.11235 − 10
log of denominator = $\bar{1}$.36315	or	9.36315 − 10
log of answer = 1.74920	or	1.74920 − 0
antilog (or answer) = 56.13		

b. TO RAISE TO A POWER OR EXTRACT A ROOT. These operations are often necessary in analytical calculations dealing with ionization constants, solubility products, and pH values. To raise a given number to some power, the logarithm is multiplied by the power desired. In this operation look up the logarithm of the number, multiply by the desired power, and then look up the antilog. For example, calculate the value of $(13.5)^5$ and the value of $(0.00246)^4$.

$$\begin{aligned} \text{log of } 13.5 &= 1.13033 \\ 5 \text{ times log } 13.5 &= 5.65165 \\ \text{antilog} &= 448{,}380 \end{aligned}$$

or, expressed in the usual exponential way to three significant figures, 4.48×10^5.

For numbers less than 1, the operation is the same, but the student must remember that the characteristic is negative and the mantissa is positive.

$$\begin{aligned} \text{log of } 0.00246 &= \bar{3}.39094 \quad \text{or} \quad 7.39094 - 10 \\ 4 \text{ times log } 0.00246 &= \overline{11}.56376 \quad \text{or} \quad 29.56376 - 40 \quad \text{or} \quad 9.56376 - 20 \\ \text{antilog} &= 0.000000000036623 \end{aligned}$$

or, expressed in the usual exponential way to three significant figures, 3.66×10^{-11}.

To extract a root, the reverse process is used. To extract the third root of 3493 and the fifth root of 0.002638, the following manipulations are performed:

$$\begin{aligned} \text{log of } 3493 &= 3.54320 \\ \text{log of } 3493 \div 3 &= 1.18107 \\ \text{antilog (or number)} &= 15.173 \end{aligned}$$

$$\begin{aligned} \text{log of } 0.002638 &= \bar{3}.42127 \quad \text{or} \quad 7.42127 - 10 \quad \text{or} \quad 47.42127 - 50 \\ \text{log of } 0.002638 \div 5 &= \bar{1}.48425 \quad \text{or} \quad 9.48425 - 10 \\ \text{antilog (or number)} &= 0.30496 \end{aligned}$$

When extracting a root of a number less than 1, the logarithm should be expressed in the plus 10 minus 10 method but should use multiples of 10 by the root to be extracted instead of 10 itself. In the above example, 50 is used in place of 10 since the fifth root is to be extracted.

D. Significant Figures in Using Logarithms

In general, there should be the same number of significant figures in the mantissa of the logarithm as there are significant figures in the number being expressed. Thus the log of 1.2300 is 0.08991, the log of 1.230 is 0.0899, the log of 1.23 is 0.090, and the log of 1.2 is 0.08. This rule has not been followed in the above examples since the primary purpose of the examples is to emphasize the manipulations involved in the use of logarithms.

E. Negative Logarithms

Some of the concepts used in analytical chemistry are defined in terms of negative logarithms. Examples are $p\text{H} = -\log(\text{H}^+)$, $p\text{K} = -\log K$, $A = -\log T$, and $p_c\text{M} = -\log[\text{M}^{n+}]$. In using negative logarithms, it must be remembered that the decimal part of the negative log is negative, while the decimal part (the mantissa) of a regular logarithm is positive and that the values found in log tables are thus positive. Consequently it is not possible to treat the decimal part of a negative logarithm as a mantissa for location in a log table. To express a number as a negative logarithm, it is usually advisable to write the number in the usual exponential form of a number between 1 and 10 multiplied by 10 with a whole number exponent. The mantissa of the integer sequence is then looked up and the number expressed as 10, with the mantissa as the exponent. The exponents are then added algebraically and expressed as 10 raised to the proper power. The negative logarithm of this number is simply the exponent with the sign changed. Examples are shown below to illustrate the manipulations involved.

1. If (H^+) is 4.5×10^{-6}, what is the pH?

$$4.5 \times 10^{-6} = 10^{0.65} \times 10^{-6} = 10^{-5.35}$$
$$p\text{H} = -\log(\text{H}^+) = -\log 10^{-5.35} = 5.35$$

Negative exponents of this type are usually expressed to either two or three decimal places depending upon the number of significant figures in the number being converted. The rule to be followed is to use the same number of figures in the decimal part of the logarithm as is used in the number itself. In the above case the mantissa for 4.5 is 0.65321, which rounds off to 0.65 to give the proper number of significant figures.

2. If pH is -0.36, what is the hydrogen ion activity? Since

$$p\text{H} = -\log(\text{H}^+) = -\log 10^{-0.36} = \log 10^{0.36}$$
$$(\text{H}^+) = 10^{0.36} = 2.3 \times 10^0 = 2.3$$

In this case, the number corresponding to the mantissa 0.36 is 2.2902, which is rounded to 2.3 according to the significant figure rule.

3. If the absorbance of a solution is 0.356, what is the per cent transmittance of the solution?

$$T = 10^{-0.356} = 10^{0.644} \times 10^{-1} = 4.41 \times 10^{-1} = 0.441$$
$$\%T = 100T = 44.1\%$$

In this case the number corresponding to the mantissa 0.644 is 4.406, which is rounded off to 4.41 by the significant figure rule.

4. What is the p_cCu of a 0.1063 *M* solution of $CuCl_2$?

$$0.1063 = 1.063 \times 10^{-1} = 10^{0.0265} \times 10^{-1} = 10^{-0.9735}$$
$$p_c\text{Cu} = -\log 10^{-0.9735} = \log 10^{0.9735} = 0.9735$$

In this case, four places are carried in the mantissa since there are four significant figures in the number.

In the above examples, enclosure of a symbol in parentheses refers to active concentration or activity, and enclosure in brackets represents molar concentration. The subscript *c* is used in the *p* notation to indicate that molar instead of active concentrations are being used.

APPENDIX 11

Five-Place Logarithms

N	0	1	2	3	4	5	6	7	8	9	Proportional parts			
100	00 000	043	087	130	173	217	260	303	346	389		**44**	**43**	**42**
101	432	475	518	561	604	647	689	732	775	817	1	4,4	4,3	4,2
102	860	903	945	988	*030	*072	*115	*157	*199	*242	2	8,8	8,6	8,4
103	01 284	326	368	410	452	494	536	578	620	662	3	13,2	12,9	12,6
104	703	745	787	828	870	912	953	995	*036	*078	4	17,6	17,2	16,8
105	02 119	160	202	243	284	325	366	407	449	490	5	22,0	21,5	21,0
106	531	572	612	653	694	735	776	816	857	898	6	26,4	25,8	25,2
107	938	979	*019	*060	*100	*141	*181	*222	*262	*302	7	30,8	30,1	29,4
108	03 342	383	423	463	503	543	583	623	663	703	8	35,2	34,4	33,6
109	743	782	822	862	902	941	981	*021	*060	*100	9	39,6	38,7	37,8
110	04 139	179	218	258	297	336	376	415	454	493		**41**	**40**	**39**
111	532	571	610	650	689	727	766	805	844	883	1	4,1	4,0	3,9
112	922	961	999	*038	*077	*115	*154	*192	*231	*269	2	8,2	8,0	7,8
113	05 308	346	385	423	461	500	538	576	614	652	3	12,3	12,0	11,7
114	690	729	767	805	843	881	918	956	994	*032	4	16,4	16,0	15,6
115	06 070	108	145	183	221	258	296	333	371	408	5	20,5	20,0	19,5
116	446	483	521	558	595	633	670	707	744	781	6	24,6	24,0	23,4
117	819	856	893	930	967	*004	*041	*078	*115	*151	7	28,7	28,0	27,3
118	07 188	225	262	298	335	372	408	445	482	518	8	32,8	32,0	32,2
119	555	591	628	664	700	737	773	809	846	882	9	36,9	36,0	35,1
120	918	954	990	*027	*063	*099	*135	*171	*207	*243		**38**	**37**	**36**
121	08 279	314	350	386	422	458	493	529	565	600	1	3,8	3,7	3,6
122	636	672	707	743	778	814	849	884	920	955	2	7,6	7,4	7,2
123	991	*026	*061	*096	*132	*167	*202	*237	*272	*307	3	11,4	11,1	10,8
124	09 342	377	412	447	482	517	552	587	621	656	4	15,2	14,8	14,4
125	691	726	760	795	830	864	899	934	968	*003	5	19,0	18,5	18,0
126	10 037	072	106	140	175	209	243	278	312	346	6	22,8	22,2	21,6
127	380	415	449	483	517	551	585	619	653	687	7	26,6	25,9	25,2
128	721	755	789	823	857	890	924	958	992	*025	8	30,4	29,6	28,8
129	11 059	093	126	160	193	227	261	294	327	361	9	34,2	33,3	32,4
130	394	428	461	494	528	561	594	628	661	694		**35**	**34**	**33**
131	727	760	793	826	860	893	926	959	992	*024	1	3,5	3,4	3,3
132	12 057	090	123	156	189	222	254	287	320	352	2	7,0	6,8	6,6
133	385	418	450	483	516	548	581	613	646	678	3	10,5	10,2	9,9
134	710	743	775	808	840	872	905	937	969	*001	4	14,0	13,6	13,2
135	13 033	066	098	130	162	194	226	258	290	322	5	17,5	17,0	16,5
136	354	386	418	450	481	513	545	577	609	640	6	21,0	20,4	19,8
137	672	704	735	767	799	830	862	893	925	956	7	24,5	23,8	23,1
138	988	*019	*051	*082	*114	*145	*176	*208	*239	*270	8	28,0	27,2	26,4
139	14 301	333	364	395	426	457	489	520	551	582	9	31,5	30,6	29,7
140	613	644	675	706	737	768	799	829	860	891		**32**	**31**	**30**
141	922	953	983	*014	*045	*076	*106	*137	*168	*198	1	3,2	3,1	3,0
142	15 229	259	290	320	351	381	412	442	473	503	2	6,4	6,2	6,0
143	534	564	594	625	655	685	715	746	776	806	3	9,6	9,3	9,0
144	836	866	897	927	957	987	*017	*047	*077	*107	4	12,8	12,4	12,0
145	16 137	167	197	227	256	286	316	346	376	406	5	16,0	15,5	15,0
146	435	465	495	524	554	584	613	643	673	702	6	19,2	18,6	18,0
147	732	761	791	820	850	879	909	938	967	997	7	22,4	21,7	21,0
148	17 026	056	085	114	143	173	202	231	260	289	8	25,6	24,8	24,0
149	319	348	377	406	435	464	493	522	551	580	9	28,8	27,9	27,0
150	609	638	667	696	725	754	782	811	840	869				
N	0	1	2	3	4	5	6	7	8	9	Proportional parts			

N	0	1	2	3	4	5	6	7	8	9	Proportional parts		
150	17 609	638	667	696	725	754	782	811	840	869		**29**	**28**
151	898	926	955	984	*013	*041	*070	*099	*127	*156	1	2,9	2,8
152	18 184	213	241	270	298	327	355	384	412	441	2	5,8	5,6
153	469	498	526	554	583	611	639	667	696	724	3	8,7	8,4
154	752	780	808	837	865	893	921	949	977	*005	4	11,6	11,2
155	19 033	061	089	117	145	173	201	229	257	285	5	14,5	14,0
156	312	340	368	396	424	451	479	507	535	562	6	17,4	16,8
157	590	618	645	673	700	728	756	783	811	838	7	20,3	19,6
158	866	893	921	948	976	*003	*030	*058	*085	*112	8	23,2	22,4
159	20 140	167	194	222	249	276	303	330	358	385	9	26,1	25,2
160	412	439	466	493	520	548	575	602	629	656		**27**	**26**
161	683	710	737	763	790	817	844	871	898	925	1	2,7	2,6
162	952	978	*005	*032	*059	*085	*112	*139	*165	*192	2	5,4	5,2
163	21 219	245	272	299	325	352	378	405	431	458	3	8,1	7,8
164	484	511	537	564	590	617	643	669	696	722	4	10,8	10,4
165	748	775	801	827	854	880	906	932	958	985	5	13,5	13,0
166	22 011	037	063	089	115	141	167	194	220	246	6	16,2	15,6
167	272	298	324	350	376	401	427	453	479	505	7	18,9	18,2
168	531	557	583	608	634	660	686	712	737	763	8	21,6	20,8
169	789	814	840	866	891	917	943	968	994	*019	9	24,3	23,4
170	23 045	070	096	121	147	172	198	223	249	274		**25**	
171	300	325	350	376	401	426	452	477	502	528	1	2,5	
172	553	578	603	629	654	679	704	729	754	779	2	5,0	
173	805	830	855	880	905	930	955	980	*005	*030	3	7,5	
174	24 055	080	105	130	155	180	204	229	254	279	4	10,0	
175	304	329	353	378	403	428	452	477	502	527	5	12,5	
176	551	576	601	625	650	674	699	724	748	773	6	15,0	
177	797	822	846	871	895	920	944	969	993	*018	7	17,5	
178	25 042	066	091	115	139	164	188	212	237	261	8	20,0	
179	285	310	334	358	382	406	431	455	479	503	9	22,5	
180	527	551	575	600	624	648	672	696	720	744		**24**	**23**
181	768	792	816	840	864	888	912	935	959	983	1	2,4	2,3
182	26 007	031	055	079	102	126	150	174	198	221	2	4,8	4,6
183	245	269	293	316	340	364	387	411	435	458	3	7,2	6,9
184	482	505	529	553	576	600	623	647	670	694	4	9,6	9,2
185	717	741	764	788	811	834	858	881	905	928	5	12,0	11,5
186	951	975	998	*021	*045	*068	*091	*114	*138	*161	6	14,4	13,8
187	27 184	207	231	254	277	300	323	346	370	393	7	16,8	16,1
188	416	439	462	485	508	531	554	577	600	623	8	19,2	18,4
189	646	669	692	715	738	761	784	807	830	852	9	21,6	20,7
190	875	898	921	944	967	989	*012	*035	*058	*081		**22**	**21**
191	28 103	126	149	171	194	217	240	262	285	307	1	2,2	2,1
192	330	353	375	398	421	443	466	488	511	533	2	4,4	4,2
193	556	578	601	623	646	668	691	713	735	758	3	6,6	6,3
194	780	803	825	847	870	892	914	937	959	981	4	8,8	8,4
195	29 003	026	048	070	092	115	137	159	181	203	5	11,0	10,5
196	226	248	270	292	314	336	358	380	403	425	6	13,2	12,6
197	447	469	491	513	535	557	579	601	623	645	7	15,4	14,7
198	667	688	710	732	754	776	798	820	842	863	8	17,6	16,8
199	885	907	929	951	973	994	*016	*038	*060	*081	9	19,8	18,9
200	30 103	125	146	168	190	211	233	255	276	298			
N	0	1	2	3	4	5	6	7	8	9	Proportional parts		

N	0	1	2	3	4	5	6	7	8	9	Proportional parts		
200	30 103	125	146	168	190	211	233	255	276	298		**22**	**21**
201	320	341	363	384	406	428	449	471	492	514	1	2,2	2,1
202	535	557	578	600	621	643	664	685	707	728	2	4,4	4,2
203	750	771	792	814	835	856	878	899	920	942	3	6,6	6,3
204	963	984	*006	*027	*048	*069	*091	*112	*133	*154	4	8,8	8,4
205	31 175	197	218	239	260	281	302	323	345	366	5	11,0	10,5
206	387	408	429	450	471	492	513	534	555	576	6	13,2	12,6
207	597	618	639	660	681	702	723	744	765	785	7	15,4	14,7
208	806	827	848	869	890	911	931	952	973	994	8	17,6	16,8
209	32 015	035	056	077	098	118	139	160	181	201	9	19,8	18,9
210	222	243	263	284	305	325	346	366	387	408		**20**	
211	428	449	469	490	510	531	552	572	593	613	1	2,0	
212	634	654	675	695	715	736	756	777	797	818	2	4,0	
213	838	858	879	899	919	940	960	980	*001	*021	3	6,0	
214	33 041	062	082	102	122	143	163	183	203	224	4	8,0	
215	244	264	284	304	325	345	365	385	405	425	5	10,0	
216	445	465	486	506	526	546	566	586	606	626	6	12,0	
217	646	666	686	706	726	746	766	786	806	826	7	14,0	
218	846	866	885	905	925	945	965	985	*005	*025	8	16,0	
219	34 044	064	084	104	124	143	163	183	203	223	9	18,0	
220	242	262	282	301	321	341	361	380	400	420		**19**	
221	439	459	479	498	518	537	557	577	596	616	1	1,9	
222	635	655	674	694	713	733	753	772	792	811	2	3,8	
223	830	850	869	889	908	928	947	967	986	*005	3	5,7	
224	35 025	044	064	083	102	122	141	160	180	199	4	7,6	
225	218	238	257	276	295	315	334	353	372	392	5	9,5	
226	411	430	449	468	488	507	526	545	564	583	6	11,4	
227	603	622	641	660	679	698	717	736	755	774	7	13,3	
228	793	813	832	851	870	889	908	927	946	965	8	15,2	
229	984	*003	*021	*040	*059	*078	*097	*116	*135	*154	9	17,1	
230	36 173	192	211	229	248	267	286	305	324	342		**18**	
231	361	380	399	418	436	455	474	493	511	530	1	1,8	
232	549	568	586	605	624	642	661	680	698	717	2	3,6	
233	736	754	773	791	810	829	847	866	884	903	3	5,4	
234	922	940	959	977	996	*014	*033	*051	*070	*088	4	7,2	
235	37 107	125	144	162	181	199	218	236	254	273	5	9,0	
236	291	310	328	346	365	383	401	420	438	457	6	10,8	
237	475	493	511	530	548	566	585	603	621	639	7	12,6	
238	658	676	694	712	731	749	767	785	803	822	8	14,4	
239	840	858	876	894	912	931	949	967	985	*003	9	16,2	
240	38 021	039	057	075	093	112	130	148	166	184		**17**	
241	202	220	238	256	274	292	310	328	346	364	1	1,7	
242	382	399	417	435	453	471	489	507	525	543	2	3,4	
243	561	578	596	614	632	650	668	686	703	721	3	5,1	
244	739	757	775	792	810	828	846	863	881	899	4	6,8	
245	917	934	952	970	987	*005	*023	*041	*058	*076	5	8,5	
246	39 094	111	129	146	164	182	199	217	235	252	6	10,2	
247	270	287	305	322	340	358	375	393	410	428	7	11,9	
248	445	463	480	498	515	533	550	568	585	602	8	13,6	
249	620	637	655	672	690	707	724	742	759	777	9	15,3	
250	794	811	829	846	863	881	898	915	933	950			
N	0	1	2	3	4	5	6	7	8	9	Proportional parts		

N	0	1	2	3	4	5	6	7	8	9	Proportional parts
250	39 794	811	829	846	863	881	898	829	933	950	**18**
251	967	985	*002	*019	*037	*054	*071	*088	*106	*123	1 1,8
252	40 140	157	175	192	209	226	243	261	278	295	2 3,6
253	312	329	346	364	381	398	415	432	449	466	3 5,4
254	483	500	518	535	552	569	586	603	620	637	4 7,2
255	654	671	688	705	722	739	756	773	790	807	5 9,0
256	824	841	858	875	892	909	926	943	960	976	6 10,8
257	993	*010	*027	*044	*061	*078	*095	*111	*128	*145	7 12,6
258	41 162	179	196	212	229	246	263	280	296	313	8 14,4
259	330	347	363	380	397	414	430	447	464	481	9 16,2
260	497	514	531	547	564	581	597	614	631	647	**17**
261	664	681	697	714	731	747	764	780	797	814	1 1,7
262	830	847	863	880	896	913	929	946	963	979	2 3,4
263	996	*012	*029	*045	*062	*078	*095	*111	*127	*144	3 5,1
264	42 160	177	193	210	226	243	259	275	292	308	4 6,8
265	325	341	357	374	390	406	423	439	455	472	5 8,5
266	488	504	521	537	553	570	586	602	619	635	6 10,2
267	651	667	684	700	716	732	749	765	781	797	7 11,9
268	813	830	846	862	878	894	911	927	943	959	8 13,6
269	975	991	*008	*024	*040	*056	*072	*088	*104	*120	9 15,3
270	43 136	152	169	185	201	217	233	249	265	281	**16**
271	297	313	329	345	361	377	393	409	425	441	1 1,6
272	457	473	489	505	521	537	553	569	584	600	2 3,2
273	616	632	648	664	680	696	712	727	743	759	3 4,8
274	775	791	807	823	838	854	870	886	902	917	4 6,4
275	933	949	965	981	996	*012	*028	*044	*059	*075	5 8,0
276	44 091	107	122	138	154	170	185	201	217	232	6 9,6
277	248	264	279	295	311	326	342	358	373	389	7 11,2
278	404	420	436	451	467	483	498	514	529	545	8 12,8
279	560	576	592	607	623	638	654	669	685	700	9 14,4
280	716	731	747	762	778	793	809	824	840	855	**15**
281	871	886	902	917	932	948	963	979	994	*010	1 1,5
282	45 025	040	056	071	086	102	117	133	148	163	2 3,0
283	179	194	209	225	240	255	271	286	301	317	3 4,5
284	332	347	362	378	393	408	423	439	454	469	4 6,0
285	484	500	515	530	545	561	576	591	606	621	5 7,5
286	637	652	667	682	697	712	728	743	758	773	6 9,0
287	788	803	818	834	849	864	879	894	909	924	7 10,5
288	939	954	969	984	*000	*015	*030	*045	*060	*075	8 12,0
289	46 090	105	120	135	150	165	180	195	210	225	9 13,5
290	240	255	270	285	300	315	330	345	359	374	**14**
291	389	404	419	434	449	464	479	494	509	523	1 1,4
292	538	553	568	583	598	613	627	642	657	672	2 2,8
293	687	702	716	731	746	761	776	790	805	820	3 4,2
294	835	850	864	879	894	909	923	938	953	967	4 5,6
295	982	997	*012	*026	*041	*056	*070	*085	*100	*114	5 7,0
296	47 129	144	159	173	188	202	217	232	246	261	6 8,4
297	276	290	305	319	334	349	363	378	392	407	7 9,8
298	422	436	451	465	480	494	509	524	538	553	8 11,2
299	567	582	596	611	625	640	654	669	683	698	9 12,6
300	712	727	741	756	770	784	799	813	828	842	
N	0	1	2	3	4	5	6	7	8	9	Proportional parts

N	0	1	2	3	4	5	6	7	8	9	Proportional parts
300	47 712	727	741	756	770	784	799	813	828	842	
301	857	871	885	900	914	929	943	958	972	986	
302	48 001	015	029	044	058	073	087	101	116	130	
303	144	159	173	187	202	216	230	244	259	273	**15**
304	287	302	316	330	344	359	373	387	401	416	1 1,5
305	430	444	458	473	487	501	515	530	544	558	2 3,0
306	572	586	601	615	629	643	657	671	686	700	3 4,5
307	714	728	742	756	770	785	799	813	827	841	4 6,0
308	855	869	883	897	911	926	940	954	968	982	5 7,5
309	996	*010	*024	*038	*052	*066	*080	*094	*108	*122	6 9,0
310	49 136	150	164	178	192	206	220	234	248	262	7 10,5
311	276	290	304	318	332	346	360	374	388	402	8 12,0
312	415	429	443	457	471	485	499	513	527	541	9 13,5
313	554	568	582	596	610	624	638	651	665	679	
314	693	707	721	734	748	762	776	790	803	817	
315	831	845	859	872	886	900	914	927	941	955	**14**
316	969	982	996	*010	*024	*037	*051	*065	*079	*092	1 1,4
317	50 106	120	133	147	161	174	188	202	215	229	2 2,8
318	243	256	270	284	297	311	325	338	352	365	3 4,2
319	379	393	406	420	433	447	461	474	488	501	4 5,6
320	515	529	542	556	569	583	596	610	623	637	5 7,0
321	651	664	678	691	705	718	732	745	759	772	6 8,4
322	786	799	813	826	840	853	866	880	893	907	7 9,8
323	920	934	947	961	974	987	*001	*014	*028	*041	8 11,2
324	51 055	068	081	095	108	121	135	148	162	175	9 12,6
325	188	202	215	228	242	255	268	282	295	308	
326	322	335	348	362	375	388	402	415	428	441	
327	455	468	481	495	508	521	534	548	561	574	**13**
328	587	601	614	627	640	654	667	680	693	706	1 1,3
329	720	733	746	759	772	786	799	812	825	838	2 2,6
330	851	865	878	891	904	917	930	943	957	970	3 3,9
331	983	996	*009	*022	*035	*048	*061	*075	*088	*101	4 5,2
332	52 114	127	140	153	166	179	192	205	218	231	5 6,5
333	244	257	270	284	297	310	323	336	349	362	6 7,8
334	375	388	401	414	427	440	453	466	479	492	7 9,1
335	504	517	530	543	556	569	582	595	608	621	8 10,4
336	634	647	660	673	686	699	711	724	737	750	9 11,7
337	763	776	789	802	815	827	840	853	866	879	
338	892	905	917	930	943	956	969	982	994	*007	
339	53 020	033	046	058	071	084	097	110	122	135	**22**
340	148	161	173	186	199	212	224	237	250	263	1 1,2
341	275	288	301	314	326	339	352	364	377	390	2 2,4
342	403	415	428	441	453	466	479	491	504	517	3 3,6
343	529	542	555	567	580	593	605	618	631	643	4 4,8
344	656	668	681	694	706	719	732	744	757	769	5 6,0
345	782	794	807	820	832	845	857	870	882	895	6 7,2
346	908	920	933	945	958	970	983	995	*008	*020	7 8,4
347	54 033	045	058	070	083	095	108	120	133	145	8 9,6
348	158	170	183	095	208	220	233	245	258	270	9 10,8
349	283	295	307	320	332	345	357	370	382	394	
350	407	419	432	444	456	469	481	494	506	518	
N	0	1	2	3	4	5	6	7	8	9	Proportional parts

N	0	1	2	3	4	5	6	7	8	9	Proportional parts
350	54 407	419	432	444	456	469	481	494	506	518	
351	531	543	555	568	580	593	605	617	630	642	
352	654	667	679	691	704	716	728	741	753	765	
353	777	790	802	814	827	839	851	864	876	888	**13**
354	900	913	925	937	949	962	974	986	998	*011	1 1,3
355	55 023	035	047	060	072	084	096	108	121	133	2 2,6
356	145	157	169	182	194	206	218	230	242	255	3 3,9
357	267	279	291	303	315	328	340	352	364	376	4 5,2
358	388	400	413	425	437	449	461	473	485	497	5 6,5
359	509	522	534	546	558	570	582	594	606	618	6 7,8
360	630	642	654	666	678	691	703	715	727	739	7 9,1
361	751	763	775	787	799	811	823	835	847	859	8 10,4
362	871	883	895	907	919	931	943	955	967	979	9 11,7
363	991	*003	*015	*027	*038	*050	*062	*074	*086	*098	
364	56 110	122	134	146	158	170	182	194	205	217	
365	229	241	253	265	277	289	301	312	324	336	**12**
366	348	360	372	384	396	407	419	431	443	455	1 1,2
367	467	478	490	502	514	526	538	549	561	573	2 2,4
368	585	597	608	620	632	644	656	667	679	691	3 3,6
369	703	714	726	738	750	761	773	785	797	808	4 4,8
370	820	832	844	855	867	879	891	902	914	926	5 6,0
371	937	949	961	972	984	996	*008	*019	*031	*043	6 7,2
372	57 054	066	078	089	101	113	124	136	148	159	7 8,4
373	171	183	194	206	217	229	241	252	264	276	8 9,6
374	287	299	310	322	334	345	357	368	380	392	9 10,8
375	403	415	426	438	449	461	473	484	496	507	
376	519	530	542	553	565	576	588	600	611	623	
377	634	646	657	669	680	692	703	715	726	738	**11**
378	749	761	772	784	795	807	818	830	841	852	1 1,1
379	864	875	887	898	910	921	933	944	955	967	2 2,2
380	978	990	*001	*013	*024	*035	*047	*058	*070	*081	3 3,3
381	58 092	104	115	127	138	149	161	172	184	195	4 4,4
382	206	218	229	240	252	263	274	286	297	309	5 5,5
383	320	331	343	354	365	377	388	399	410	422	6 6,6
384	433	444	456	467	478	490	501	512	524	535	7 7,7
385	546	557	569	580	591	602	614	625	636	647	8 8,8
386	659	670	681	692	704	715	726	737	749	760	9 9,9
387	771	782	794	805	816	827	838	850	861	872	
388	883	894	906	917	928	939	950	961	973	984	
389	995	*006	*017	*028	*040	*051	*062	*073	*084	*095	**10**
390	59 106	118	129	140	151	162	173	184	195	207	1 1,0
391	218	229	240	251	262	273	284	295	306	318	2 2,0
392	329	340	351	362	373	384	395	406	417	428	3 3,0
393	439	450	461	472	483	494	506	517	528	539	4 4,0
394	550	561	572	583	594	605	616	627	638	649	5 5,0
395	660	671	682	693	704	715	726	737	748	759	6 6,0
396	770	780	791	802	813	824	835	846	857	868	7 7,0
397	879	890	901	912	923	934	945	956	966	977	8 8,0
398	988	999	*010	*021	*032	*043	*054	*065	*076	*086	9 9,0
399	60 097	108	119	130	141	152	163	173	184	195	
400	206	217	228	239	249	260	271	282	293	304	
N	0	1	2	3	4	5	6	7	8	9	Proportional parts

N		0	1	2	3	4	5	6	7	8	9	Proportional parts	
400	60	206	217	228	239	249	260	271	282	293	304		
401		314	325	336	347	358	369	379	390	401	412		
402		423	433	444	455	466	477	487	498	509	520		
403		531	541	552	563	574	584	595	606	617	627		
404		638	649	660	670	681	692	703	713	724	735		
405		746	756	767	778	788	799	810	821	831	842		
406		853	863	874	885	895	906	917	927	938	949		
407		959	970	981	991	*002	*013	*023	*034	*045	*055		**11**
408	61	066	077	087	098	109	119	130	140	151	162	1	1,1
409		172	183	194	204	215	225	236	247	257	268	2	2,2
410		278	289	300	310	321	331	342	352	363	374	3	3,3
411		384	395	405	416	426	437	348	458	469	479	4	4,4
412		490	500	511	521	532	542	553	563	574	584	5	5,5
413		595	606	616	627	637	648	658	669	679	690	6	6,6
414		700	711	721	731	742	752	763	773	784	794	7	7,7
415		805	815	826	836	847	857	868	878	888	899	8	8,8
416		909	920	930	941	951	962	972	982	993	*003	9	9,9
417	62	014	024	034	045	055	066	076	086	097	107		
418		118	128	138	149	159	170	180	190	201	211		
419		221	232	242	252	263	273	284	294	304	315		
420		325	335	346	356	366	377	387	397	408	418		**10**
421		428	439	449	459	469	480	490	500	511	521	1	1,0
422		531	542	552	562	572	583	593	603	613	624	2	2,0
423		634	644	655	665	675	685	696	706	716	726	3	3,0
424		737	747	757	767	778	788	798	808	818	829	4	4,0
425		839	849	859	870	880	890	900	910	921	931	5	5,0
426		941	951	961	972	982	992	*002	*012	*022	*033	6	6,0
427	63	043	053	063	073	083	094	104	114	124	134	7	7,0
428		144	155	165	175	185	195	205	215	225	236	8	8,0
429		246	256	266	276	286	296	306	317	327	337	9	9,0
430		347	357	367	377	387	397	407	417	428	438		
431		448	458	468	478	488	498	508	518	528	538		
432		548	558	568	579	589	599	609	619	629	639		
433		649	659	669	679	689	699	709	719	729	739		
434		749	759	769	779	789	799	809	819	829	839		
435		849	859	869	879	889	899	909	919	929	939		**9**
436		949	959	969	979	988	998	*008	*018	*028	*038	1	0,9
437	64	048	058	068	078	088	098	108	118	128	137	2	1,8
438		147	157	167	177	187	197	207	217	227	237	3	2,7
439		246	256	266	276	286	296	306	316	326	335	4	3,6
440		345	355	365	375	385	395	404	414	424	434	5	4,5
441		444	454	464	473	483	493	503	513	523	532	6	5,4
442		542	552	562	572	582	591	601	611	621	631	7	6,3
443		640	650	660	670	680	689	699	709	719	729	8	7,2
444		738	748	758	768	777	787	797	807	816	826	9	8,1
445		836	846	856	865	875	885	895	904	914	924		
446		933	943	953	963	972	982	992	*002	*011	*021		
447	65	031	040	050	060	070	079	089	099	108	118		
448		128	137	147	157	167	176	186	196	205	215		
449		225	234	244	254	263	273	283	292	302	312		
450		321	331	341	350	360	369	379	389	398	408		
N		0	1	2	3	4	5	6	7	8	9	Proportional parts	

N	0	1	2	3	4	5	6	7	8	9	Proportional parts	
450	65 321	331	341	350	360	369	379	389	398	408		
451	418	427	437	447	456	466	475	485	495	504		
452	514	523	533	543	552	562	571	581	591	600		
453	610	619	629	639	648	658	667	677	686	696		
454	706	715	725	734	744	753	763	772	782	792		
455	801	811	820	830	839	849	858	868	877	887		
456	896	906	916	925	935	944	954	963	973	982		
457	992	*001	*011	*020	*030	*039	*049	*058	*068	*077		**10**
458	66 087	096	106	115	124	134	143	153	162	172	1	1,0
459	181	191	200	210	219	229	238	247	257	266	2	2,0
460	276	285	295	304	314	323	332	342	351	361	3	3,0
461	370	380	389	398	408	417	427	436	445	455	4	4,0
462	464	474	483	492	502	511	521	530	539	549	5	5 0
463	558	567	577	586	596	605	614	624	633	642	6	6 0
464	652	661	671	680	689	699	708	717	727	736	7	7,0
465	745	755	764	773	783	792	801	811	820	829	8	8,0
466	839	848	857	867	876	885	894	904	913	922	9	9,0
467	932	941	950	960	969	978	987	997	*006	*015		
468	67 025	034	043	052	062	071	080	089	099	108		
469	117	127	136	145	154	164	173	182	191	201		
470	210	219	228	237	247	256	265	274	284	293		
471	302	311	321	330	339	348	357	367	376	385		**9**
472	394	403	413	422	431	440	449	459	468	477	1	0,9
473	486	495	504	514	523	532	541	550	560	569	2	1,8
474	578	587	596	605	614	624	633	642	651	660	3	2,7
475	669	679	688	697	706	715	724	733	742	752	4	3,6
476	761	770	779	788	797	806	815	825	834	843	5	4,5
477	852	861	870	879	888	897	906	916	925	934	6	5,4
478	943	952	961	970	979	988	997	*006	*015	*024	7	6,3
479	68 034	043	052	061	070	079	088	097	106	115	8	7,2
480	124	133	142	151	160	169	178	187	196	205	9	8,1
481	215	224	233	242	251	260	269	278	287	296		
482	305	314	323	332	341	350	359	368	377	386		
483	395	404	413	422	431	440	449	458	467	476		
484	485	494	502	511	520	529	538	547	556	565		
485	574	583	592	601	610	619	628	637	646	655		**8**
486	664	673	681	690	699	708	717	726	735	744	1	0,8
487	753	762	771	780	789	797	806	815	824	833	2	1,6
488	842	851	860	869	878	886	895	904	913	922	3	2,4
489	931	940	949	958	966	975	984	993	*002	*011	4	3,2
490	69 020	028	037	046	055	064	073	082	090	099	5	4,0
491	108	117	126	135	144	152	161	170	179	188	6	4,8
492	197	205	214	223	232	241	249	258	267	276	7	5,6
493	285	294	302	311	320	329	338	346	355	364	8	6,4
494	373	381	390	399	408	417	425	434	443	452	9	7,2
495	461	469	478	487	496	504	513	522	531	539		
496	548	557	566	574	583	592	601	609	618	627		
497	636	644	653	662	671	679	688	697	705	714		
498	723	732	740	749	758	767	775	784	793	801		
499	810	819	827	836	845	854	862	871	880	888		
500	897	906	914	923	932	940	949	958	966	975		
N	0	1	2	3	4	5	6	7	8	9	Proportional parts	

N	0	1	2	3	4	5	6	7	8	9	Proportional parts
500	69 897	906	914	923	932	940	949	958	966	975	
501	984	992	001	*010	*018	*027	*036	*044	*053	*062	
502	70 070	079	088	096	105	114	122	131	140	148	
503	157	165	174	183	191	200	209	217	226	234	
504	243	252	260	269	278	286	295	303	312	321	
505	329	338	346	355	364	372	381	389	398	406	
506	415	424	432	441	449	458	467	475	484	492	
507	501	509	518	526	535	544	552	561	569	578	**9**
508	586	595	603	612	621	629	638	646	655	663	1 | 0,9
509	672	680	689	697	706	714	723	731	740	749	2 | 1,8
510	757	766	774	783	791	800	808	817	825	834	3 | 2,7
511	842	851	859	868	876	885	893	902	910	919	4 | 3,6
512	927	935	944	952	961	969	978	986	995	*003	5 | 4,5
513	71 012	020	029	037	046	054	063	071	079	088	6 | 5,4
514	096	105	113	122	130	139	147	155	164	172	7 | 6,3
515	181	189	198	206	214	223	231	240	248	257	8 | 7,2
516	265	273	282	290	299	307	315	324	332	341	9 | 8,1
517	349	357	366	374	383	391	399	408	416	425	
518	433	441	450	458	466	475	483	492	500	508	
519	517	525	533	542	550	559	567	575	584	592	
520	600	609	617	625	634	642	650	659	667	675	**8**
521	684	692	700	709	717	725	734	742	750	759	1 | 0,8
522	767	775	784	792	800	809	817	825	834	842	2 | 1,6
523	850	858	867	875	883	892	900	908	917	925	3 | 2,4
524	933	941	950	958	966	975	983	991	999	*008	4 | 3,2
525	72 016	024	032	041	049	057	066	074	082	090	5 | 4,0
526	099	107	115	123	132	140	148	156	165	173	6 | 4,8
527	181	189	198	206	214	222	230	239	247	255	7 | 5,6
528	263	272	280	288	296	304	313	321	329	337	8 | 6,4
529	346	354	362	370	378	387	395	403	411	419	9 | 7,2
530	428	436	444	452	460	469	477	485	493	501	
531	509	518	526	534	542	550	558	567	575	583	
532	591	599	607	616	624	632	640	648	656	665	
533	673	681	689	697	705	713	722	730	738	746	
534	754	762	770	779	787	795	803	811	819	827	
535	835	843	852	860	868	876	884	892	900	908	**7**
536	916	925	933	941	949	957	965	973	981	989	1 | 0,7
537	997	*006	*014	*022	*030*	*038	*046	*054	*062	*070	2 | 1,4
538	73 078	086	094	102	111	119	127	135	143	151	3 | 2,1
539	159	167	175	183	191	199	207	215	223	231	4 | 2,8
540	239	247	255	263	272	280	288	296	304	312	5 | 3,5
541	320	328	336	344	352	360	368	376	384	392	6 | 4,2
542	400	408	416	424	432	440	448	456	464	472	7 | 4,9
543	480	488	496	504	512	520	528	536	544	552	8 | 5,6
544	560	568	576	584	592	600	608	616	624	632	9 | 6,3
545	640	648	656	664	672	679	687	695	703	711	
546	719	727	735	743	751	759	767	775	783	791	
547	799	807	815	823	830	838	846	854	862	870	
548	878	886	894	902	910	918	926	933	941	949	
549	957	965	973	981	989	997	*005	*013	*020	*028	
550	74 036	044	052	060	068	076	084	092	099	107	
N	0	1	2	3	4	5	6	7	8	9	Proportional parts

N	0	1	2	3	4	5	6	7	8	9	Proportional parts
550	74 036	044	052	060	068	076	084	092	099	107	
551	115	123	131	139	147	155	162	170	178	186	
552	194	202	210	218	225	233	241	249	257	265	
553	273	280	288	296	304	312	320	327	335	343	
554	351	359	367	374	382	390	398	406	414	421	
555	429	437	445	453	461	468	476	484	492	500	
556	507	515	523	531	539	547	554	562	570	578	
557	586	593	601	609	617	624	632	640	648	656	
558	663	671	679	687	695	702	710	718	726	733	
559	741	749	757	764	772	780	788	796	803	811	
560	819	827	834	842	850	858	865	873	881	889	**8**
561	896	904	912	920	927	935	943	950	958	966	1 | 0,8
562	974	981	989	997	*005	*012	*020	*028	*035	*043	2 | 1,6
563	75 051	059	066	074	082	089	097	105	113	120	3 | 2,4
564	128	136	143	151	159	166	174	182	189	197	4 | 3,2
565	205	213	220	228	236	243	251	259	266	274	5 | 4,0
566	282	289	297	305	312	320	328	335	343	351	6 | 4,8
567	358	366	374	381	389	397	404	412	420	427	7 | 5,6
568	435	442	450	458	465	473	481	488	496	504	8 | 6,4
569	511	519	526	534	542	549	557	565	572	580	9 | 7,2
570	587	595	603	610	618	626	633	641	648	656	
571	664	671	679	686	694	702	709	717	724	732	
572	740	747	755	762	770	778	785	793	800	808	
573	815	823	831	838	846	853	861	868	876	884	
574	891	899	906	914	921	929	937	944	952	959	
575	967	974	982	989	997	*005	*012	*020	*027	*035	
576	76 042	050	057	065	072	080	087	095	103	110	
577	118	125	133	140	148	155	163	170	178	185	
587	193	200	208	215	223	230	238	245	253	260	
579	268	275	283	290	298	305	313	320	328	335	
580	343	350	358	365	373	380	388	395	403	410	**7**
581	418	425	433	440	448	455	462	470	477	485	1 | 0,7
582	492	500	507	515	522	530	537	545	552	559	2 | 1,4
583	567	574	582	589	597	604	612	619	626	634	3 | 2,1
584	641	649	656	664	671	678	686	693	701	708	4 | 2,8
585	716	723	730	738	745	753	760	768	775	782	5 | 3,5
586	790	797	805	812	819	827	834	842	849	856	6 | 4,2
587	864	871	879	886	893	901	908	916	923	930	7 | 4,9
588	938	945	953	960	967	975	982	989	997	*004	8 | 5,6
589	77 012	019	026	034	041	048	056	063	070	078	9 | 6,3
590	085	093	100	107	115	122	129	137	144	151	
591	159	166	173	181	188	195	203	210	217	225	
592	232	240	247	254	262	269	276	283	291	298	
593	305	313	320	327	335	342	349	357	364	371	
594	379	386	393	401	408	415	422	430	437	444	
595	452	459	466	474	481	488	495	503	510	517	
596	525	532	539	546	554	561	568	576	583	590	
597	597	605	612	619	627	634	641	648	656	663	
598	670	677	685	692	699	706	714	721	728	735	
599	743	750	757	764	772	779	786	793	801	808	
600	815	822	830	837	844	851	859	866	873	880	
N	0	1	2	3	4	5	6	7	8	9	Proportional parts

N	0	1	2	3	4	5	6	7	8	9	Proportional parts	
600	77 815	822	830	837	844	851	859	866	873	880		
601	887	895	902	909	916	924	931	938	945	952		
602	960	967	974	981	988	996	*003	*010	*017	*025		
603	78 032	039	046	053	061	068	075	082	089	097		
604	104	111	118	125	132	140	147	154	161	168		
605	176	183	190	197	204	211	219	226	233	240		
606	247	254	262	269	276	283	290	297	305	312		
607	319	326	333	340	347	355	362	369	376	383		**8**
608	390	398	405	412	419	426	433	440	447	455	1	0,8
609	462	469	476	483	490	497	504	512	519	526	2	1,6
610	533	540	547	554	561	569	576	583	590	597	3	2,4
611	604	611	618	625	633	640	647	654	661	668	4	3,2
612	675	682	689	696	704	711	718	725	732	739	5	4,0
613	746	753	760	767	774	781	789	796	803	810	6	4,8
614	817	824	831	838	845	852	859	866	873	880	7	5,6
615	888	895	902	909	916	923	930	937	944	951	8	6,4
616	958	965	972	979	986	993	*000	*007	*014	*021	9	7,2
617	79 029	036	043	050	057	064	071	078	085	092		
618	099	106	113	120	127	134	141	148	155	162		
619	169	176	183	190	197	204	211	218	225	232		
620	239	246	253	260	267	274	281	288	295	302		**7**
621	309	316	323	330	337	344	351	358	365	372	1	0,7
622	379	386	393	400	407	414	421	428	435	442	2	1,4
623	449	456	463	470	477	484	491	498	505	511	3	2,1
624	518	525	532	539	546	553	560	567	574	581	4	2,8
625	588	595	602	609	616	623	630	637	644	650	5	3,5
626	657	664	671	678	685	692	699	706	713	720	6	4,2
627	727	734	741	748	754	761	768	775	782	789	7	4,9
628	796	803	810	817	824	831	837	844	851	858	8	5,6
629	865	872	879	886	893	900	906	913	920	927	9	6,3
630	934	941	948	955	962	969	975	982	989	996		
631	80 003	010	017	024	030	037	044	051	058	065		
632	072	079	085	092	099	106	113	120	127	134		
633	140	147	154	161	168	175	182	188	195	202		
634	209	216	223	229	236	243	250	257	264	271		
635	277	284	291	298	305	312	318	325	332	339		**6**
636	346	353	359	366	373	380	387	393	400	407	1	0,6
637	414	421	428	434	441	448	455	462	468	475	2	1,2
638	482	489	496	502	509	516	523	530	536	543	3	1,8
639	550	557	564	570	577	584	591	598	604	611	4	2,4
640	618	625	632	638	645	652	659	665	672	679	5	3,0
641	686	693	699	706	713	720	726	733	740	747	6	3,6
642	754	760	767	774	781	787	794	801	808	814	7	4,2
643	821	828	835	841	848	855	862	868	875	882	8	4,8
644	889	895	902	909	916	922	929	936	943	949	9	5,4
645	956	963	969	976	983	990	996	*003	*010	*017		
646	81 023	030	037	043	050	057	064	070	077	084		
647	090	097	104	111	117	124	131	137	144	151		
648	158	164	171	178	184	191	198	204	211	218		
649	224	231	238	245	251	258	265	271	278	285		
650	291	298	305	311	318	325	331	338	345	351		
N	0	1	2	3	4	5	6	7	8	9	Proportional parts	

N	0	1	2	3	4	5	6	7	8	9	Proportional parts
650	81 291	298	305	311	318	325	331	338	345	351	
651	358	365	371	378	385	391	398	405	411	418	
652	425	431	438	445	451	458	465	471	478	485	
653	491	498	505	511	518	525	531	538	544	551	
654	558	564	571	578	584	591	598	604	611	617	
655	624	631	637	644	651	657	664	671	677	684	
656	690	697	704	710	717	723	730	737	743	750	
657	757	763	670	776	783	790	796	803	809	816	
658	823	829	836	842	849	856	862	869	875	882	
659	889	895	902	908	915	921	928	935	941	948	
660	954	961	968	974	981	987	994	*000	*007	*014	**7**
661	82 020	033	033	040	046	053	060	066	073	079	1 | 0,7
662	086	092	099	105	112	119	125	132	138	145	2 | 1,4
663	151	158	164	171	178	184	191	197	204	210	3 | 2,1
664	217	223	230	236	243	249	256	263	269	276	4 | 2,8
665	282	289	295	302	308	315	321	328	334	341	5 | 3,5
666	347	354	360	367	373	380	387	393	400	406	6 | 4,2
667	413	419	426	432	439	445	452	458	465	471	7 | 4,9
668	478	484	491	497	504	510	517	523	530	536	8 | 5,6
669	543	549	556	562	569	575	582	588	595	601	9 | 6,3
670	607	614	620	627	633	640	646	653	659	666	
671	672	679	685	792	698	705	711	718	724	730	
672	737	743	750	756	763	769	776	782	789	795	
673	802	808	814	821	827	834	840	847	853	860	
674	866	872	879	885	892	898	905	911	918	924	
675	930	937	943	950	956	963	969	975	982	988	
676	995	*001	*008	*014	*020	*027	*033	*040	*046	*052	
677	83 059	065	072	078	085	091	097	104	110	117	
678	123	129	136	142	149	155	161	168	174	181	
679	187	193	200	206	213	219	225	232	238	245	
680	251	257	264	270	276	283	289	296	302	308	**6**
681	315	321	327	334	340	347	353	359	366	372	1 | 0,6
682	378	385	391	398	404	410	417	423	429	436	2 | 1,2
683	442	448	455	461	467	474	480	487	493	499	3 | 1,8
684	506	512	518	525	531	537	544	550	556	563	4 | 2,4
685	569	575	582	588	594	601	607	613	620	626	5 | 3,0
686	632	639	645	651	658	664	670	677	683	689	6 | 3,6
687	696	702	708	715	721	727	734	740	746	753	7 | 4,2
688	759	765	771	778	784	790	797	803	809	816	8 | 4,8
689	822	828	835	841	847	853	860	866	872	879	9 | 5,4
690	885	891	897	904	910	916	923	929	935	942	
691	948	954	960	967	973	979	985	992	998	*004	
692	84 011	017	023	029	036	042	048	055	061	067	
693	073	080	086	092	098	105	111	117	123	130	
694	136	142	148	155	161	167	173	180	186	192	
695	198	205	211	217	223	230	236	242	248	255	
696	261	267	273	280	286	292	298	305	311	317	
697	323	330	336	342	348	354	361	367	373	379	
698	386	392	398	404	410	417	423	429	435	442	
699	448	454	460	466	473	479	485	491	497	504	
700	510	516	522	528	535	541	547	553	559	566	
N	0	1	2	3	4	5	6	7	8	9	Proportional parts

N	0	1	2	3	4	5	6	7	8	9	Proportional parts
700	84 510	516	522	528	535	541	547	553	559	566	
701	572	578	584	590	597	603	609	615	621	628	
702	634	640	646	652	658	665	671	677	683	689	
703	696	702	708	714	720	726	733	739	745	751	
704	757	763	770	776	782	788	794	800	807	813	
705	819	825	831	837	844	850	856	862	868	874	
706	880	887	893	899	905	911	917	924	930	936	
707	942	948	954	960	967	973	979	985	991	997	**7**
708	85 003	009	016	022	028	034	040	046	052	058	1 | 0,7
709	065	071	077	083	089	095	101	107	114	120	2 | 1,4
710	126	132	138	144	150	156	163	169	175	181	3 | 2,1
711	187	193	199	205	211	217	224	230	236	242	4 | 2,8
712	248	254	260	266	272	278	285	291	297	303	5 | 3,5
713	309	315	321	327	333	339	345	352	358	364	6 | 4,2
714	370	376	382	388	394	400	406	412	418	425	7 | 4,9
715	431	437	443	449	455	461	467	473	479	485	8 | 5,6
716	491	497	503	509	516	522	528	534	540	546	9 | 6,3
717	552	558	564	570	576	582	588	594	600	606	
718	612	618	625	631	637	643	649	655	661	667	
719	673	679	685	691	697	703	709	715	721	727	
720	733	739	745	751	757	763	769	775	781	788	
721	794	800	806	812	818	824	830	836	842	848	**6**
722	854	860	866	872	878	884	890	896	902	908	1 | 0,6
723	914	920	926	932	938	944	950	956	962	968	2 | 1,2
724	974	980	986	992	998	*004	*010	*016	*022	*028	3 | 1,8
725	86 034	040	046	052	058	064	070	076	082	088	4 | 2,4
726	094	100	106	112	118	124	130	136	141	147	5 | 3,0
727	153	159	165	171	177	183	189	195	201	207	6 | 3,6
728	213	219	225	231	237	243	249	255	261	267	7 | 4,2
729	273	279	285	291	297	303	308	314	320	326	8 | 4,8
730	332	338	344	350	356	362	368	374	380	386	9 | 5,4
731	392	398	404	410	415	421	427	433	439	445	
732	451	457	463	469	475	481	487	493	499	504	
733	510	516	522	528	534	540	546	552	558	564	
734	570	576	581	587	593	599	605	611	617	623	
735	629	635	641	646	652	658	664	670	676	682	**5**
736	688	694	700	705	711	717	723	729	735	741	1 | 0,5
737	747	753	759	764	770	776	782	688	794	800	2 | 1,0
738	806	812	817	823	829	835	841	847	853	859	3 | 1,5
739	864	870	876	882	888	894	900	906	911	917	4 | 2,0
740	923	929	935	941	947	953	958	964	970	976	5 | 2,5
741	982	988	994	999	*005	*011	*017	*023	*029	*035	6 | 3,0
742	87 040	046	052	058	064	070	075	081	087	093	7 | 3,5
743	099	105	111	116	122	128	134	140	146	151	8 | 4,0
744	157	163	169	175	181	186	192	198	204	210	9 | 4,5
745	216	221	227	233	239	245	251	256	262	268	
746	274	280	286	291	297	303	309	315	320	326	
747	332	338	344	349	355	361	367	373	379	384	
748	390	396	402	408	413	419	425	431	437	442	
749	448	454	460	466	471	377	483	489	495	500	
750	506	512	518	523	529	535	541	547	552	558	
N	0	1	2	3	4	5	6	7	8	9	Proportional parts

N	0	1	2	3	4	5	6	7	8	9	Proportional parts
750	87 506	512	518	523	529	535	541	547	552	558	
751	564	570	576	581	587	593	599	604	610	616	
752	622	628	633	639	645	651	656	662	668	674	
753	679	685	691	697	703	708	714	720	726	731	
754	737	743	749	754	760	766	772	777	783	789	
755	795	800	806	812	818	823	829	835	841	846	
756	852	858	864	869	875	881	887	892	898	904	
757	910	915	921	927	933	938	944	950	955	961	
758	967	973	978	984	990	996	*001	*007	*013	*018	
759	88 024	030	036	041	047	053	058	064	070	076	
760	081	087	093	098	104	110	116	121	127	133	**6**
761	138	144	150	156	161	167	173	178	184	190	1 | 0,6
762	195	201	207	213	218	224	230	235	241	247	2 | 1,2
763	252	258	264	270	275	281	287	292	298	304	3 | 1,8
764	309	315	321	326	332	338	343	349	355	360	4 | 2,4
765	366	372	377	383	389	395	400	406	412	417	5 | 3,0
766	423	429	434	440	446	451	457	463	468	474	6 | 3,6
767	480	485	491	497	502	508	513	519	525	530	7 | 4,2
768	536	542	547	553	559	564	570	576	581	587	8 | 4,8
769	593	598	604	610	615	621	627	632	638	643	9 | 5,4
770	649	655	660	666	672	677	683	689	694	700	
771	705	711	717	722	728	734	739	745	750	756	
772	762	767	773	779	784	790	795	801	807	812	
773	818	824	829	835	840	846	852	857	863	868	
774	874	880	885	891	897	902	908	913	919	925	
775	930	936	941	947	953	958	964	969	975	981	
776	986	992	997	*003	*009	*014	*020	*025	*031	*037	
777	89 042	048	053	059	064	070	076	081	087	092	
778	098	104	109	115	120	126	131	137	143	148	
779	154	159	165	170	176	182	187	193	198	204	
780	209	215	221	226	232	237	243	248	254	260	**5**
781	265	271	276	282	287	293	298	304	310	315	1 | 0,5
782	321	326	332	337	343	348	354	360	365	371	2 | 1,0
783	376	382	387	393	398	404	409	415	421	426	3 | 1,5
784	432	437	443	448	454	459	465	470	476	481	4 | 2,0
785	487	492	498	504	509	515	520	526	531	537	5 | 2,5
786	542	548	553	559	564	570	575	581	586	592	6 | 3,0
787	597	603	609	614	620	625	631	636	642	647	7 | 3,5
788	653	658	664	669	675	680	686	691	697	702	8 | 4,0
789	708	713	719	724	730	735	741	746	752	757	9 | 4,5
790	763	768	774	779	785	790	796	801	807	812	
791	818	823	829	834	840	845	851	856	862	867	
792	873	878	883	889	894	900	905	911	916	922	
793	927	933	938	944	949	955	960	966	971	977	
794	982	988	993	998	*004	*009	*015	*020	*026	*031	
795	90 037	042	048	053	059	064	069	075	080	086	
796	091	097	102	108	113	119	124	129	135	140	
797	146	151	157	162	168	173	179	184	189	195	
798	200	206	211	217	222	227	233	238	244	249	
799	255	260	266	271	276	282	287	293	298	304	
800	309	314	320	325	331	336	342	347	352	358	
N	0	1	2	3	4	5	6	7	8	9	Proportional parts

N		0	1	2	3	4	5	6	7	8	9	Proportional parts	
800	90	309	314	320	325	331	336	342	347	352	358		
801		363	369	374	380	385	390	396	401	407	412		
802		417	423	428	434	439	445	450	455	461	466		
803		472	477	382	488	493	499	504	509	515	520		
804		526	531	536	542	547	553	558	563	569	574		
805		580	585	590	596	601	607	612	617	623	628		
806		634	639	644	650	655	660	666	671	677	682		
807		687	693	698	703	709	714	720	725	730	736		
808		741	747	752	757	763	768	773	779	784	789		
809		795	800	806	811	816	822	827	832	838	843		
810		849	854	859	865	870	875	881	886	891	897		**6**
811		902	907	913	918	924	929	934	940	945	950	1	0,6
812		956	961	966	972	977	982	988	993	998	*004	2	1,2
813	91	009	014	020	025	030	036	041	046	052	057	3	1,8
814		062	068	073	078	084	089	094	100	105	110	4	2,4
815		116	121	126	132	137	142	148	153	158	164	5	3,0
816		169	174	180	185	190	196	201	206	212	217	6	3,6
817		222	228	233	238	243	249	254	259	265	270	7	4,2
818		275	281	286	291	297	302	307	312	318	323	8	4,8
819		328	334	339	344	350	355	360	365	371	376	9	5,4
820		381	387	392	397	403	408	413	418	424	429		
821		434	440	445	450	455	461	466	471	477	482		
822		487	492	498	503	508	514	519	524	529	535		
823		540	547	551	556	561	566	572	577	582	587		
824		593	598	603	609	614	619	624	630	635	640		
825		645	651	656	661	666	672	677	682	687	693		
826		698	703	709	714	719	724	730	735	740	745		
827		751	756	761	766	772	777	782	787	793	798		
828		803	808	814	819	824	829	834	840	845	850		
829		855	861	866	871	876	882	887	892	897	903		
830		908	913	918	924	929	934	939	944	950	955		**5**
831		960	965	971	976	981	986	991	997	*002	*007	1	0,5
832	92	012	018	023	028	033	038	044	049	054	059	2	1,0
833		065	070	075	080	085	091	096	101	106	111	3	1,5
834		117	122	127	132	137	143	148	153	158	163	4	2,0
835		169	174	179	184	189	195	200	205	210	215	5	2,5
836		221	226	231	236	241	247	252	257	262	267	6	3,0
837		273	278	283	288	293	298	304	309	314	319	7	3,5
838		324	330	335	340	345	350	355	361	366	371	8	4,0
839		376	381	387	392	397	402	407	412	418	423	9	4,5
840		428	433	438	443	449	454	459	464	469	474		
841		480	485	490	495	500	505	511	516	521	526		
842		531	536	542	547	552	557	562	567	572	578		
843		583	588	593	598	603	609	614	619	624	629		
844		634	639	645	650	655	660	665	670	675	681		
845		686	691	696	701	706	711	716	722	727	732		
846		737	742	747	752	758	763	768	773	778	783		
847		788	793	799	804	809	814	819	824	829	834		
848		840	845	850	855	860	865	870	875	881	886		
849		891	896	901	906	911	916	921	927	932	937		
850		942	947	952	957	962	967	973	978	983	988		
N		0	1	2	3	4	5	6	7	8	9	Proportional parts	

N	0	1	2	3	4	5	6	7	8	9	Proportional parts
850	92 942	947	952	957	962	967	973	978	983	988	
851	993	998	*003	*008	*013	*018	*024	*029	*034	*039	
852	93 044	049	054	059	064	069	075	080	085	090	
853	095	100	105	110	115	120	125	131	136	141	
854	146	151	156	161	166	171	176	181	186	192	
855	197	202	207	212	217	222	227	232	237	242	
856	247	252	258	263	268	273	278	283	288	293	**6**
857	298	303	308	313	318	323	328	334	339	344	1 0,6
858	349	354	359	364	369	374	379	384	389	394	2 1,2
859	399	404	409	414	420	425	430	435	440	445	3 1,8
860	450	455	460	465	470	475	480	485	490	495	4 2,4
861	500	505	510	515	520	526	531	536	541	546	5 3,0
862	551	556	561	566	571	576	581	586	591	596	6 3,6
863	601	606	611	616	621	626	631	636	641	646	7 4,2
864	651	656	661	666	671	676	682	687	692	697	8 4,8
865	702	707	712	717	722	727	732	737	742	747	9 5,4
866	752	757	762	767	772	777	782	787	792	797	
867	802	807	812	817	822	827	832	837	842	847	
868	852	857	862	867	872	877	882	887	892	897	
869	902	907	912	917	922	927	932	937	942	947	
870	952	957	962	967	972	977	982	987	992	997	**5**
871	94 002	007	012	017	022	027	032	037	042	047	1 0,5
872	052	057	062	067	072	077	082	086	091	096	2 1,0
873	101	106	111	116	121	126	131	136	141	146	3 1,5
874	151	156	161	166	171	176	181	186	191	196	4 2,0
875	201	206	211	216	221	226	231	236	240	245	5 2,5
876	250	255	260	265	270	275	280	285	290	295	6 3,0
877	300	305	310	315	320	325	330	335	340	345	7 3,5
878	349	354	359	364	369	374	379	384	389	394	8 4,0
879	399	404	409	414	419	424	429	433	438	443	9 4,5
880	448	453	458	463	468	473	478	483	488	493	
881	498	503	507	512	517	522	527	532	537	542	
882	547	552	557	562	567	571	576	581	586	591	
883	596	601	606	611	616	621	626	630	635	640	
884	645	650	655	660	665	670	675	680	685	689	**4**
885	694	699	704	709	714	719	724	729	734	738	1 0,4
886	743	748	753	758	763	768	773	778	783	787	2 0,8
887	792	797	802	807	812	817	822	827	832	836	3 1,2
888	841	846	851	856	861	866	871	876	880	885	4 1,6
889	890	895	900	905	910	915	919	924	929	934	5 2,0
890	939	944	949	954	959	963	968	973	978	983	6 2,4
891	988	993	998	*002	*007	*012	*017	*022	*027	*032	7 2,8
892	95 036	041	046	051	056	061	066	071	075	080	8 3,2
893	085	090	095	100	105	109	114	119	124	129	9 3,6
894	134	139	143	148	153	158	163	168	173	177	
895	182	187	192	197	202	207	211	216	221	226	
896	231	236	240	245	250	255	260	265	270	274	
897	279	284	289	294	299	303	308	313	318	323	
898	328	332	337	342	347	352	357	361	366	371	
899	376	381	386	390	395	400	405	410	415	419	
900	424	429	434	439	444	448	453	458	463	468	
N	0	1	2	3	4	5	6	7	8	9	Proportional parts

N	0	1	2	3	4	5	6	7	8	9	Proportional parts
900	95 424	429	434	439	444	448	453	458	463	468	
901	472	477	482	487	492	497	501	506	511	516	
902	521	525	530	535	540	545	550	554	559	564	
903	569	574	578	583	588	593	598	602	607	612	
904	617	622	626	631	636	641	646	650	655	660	
905	665	670	674	679	684	689	694	698	703	708	
906	713	718	722	727	732	737	742	746	751	756	
907	761	766	770	775	780	785	789	794	799	804	
908	809	813	818	823	828	832	837	842	847	852	
909	856	861	866	871	875	880	885	890	895	899	
910	904	909	914	918	923	928	933	938	942	947	**5**
911	952	957	961	966	971	976	980	985	990	995	1 | 0,5
912	999	*004	*009	*014	*019	*023	*028	*033	*038	*042	2 | 1,0
913	96 047	052	057	061	066	071	076	080	085	090	3 | 1,5
914	095	099	104	109	114	118	123	128	133	137	4 | 2,0
915	142	147	152	156	161	166	171	175	180	185	5 | 2,5
916	190	194	199	204	209	213	218	223	227	232	6 | 3,0
917	237	242	246	251	256	261	265	270	275	280	7 | 3,5
918	284	289	294	298	303	308	313	317	322	327	8 | 4,0
919	332	336	341	346	350	355	360	365	369	374	9 | 4,5
920	379	384	388	393	398	402	407	412	417	421	
921	426	431	435	440	445	450	454	459	464	468	
922	473	478	483	487	492	497	501	506	511	515	
923	520	525	530	534	539	544	548	553	558	562	
924	567	572	577	581	586	591	595	600	605	609	
925	614	619	624	628	633	638	642	647	652	656	
926	661	666	670	675	680	685	689	694	699	703	
927	708	713	717	722	727	731	736	741	745	750	
928	755	759	764	769	774	778	783	788	792	797	
929	802	806	811	816	820	825	830	834	839	844	
930	848	853	858	862	867	872	876	881	886	890	**4**
931	895	900	904	909	914	918	923	928	932	937	1 | 0,4
932	942	946	951	956	960	965	970	974	979	984	2 | 0,8
933	988	993	997	*002	*007	*011	*016	*021	*025	*030	3 | 1,2
934	97 035	039	044	049	053	058	063	067	072	077	4 | 1,6
935	081	086	090	095	100	104	109	114	118	123	5 | 2,0
936	128	132	137	142	146	151	155	160	165	169	6 | 2,4
937	174	179	183	188	192	197	202	206	211	216	7 | 2,8
938	220	225	230	234	239	243	248	253	257	262	8 | 3,2
939	267	271	276	280	285	290	294	299	304	308	9 | 3,6
940	313	317	322	327	331	336	340	345	350	354	
941	359	364	368	373	377	382	387	391	396	400	
942	405	410	414	419	424	428	433	437	442	447	
943	451	456	460	465	470	474	479	483	488	493	
944	497	502	506	511	516	520	525	529	534	539	
945	543	548	552	557	562	566	571	575	580	585	
946	589	594	598	603	607	612	617	621	626	630	
947	635	640	644	649	653	658	663	667	672	676	
948	681	685	690	695	699	704	708	713	717	722	
949	727	731	736	740	745	749	754	759	763	768	
950	772	777	782	786	791	795	800	804	809	813	
N	0	1	2	3	4	5	6	7	8	9	Proportional parts

N	0	1	2	3	4	5	6	7	8	9	Proportional parts	
950	97 772	777	782	786	791	795	800	804	809	813		
951	818	823	827	832	836	841	845	850	855	859		
952	864	868	873	877	882	886	891	896	900	905		
953	909	914	918	923	928	932	937	941	946	950		
954	955	959	964	968	973	978	982	987	991	996		
955	98 000	005	009	014	019	023	028	032	037	041		
956	046	050	055	059	064	068	073	078	082	087		
957	091	096	100	105	109	114	118	123	127	132		
958	137	141	146	150	155	159	164	168	173	177		
959	182	186	191	195	200	204	209	214	218	223		
960	227	232	236	241	245	250	254	259	263	268		**5**
961	272	277	281	286	290	295	299	304	308	313	1	0,5
962	318	322	327	331	336	340	345	349	354	358	2	1,0
963	363	367	372	376	381	385	390	394	399	403	3	1,5
964	408	412	417	421	426	430	435	439	444	448	4	2,0
965	453	457	462	466	471	475	480	484	489	493	5	2,5
966	498	502	507	511	516	520	525	529	534	538	6	3,0
967	543	547	552	556	561	565	570	574	579	583	7	3,5
968	588	592	597	601	605	610	614	619	623	628	8	4,0
969	632	637	641	646	650	655	659	664	668	673	9	4,5
970	677	682	686	691	695	700	704	709	713	717		
971	722	726	731	735	740	744	749	753	758	762		
972	767	771	776	780	784	789	793	798	802	807		
973	811	816	820	825	829	834	838	843	847	851		
974	856	860	865	869	874	878	883	887	892	896		
975	900	905	909	914	918	923	927	932	936	941		
976	945	949	954	958	963	967	972	976	981	985		
977	989	994	998	*003	*007	*012	*016	*021	*025	*029		
978	99 034	038	043	047	052	056	061	065	069	074		
979	078	083	087	092	096	100	105	109	114	118		
980	123	127	131	136	140	145	149	154	158	162		**4**
981	167	171	176	180	185	189	193	198	202	207	1	0,4
982	211	216	220	224	229	233	238	242	247	251	2	0,8
983	255	260	264	269	273	277	282	286	291	295	3	1,2
984	300	304	308	313	317	322	326	330	335	339	4	1,6
985	344	348	352	357	361	366	370	373	379	383	5	2,0
986	388	392	396	401	405	410	414	419	423	427	6	2,4
987	432	436	441	445	449	454	458	463	467	471	7	2,8
988	476	480	484	489	493	498	502	506	511	515	8	3,2
989	520	524	528	533	537	542	546	550	555	559	9	3,6
990	564	568	572	577	581	585	590	594	599	603		
991	607	612	616	621	625	629	634	638	642	647		
992	651	656	660	664	669	673	677	682	686	691		
993	695	699	704	708	712	717	721	726	730	734		
994	739	743	747	752	756	760	765	769	774	778		
995	782	787	791	795	800	804	808	813	817	822		
996	826	830	835	839	843	848	852	856	861	865		
997	870	874	878	883	887	891	896	900	904	909		
998	913	917	922	926	930	935	939	944	948	952		
999	957	961	965	970	974	978	983	987	991	996		
1000	00 000	004	009	013	017	022	026	030	035	39		
N	0	1	2	3	4	5	6	7	8	9	Proportional parts	

INDEX

H_2SO_3	82.08
H_2SO_4	98.08
HVO_3	99.95
HgO	216.59
HgS	232.65
Hg_2Br_2	561.00
Hg_2Cl_2	472.08
Hg_2I_2	654.99
$KB(C_6H_5)_4$	358.34
KBF_4	125.82
KBr	119.01
$KBrO_3$	167.01
KCN	65.12
K_2CO_3	138.22
KCl	74.56
$KClO_3$	122.55
$KClO_4$	138.55
K_2CrO_4	194.20
$K_2Cr_2O_7$	294.20
$K_3Fe(CN)_6$	329.26
$K_4Fe(CN)_6$	368.36
$KHCO_3$	100.12
KHC_2O_4	128.13
$KHC_2O_4 \cdot H_2C_2O_4 \cdot 2H_2O$	254.20
$KHC_4H_4O_6$ (tartrate)	188.19
$KHC_8H_4O_4$ (phthalate)	204.23
$KH(IO_3)_2$	389.92
KH_2PO_4	136.09
K_2HPO_4	174.19
KI	166.00
KIO_3	214.00
KIO_4	230.00
$KMnO_4$	158.04
KNO_2	85.11
KNO_3	101.11
K_2O	94.20
KOH	56.11
K_3PO_4	212.01
K_2PtCl_6	486.01
KSCN	97.16
K_2SO_4	174.26
$K_2S_2O_8$	270.32
LiBr	86.86
LiCl	42.39
Li_2SO_4	109.94
$MgCO_3$	84.32
MgC_2O_4	112.33
$Mg(C_9H_6ON)_2$ (oxinate)	312.62
$MgCl_2$	95.22
$Mg(ClO_4)_2$	223.22
$MgNH_4PO_4$	137.33
MgI_2	278.12
MgO	40.31
$Mg(OH)_2$	58.33
$Mg_2P_2O_7$	222.56
$MgSO_4$	120.37
MnO_2	86.94
Mn_3O_4	228.81
$Mn(OH)_2$	88.96
MnS	87.00
MoO_3	143.94
Mo_2O_3	239.88
$NaBO_2$	65.80
$Na_2B_4O_7 \cdot 10H_2O$ (borax)	381.38
NaBr	102.90
$NaCHO_2$ (formate)	68.01
$NaC_2H_3O_2$ (acetate)	82.03
$NaC_7H_5O_2$ (benzoate)	144.11
NaCN	49.01
Na_2CO_3	105.99
$Na_2C_2O_4$	134.00
NaCl	58.44
NaClO	74.44
$NaClO_2$	90.44
Na_2HAsO_4	185.91
$NaHCO_3$	84.01
NaH_2PO_4	119.99
Na_2HPO_4	141.98
NaI	149.89
NaN_3	65.02
$NaNO_2$	69.00
$NaNO_3$	85.00
Na_2O	61.98
Na_2O_2	77.98
NaOH	40.00
Na_3PO_4	163.95
Na_2S	78.04
NaSCN	81.07
Na_2SO_3	126.04
Na_2SO_4	142.04
$Na_2S_2O_3 \cdot 5H_2O$	248.18
NH_3	17.03
$(NH_4)_2CO_3$	96.09
$(NH_4)_2C_2O_4$	124.10
NH_4Cl	53.49
NH_2OH	33.03
NH_4OH	35.05
$(NH_4)_3P(Mo_3O_{10})_4$	1876.36
$(NH_4)_2PtCl_6$	443.89
$(NH_4)_2SO_4$	132.13
$NiCl_2$	129.62
P_2O_5	141.95